Art Scents

THINKING ART

Series Editors
Noël Carroll and Jesse Prinz, CUNY Graduate Center

Thinking Art fills an important gap in contemporary philosophy of art, focusing on cutting edge ideas and approaches to the subject.

PUBLISHED IN THE SERIES:

Attentional Agency:
A Perceptual Theory of The Arts
William P. Seeley

Games: Agency as Art
C. Thi Nguyen

Art Scents

Exploring the Aesthetics of Smell and the Olfactory Arts

LARRY SHINER

Oxford University Press is a department of the University of Oxford. It furthers
the University's objective of excellence in research, scholarship, and education
by publishing worldwide. Oxford is a registered trade mark of Oxford University
Press in the UK and certain other countries.

Published in the United States of America by Oxford University Press
198 Madison Avenue, New York, NY 10016, United States of America.

© Oxford University Press 2020

All rights reserved. No part of this publication may be reproduced, stored in
a retrieval system, or transmitted, in any form or by any means, without the
prior permission in writing of Oxford University Press, or as expressly permitted
by law, by license, or under terms agreed with the appropriate reproduction
rights organization. Inquiries concerning reproduction outside the scope of the
above should be sent to the Rights Department, Oxford University Press, at the
address above.

You must not circulate this work in any other form
and you must impose this same condition on any acquirer.

Library of Congress Cataloging-in-Publication Data
Names: Shiner, L. E. (Larry E.), 1934– author.
Title: Art scents : exploring the aesthetics of smell and the olfactory arts / Larry Shiner.
Description: New York, NY, United States of America. : Oxford University Press, 2020. |
Series: Thinking art | Includes bibliographical references.
Identifiers: LCCN 2019037447 (print) | LCCN 2019037448 (ebook) |
ISBN 9780190089818 (hardback) | ISBN 9780190089825 (updf) |
ISBN 9780190089832 (epub) | ISBN 9780190089849 (online)
Subjects: LCSH: Aesthetics. | Odors.
Classification: LCC BH301.O35 S55 2020 (print) |
LCC BH301.O35 (ebook) | DDC 152.1/66—dc2 3
LC record available at https://lccn.loc.gov/2019037447
LC ebook record available at https://lccn.loc.gov/2019037448

Contents

Acknowledgments ix
Introduction 1
   Organization 3
   The Argument 4

I. WHAT CAN THE NOSE KNOW?

Overview: The Challenge of the Olfactory Arts 13
Prelude: Nietzsche's Nose 17
1. The Fear of Smell 19
   The Historical Prejudice against Smell 19
   The Case against Odors 21
   The Case against the Sense of Smell 23
      Smell Is Disreputable 23
      Smell Is Deficient 24
      Smell Is Deceptive 25
      Smell Is Dispensable 28
Interlude: Calvino's "The Name, the Nose" 35
2. The Neuroscience and Psychology of Smell I: What the Nose Can Do 37
   Odors and the Human Olfactory System 37
   The "Odor Object" Theory 41
   Detection, Discrimination, Learning 43
   Social Communication 45
Interlude: The Pheromone Myth 51
3. The Neuroscience and Psychology of Smell II: What the Nose Can't Do 54
   Four Counts against Smell's Cognitive Adequacy 54
   What Is Smell's Dominant Characteristic? 56
   Measuring the Abilities of Olfactory Experts 58
   Interpreting Olfactory Experts' Abilities 63

4. Smell, Emotion, and Aesthetics — 68
   The Intelligence of the Emotions — 68
   Aesthetics and Emotion — 71

II. SMELL REDEEMED: LANGUAGE, CULTURE, AND MEMORY

Overview: A Biocultural Approach — 79

Prelude: Darwin, Smell, and Evolution — 82

5. The Dialectic of Deodorization: Smell in Western History — 87
   Our Forgotten Olfactory Past — 88
   The "Deodorization" of Western Societies — 91
   Philosophical Implications — 94

Interlude: Fragrant Asia — 98

6. Language, Culture, and Smell — 103
   Learning, Labeling, and Classification — 104
   Odor Terms in Western Languages — 108
   Smell in Non-Western Cultures and Languages — 110

7. Writing Smell — 117
   Poetry — 117
   Novelistic Devices — 120
   Character — 122

8. Odor, Memory, and Proust — 127
   Voluntary versus Involuntary Memories — 127
   Proust, Psychology, and Transcendence — 129

Postlude: Is an Olfactory Aesthetics Possible? — 138

III. DISCOVERING THE OLFACTORY ARTS

Overview: What Is Olfactory Art? — 145

Prelude: Picturing Smell — 152

9. Toward a Total Work of Art: Smell in Theater, Film, and Music — 158
   Theater — 158
   Film — 165
   Music — 170

Interlude: *Smeller 2.0* and the Osmodrama — 178

10. Sublime Stenches: Contemporary Olfactory Art	182
Some Types of Olfactory or Scent Art	183
Scent Sculptures	184
Installations	184
Performance	185
Participation	185
Perfumes	186
Atmospheres	186
Olfactory/Scent Art and Sound Art as Art Forms	186
The History of Olfactory/Scent Art	188
Some Self-Identified Scent or Olfactory Artists	190
Exhibition, Conservation, and Ontology	191
The Interpretation of Olfactory Art	193
Olfactory Art and Aesthetic Appreciation	196
Interlude: Kodo, an Art of Incense	201
What Kodo Is	201
Art Form or Aesthetic Practice?	204
11. Beautiful Fragrances: Is Perfume a Fine Art?	209
An Aesthetic Case for Perfume as Fine Art	211
A Contextual Case against Perfumes as Fine Art	217
An Impasse	224
12. Perfume between Art and Design	229
From "Art" to Art	229
Ways Out of the Impasse between Fine Art and Design	231
Combining Aesthetic and Contextual Theories	231
Local Analogies and the Possibility of "Art Perfumes"	234
Art Perfumes, Scent Art, and Standard Perfumes	240
Postlude: Free Art versus Design Art	243

IV. THE AESTHETICS AND ETHICS OF SCENTING

Overview: Varieties of Aesthetic Experience	255
Prelude: Two Cautionary Tales	261
13. The Meanings and Morality of Scenting the Body	265
Greek Philosophers, Roman Moralists, and Church Fathers	266
Perfume Pleasures in Asia and the Middle East	270
Contemporary Meanings	271

14. Ambient Scenting, Architecture, and the City	278
Smellwalking and Scent Arts in the City	279
Smell and Unban Design	281
Aromatic Architecture	283
The Ethics of Ambient Scenting	286
Aromatherapy versus Fragrance Bans	290
15. Enhancing Flavors with Scents in Contemporary Cuisine	296
Avant-Garde Cuisine: Fine Art or Design Art?	299
The Everyday Aesthetics of Aromas	301
The Ethics of Aromas and Flavors in Fast Food	302
Postlude: Wilderness, Gardens, and Paradise	307
16. An Invitation to Discovery	315
Bibliography	319
Index	341

Acknowledgments

For many years I taught a course on aesthetics for graduate students in arts management at the University of Illinois at Springfield. One day a student from Russia, Yulia Kriskovets, walked into my office and asked if I would be her thesis adviser. "Depends on the topic," I said. "Olfactory art and how to exhibit it." I was dumbfounded. I thought I knew something about contemporary art, but I didn't know what "olfactory art" was. Thanks to Yulia, I embarked on a crash course on smell in the arts a dozen years ago, and after she finished her thesis, we collaborated on an article published in the *Journal of Aesthetics and Art Criticism*.

That article led to an invitation from Marta Tafalla in 2013 to participate in a colloquium she organized in Barcelona on smell and aesthetics that was the impetus to begin writing this book. In addition to the stimulating exchanges I have had since then with Marta, my research has been deeply informed by the presentations of the other participants in the colloquium as well as by their other writings: Emily Brady, Jim Drobnick, Victoria Henshaw, Laura López-Mascaraque, Barry C. Smith, and Cain Todd. Since then, many others have helped me.

A number of friends who are part of a local interdisciplinary study group have read many of the chapters as the book developed, and their frank criticism and encouragement were a constant spur to keep going: José Arce, Harry Berman, Pyotr Boltuc, Gene Brodland, Meredith Cargill, Cynthia Cochran, Bernd Estabrook, Robert Kunath, Rosina Nejinsky, Lynn Pardie, Bill Underwood, and Peter Wenz. Numerous colleagues from the American Society for Aesthetics have offered encouragement and advice on a number of issues. I owe a special debt to Kevin Sweeney for stimulating discussions and above all for reading the entire manuscript at a crucial stage and offering valuable suggestions. Other ASA colleagues whose conversation has helped me with various issues include Arnold Berleant, Allen Carlson, John Carvahlo, Donald Crawford, Susan Feagin, Cynthia Freeland, Ivan Gaskell, David Goldblatt, Karen Gover, Sherri Irvin, Gary Iseminger, Jennifer Judkins, Carolyn Korsmeyer, Thomas Leddy, Sheila Lintott, Dominic Lopes, Jonathan Maskit, Mara Miller, Glenn Parsons, Monique Roelofs, Stephanie Ross, Yuriko Saito, Elisabeth Schellekens, Robert Stecker, and Mary Wiseman.

Mădălina Diaconu, of the University of Vienna, who has written perceptively on olfaction and aesthetics, has offered helpful suggestions on a number of points. I also owe a particular debt to Chantal Jaquet of the University of Paris at the Sorbonne, who in addition to her many distinguished publications in the history of philosophy, has written indispensable books on the philosophy of smell and on *kodo* and has edited an important volume of essays on the olfactory arts. Professor Jaquet was kind enough to meet with me in Paris in the summer of 2018, and I am grateful for a memorable and helpful conversation.

Many artists, musicians, and theater people from the University of Illinois at Springfield have given encouragement and helped in various ways. Jeff Robinson and Allison Lacher of the Visual Art Department have called my attention to important exhibitions involving olfactory art. Mike Miller, as always, has kept me open to new directions in contemporary art, and Sharon Graf has done the same for music, especially drawing on her background in ethnomusicology. I am also grateful for encouragement from Britton Bjorngaard, Jonathan Perkins, and Shane Harris in the art department as well as Eric and Missy Thibodeaux-Thompson in theater. Other colleagues at University of Illinois at Springfield who have either retired or moved on but have stayed in touch and given encouragement and advice include Cullom Davis, Judy Everson, Mike Lennon, Chuck Strozier, and Roy Wehrle. Two artists and former colleagues from my days at Cornell College in Iowa, Vivian Heywood and Hugh Lifson, have also helped me continue to think about new directions in contemporary art. As I was finishing this manuscript, I learned that Vivian had passed away and I remember with gratitude our many wonderful conversations over the years, including several about this project.

An invitation from Giorgio de Finis, Director of MACRO (Museo d'Arte Contemporanea di Roma) to deliver a lecture in September 2019 was helpful in crystalizing several ideas. I am especially grateful to Fabio Benincasa of MACRO for calling my attention to certain Italian olfactory artworks.

As always, the Brookens Library staff at the University of Illinois at Springfield has cheerfully fulfilled my innumerable requests to track down hard-to-find resources. I have also received valuable assistance from the Northwestern University Library and from the John M. Flaxman Library of the School of the Art Institute of Chicago.

I am grateful to Peter Ohlin of Oxford University Press for his encouragement and advice and his constant support for this project. Thanks are also due to Oxford University Press for allowing me to use parts of my article "Art Scents: Perfume, Design and Olfactory Art," which appeared in the *British Journal of Aesthetics* 55, no. 3 (2015): 375–92.

Most importantly, I want to thank my wife, my partner, and intellectual companion, Catherine Walters, for her constant encouragement, stimulating conversations about ideas, and the many suggestions that have immeasurably improved this work.

Art Scents

Introduction

An increasing number of artists have begun using scents in their works over the last two decades and it's time for aesthetics to take notice. Consider Otobong Nkanga's 2018 work *Anamnesis*, a long, freestanding, white wall with a dark, river-like incision running around it at nose level. She filled the incision with aromatic coffee beans, chopped tobacco leaves, cloves, and other spices of the kind that were exploited in the African colonial trade, and it gave museum visitors a palpable experience of Nkanda's anticolonial message as they walked along smelling the river of scents.[1] The year before, Christophe Laudamiel's *Over 21* offered visitors to a New York gallery a very different kind of olfactory experience. Laudamiel had placed ten canisters of synthetic scents with overt sexual references around a dining table; visitors dipped perfume blotters into a small hole in the top of each canister and inhaled scents with names like Elephant in Musth and Green Fairy in Chelsea.[2] In 2015 Basel's Tinguely Museum presented a survey of sixty olfactory artworks past and present. But museum and gallery works like these are not the only kinds of contemporary olfactory arts. The French drama *Scents of the Soul* (2012) released a dozen strategic scents from beneath theater seats, and *Green Aria: A Scent Opera* (2009) at the Guggenheim combined electronic music with abstract odors to narrate an environmental message. Meanwhile, designers have been putting odors into everything from urban streetscapes and signature scents for hotels to fabrics with embedded fragrances.[3]

The many olfactory art and design works appearing today raise challenging issues for aesthetics, and even for ethics. And despite the fact that much of the philosophical tradition from Kant and Hegel to Roger Scruton and Dennis Dutton has denied that odors and the sense of smell can be the basis for genuine artworks, this book will make a case that they can and will lay the basis for an olfactory aesthetics. Yet that case will not be purely philosophical, but also biocultural, marshaling evidence from the interdisciplinary sensory revolution of the last two decades. Led by natural and social scientists, historians, artists, and activists, this revolution is overturning long-held assumptions about the so-called lower senses of smell, taste, and touch.

Yet there are still a lot of ingrained prejudices to overcome, given the long history of neglect and disdain for smell by most Western intellectuals, including Darwin and Freud. Many Western thinkers in the modern period have considered smell the lowest and most animalistic of the senses, and have ignored

Art Scents. Larry Shiner, Oxford University Press (2020) © Oxford University Press.
DOI: 10.1093/oso/9780190089818.003.0001

perfume as a trivial luxury and incense as a religious oddity. Worse yet, odors and the nose easily become the target of jokes, from Gogol's famous story about the man who lost his nose and looked all over Moscow for it, to schoolboys snickering over farts, to John Waters's handing out scratch-and-sniff cards with the movie *Polyester*. But it's time to stop snickering and ask ourselves: *why* have we been so embarrassed by our noses and our sense of smell?

One reason, as Nietzsche suggests, is that smell is a reminder that we are embodied, that we are part of the animal kingdom, and, whether we like it or not, we breathe in and give off odors all day long. Of course, most of us hardly pay attention to the odors around us, and when people are asked which of the senses they would give up, if they had to, they often name smell. This low status is confirmed by the American Medical Association guidelines for insurance companies and courts in deciding compensation for total impairment, with smell valued at 1%–5% and vision at 85%. Indeed, in an era when we spend so much of our time staring at our cell phones, notepads, and computers, it might seem that vision and hearing are gaining an even greater hold over our daily lives at the expense of taste, touch, and smell.

Since the eighteenth century, many people in the West have lived in an increasingly deodorized culture, where cities have been sanitized and noticeable odors eliminated. As a result, one of the surprises for some Westerners visiting cities in other parts of the world is to discover their rich mix of odors. In the Marrakech Medina, for example, the narrow streets exhume a constantly changing blend of smells, aromas, and fragrances that come from piles of spices in the open air, shops roasting lamb and chicken, fruit and vegetable carts lining the narrow, twisting streets, sweaty bodies pressing around you, horses dropping manure, the exhaust of motor bikes, and all this mixed with an occasional whiff of urine from the scrawny cats running loose. In up-scale urban areas of the United States, on the other hand, even coffee roasters, despite the appeal of coffee aromas in other contexts, have run afoul of city odor ordinances, and some buildings ban the wearing of perfumes in offices and elevators.

Yet, paradoxically, most of our foods and household products are scented, and many stores and hotels now use ambient odors to create atmosphere. Indeed, some of the same international corporations that produce commercial perfumes actually gain most of their income from making artificial flavors for foods and fragrances for soaps, fabric softeners, toothpastes, shampoos, and so on. Even our digital devices may eventually be programmed to emit odors of our choosing, that is, if any of a variety of current experimental models finally gains market traction. Meanwhile, scientists and technicians have already successfully developed "electronic noses" to sniff out dangerous gases or explosives. Yet the possibility that our cell phones, notepads, and computers will one day allow us to send each other odor messages may depend less on the success of the inventors than on whether they can overcome the public's olfactory ignorance and lack of interest in exploring the world of smell. There are some promising signs here and there, as aspects of the intellectual revolution I have already mentioned are filtering out to a wider audience. In the last dozen years, a number of popular books by scientists have appeared that aim at making people more aware of the importance of the sense of smell. These have come from experts in many fields ranging from psychology (Rachel Herz and Avery Gilbert) through biology (Michael Stoddart) and chemistry (Paolo Pelosi) to dog cognition (Alexandra Horowitz).[4]

This book aims to bring together cutting-edge research on olfaction in the sciences and humanities with current thinking about the nature of art and aesthetics in philosophy. Since it is intended not only for philosophers but also for artists, designers, art critics, and readers interested in the arts or curious about the sense of smell, it will not offer a detailed history of philosophical ideas on smell or make a technical contribution to the philosophy of perception. The one area of contemporary philosophy that it will engage most closely is obviously the field of aesthetics since my central concern is how we should understand and appreciate the various olfactory arts. Fortunately, there are already excellent philosophical works on the aesthetics of taste, the other "proximal" or "chemical" sense, that provide invaluable models and insights for thinking about smell.[5] But, with the noteworthy exception of Chantal Jaquet's invaluable *Philosophie de l'odorat*, there has been no other book-length philosophical discussion of smell and aesthetics, although there are several important articles from which almost any interested reader can profit.[6]

Organization

I have organized the book into four parts, each beginning with an overview. Interspersed among the individual chapters are short preludes, interludes, and postludes that treat topics of importance and intrinsic interest but are set apart lest their length break the flow of the argument within each chapter. Although the heart of the book focuses on issues surrounding the creation and aesthetic appreciation of the many olfactory arts (Parts III and IV), I will first have to counter the myths and misrepresentations that lie behind the intellectual neglect and disparagement of the sense of smell and its artistic and aesthetic potential (Parts I and II). Those first two parts will also establish a baseline of insights into the current scientific and philosophical understanding of the sense of smell and into the profound historical and cultural importance of smell and the olfactory arts. In creating that baseline, I draw from the growing work on odors and the sense of smell in the natural and social sciences as well as in the humanities. That evidence is enriched by drawing on the works of poets and novelists who have written evocatively about smell.

Since one of the main tasks of the book will be to respond to the Western tradition that denies that odors and the sense of smell can be involved in the creation of artworks or their aesthetic appreciation, it will be necessary to enter some way into current philosophical debates about what art is and about the nature of aesthetic experience and judgment. Yet, rather than announce a specific definition of art or the aesthetic at the outset, I unfold my own pluralistic position gradually since I want to show in Parts I and II that even on a more traditionalist concept of fine art and the aesthetic, a case can be made for the possibility of an olfactory aesthetics. Once that case is made, the route will be open to gradually broaden the concepts of both art and the aesthetic in Parts III and IV. Throughout the book, the focus will remain on offering an overview of the intellectual issues raised by the use of odors and the sense of smell in many kinds of artistic and aesthetic practices, including those of everyday life.

The Argument

Part I, "What Can the Nose Know?," begins with an overview that describes the challenges for an aesthetics of the olfactory arts posed by traditional mainstream assumptions about the nature of art and the aesthetic.

A prelude, "Nietzsche's Nose," calls attention to the one well-known modern philosopher who celebrated the sense of smell.

Chapter 1, "The Fear of Smell," sets the stage for later chapters by briefly reviewing and countering some of the arguments philosophers and other intellectuals have used to justify treating the human sense of smell as cognitively null and generally of little use, namely that smell is disreputable, deficient, deceptive, and dispensable. Chapter 1 closes by arguing that the intuitive and analytic arguments of the kind I have used on behalf of the sense of smell need to be corroborated by the best current work on olfaction in the sciences.

As a transition to Chapters 2 and 3 on neuroscience and psychology, I insert a literary interlude on Italo Calvino's short story "The Name, the Nose," which exemplifies many of the issues to come, and which I will use for examples from time to time.

Chapter 2, "The Neuroscience and Psychology of Smell I: What the Nose Can Do," explores the biology of the human olfactory system and surveys contemporary research on the characteristics of smell that indicate its cognitive capacity for detection, discrimination, learning, and social communication.

The topic of communication leads naturally to another brief interlude, "The Pheromone Myth."

Chapter 3, "The Neuroscience of Smell II: What the Nose Can't Do," turns to experimental evidence that could be interpreted as suggesting our sense of smell may *lack* sufficient cognitive powers to fund reflective aesthetic judgments. I focus on psychological theories that claim our olfactory system is (1) purely emotional, (2) only capable of simplistic hedonic judgments, (3) unable to reliably identify and name odors, and (4) incurably unconscious. Although these characterizations could be seen as supporting the negative intellectual tradition on smell, recent neuroscience studies of olfactory experts suggest that the human sense of smell, despite its limitations, may indeed be able to support reflective aesthetic experience and judgments.

Chapter 4, "Smell, Emotion, and Aesthetics," responds more extensively to the widely held view that associates smell exclusively with the emotions and vision with reason. I draw not only on contemporary neuroscience, psychology, and philosophy to show that the emotions have a cognitive aspect and reason an affective dimension, but that emotion plays an essential role in aesthetics. Hence, smell's strong emotional charge is not an impediment per se to smell's participation in aesthetic judgments.

Part II, "Smell Redeemed: Language, Culture, and Memory," begins with an overview showing the need to move beyond neuroscience and psychology and draw on evidence from evolutionary theory, history, anthropology, linguistics, and literature if we are to refute the claims that the sense of smell is mute and of little use.

The prelude, "Darwin, Smell, and Evolution," calls attention to recent evolutionary theories that suggest, contrary to Darwin's dismissal of smell as an evolutionary vestige, that smell may have played an important role in our becoming human.

Chapter 5, "The Dialectic of Deodorization: Smell in Western History," begins by recalling the central role that incense and perfumes once played in many aspects of Western culture, then looks at what I call the "dialectic of deodorization" over the past two centuries. The chapter ends by suggesting that this historical turn may have exacerbated our tendency to be unaware of smells and have encouraged intellectuals to ignore the sense of smell as of little importance.

An interlude, "Fragrant Asia," explores the fascinating role that smell and incense/perfume have played and continue to play in Asian societies, offering additional evidence to refute the Kantian and Darwinian dismissal of smell as largely useless.

Chapter 6, "Language, Culture, and Smell," draws on linguistics and anthropology to show that many non-Western cultures and languages have sophisticated ways of expressing smell and that peoples of these cultures can quickly and easily identify and name odors. This suggests that psychological experiments showing Westerners' poor ability to name and describe odors may not reflect a universal human trait.

Chapter 7, "Writing Smell," suggests that if we look at poetry and the novel in the West, it turns out that many Western writers such as Baudelaire, Joyce, and Woolf have been able to articulate smell experiences forcefully and convincingly.

Chapter 8, "Odor, Memory, and Proust," draws together the previous themes of emotion and language and relates them to memory. After examining some evidence from the psychology of autobiographical memory, the chapter focuses on Proust's *Remembrance of Things Past* and ends by contrasting the Proustian literary epiphanies with the directness of two Holocaust memoirs. This shows that one need not be an olfactory expert or a literary artist to give a convincing linguistic expression to smell.

A brief postlude to Part II, "Is an Olfactory Aesthetics Possible?," draws together the threads of the philosophical arguments and empirical evidence of the first eight chapters to answer in the affirmative.

The overview to Part III, "Discovering the Olfactory Arts," begins to develop an olfactory aesthetics by defining the concept "olfactory arts." I argue that we should use the term "olfactory arts" in the plural as an umbrella term to cover any kind of artwork that makes an *intentional* and *distinctive* use of *actual* odors. I suggest that "olfactory art" (or "scent art") in the singular should be used for works that meet those same three criteria plus the proviso that they are created by artists to be shown in galleries or museums. The individual chapters of Part III deal with aesthetic issues related to three broad areas of the olfactory arts: theater, film, and music as enhanced by odors; hybrids of odors with visual art forms such as sculpture or installations; and, finally, "pure" olfactory arts such as perfumes or incense. I save a discussion of the use of scents in contemporary haute cuisine, architecture, and urban design for Part IV since those uses raise unavoidable ethical as well as aesthetic issues.

Part III begins with a prelude, "Picturing Smell," exploring historical and contemporary attempts at the representation of odors and the sense of smell in the pictorial arts.

Chapter 9, "Toward a Total Work of Art: Smell in Theater, Film, and Music," argues that given the long history of the use of odors in the theater, there is reason to view the inclusion of odors in some types of contemporary theater production with cautious optimism, although the value of adding odors to films is less certain. The discussion of music focuses on *Green Aria: A Scent Opera* presented at the Guggenheim in 2009, a work that combined narrative, odors, and an electronic music score in a way that marked a decisive step toward the successful integration of actual smells with music and narrative.

An interlude titled "*Smeller 2.0* and the 'Osmodrama'" discusses Wolfgang Georgsdorf's remarkable "scent organ" that he uses not only to accompany musical works and films, but also to play independent scent compositions.

Chapter 10, "Sublime Stenches: Contemporary Olfactory Art," is devoted to a discussion of hybrids of odors with visual art genres or materials, works that are typically presented in art galleries and museums under the rubric "olfactory art." After surveying various types of olfactory art, I consider the question of whether "olfactory art" in the singular actually names a coherent category or art form. I suggest a tentative yes, based on parallels between olfactory or scent art and contemporary "sound art," such as their parallel histories, and the fact that some artists identify themselves as olfactory artists and a few of them have even issued manifestoes promoting olfactory art. I then take up questions of ontology and interpretation.

The interlude, "Kodo, an Art of Incense," focuses on the revival of the Japanese incense ceremony called *kodo*, whose more sophisticated versions are considered by some art theorists and philosophers to be a distinctive art form.

Chapter 11, "Beautiful Fragrances: Is Perfume a Fine Art?," considers the claim that the best perfumes should be classified as fine art. I argue that from the perspective of contemporary *aesthetic* definitions of fine art, perfumes have all it takes to be fine art, including formal complexity and a capacity to represent and express. Yet, in the second half of the chapter, I argue that from the perspective of contemporary *contextual* or *historical* definitions of art, perfumes are more like design art than fine art since they lack the typical intentions, norms, and routes of circulation of most artworks. In developing that case I use a model of social practices to compare a typical practice of creating a high-end perfume with a typical practice of creating a work of installation art that involves a commissioned perfume. Chapter 11 ends in an impasse.

Chapter 12, "Perfume between Art and Design," explores two main ways out of the impasse. The first way would adopt one of the current composite or disjunctive definitions of (fine) art; the second way would abandon the quest for defining (fine) art and consider instead what it would take to promote *some* perfumery practices to the status of art perfumes, parallel to the way some kinds of photography or quilt making have become art photography or art quilts.

Since the solution I propose to the impasse would mean that only certain types of perfumes could be considered art perfumes, leaving the vast majority of most standard perfumes part of design, a postlude to Part III, "Free Art versus Design Art," answers the concern that this would demote the finest perfumes for wear to "minor art" status. I propose a pluralistic view of the nature of art that would retain the important cultural distinction between art and design without accepting the invidious hierarchical implications clinging to most traditional concepts of "fine" (or "major") art. In arguing the case for a pluralistic view that would preserve the dignity of design art (or "responsive" art) as the equal of fine art (or "free" art), I draw on the parallel case of fashion design, which some theorists and certainly some practitioners and art museum curators are eager to classify as "fine art."

The overview to "Part IV: The Aesthetics and Ethics of Scenting," argues for a pluralistic concept of aesthetic experience and judgment similar to the continuum proposed for a pluralistic concept of art. I illustrate such a pluralistic approach by drawing on three contemporary attempts to broaden the concept of the aesthetic that are also open to development of an ethical criticism: "functional beauty," "everyday aesthetics," and the "aesthetics of atmospheres."

A prelude to the discussion of the ethics of scenting the body explores two cautionary tales depicting the dangers of an obsession with scents and perfumes: Huysmans's *Against Nature* and Patrick Süskind's notorious *Perfume: The Story of a Murderer*.

Chapter 13, "The Meanings and Morality of Scenting the Body," begins with the views of Plato, Aristotle, some Roman moralists, and several early Christian theologians on the ethics of wearing perfume, views that have continued to reverberate down into the present. After briefly considering the relative absence of such moral suspicions in Asian and Arab-Islamic cultures, I examine conflicting contemporary ideas about scenting the body: on the one hand, the complaint that perfumes are primarily used for seduction or masking or involve artifice, and on the other, the affirmation of motives such as identity, pleasure, and spirituality.

Chapter 14, "Ambient Scenting, Architecture, and the City," argues that the ethos of the modern city in the developed world is marked by a conflict between a continuing tendency toward "deodorizing" and a minority view that advocates greater olfactory diversity. The chapter begins by discussing smellwalks and then moves on to the role of odors in urban design and management and discusses artworks that comment on these issues. The chapter closes by considering the ethical issues surrounding ambient scenting in both the workplace and the marketplace, as well as the conflict between the claims made for aromatherapy and the demand for fragrance bans by sufferers from multiple chemical sensitivity.

Chapter 15, "Enhancing Flavors with Scents in Contemporary Cuisine," considers the central role of both orthonasal and retronasal smell in the perception of flavor and its implications for a multisensory aesthetics of food. After discussing some avant-garde aroma/flavor experiments and some parallels between the philosophical debate over whether fine cuisine is a fine art and the debate over the art status of perfumes, the chapter closes with an analysis of the place of food aromas in the health challenges posed by "fast food."

The postlude, "Wilderness, Gardens, and Paradise," briefly addresses the issue of olfaction in the areas of environmental aesthetics dealing with wilderness and gardens, and ends by examining the role of smell in the imagination of Paradise within Christian and Islamic cultures.

Chapter 16, "An Invitation to Discovery," draws together the threads of the arguments running through the preceding chapters, briefly discusses the question of whether olfactory artworks can be "profound," and suggests that those of us interested in art and aesthetics need to cultivate both our knowledge of smell and our sense of smell if we are to appreciate to their fullest the sensory riches of our environment and the creative achievements of the olfactory arts.

Notes

1. *Anamnesis* was part of an exhibition of Nkanda's work at the Museum of Contemporary Art, Chicago, called *To Dig a Hole and Watch It Collapse Again* from March 31 to September 3, 2018. A more detailed discussion of *Anamnesis* is offered at the end of Chapter 10.
2. The exhibition was held at the Dillon and Lee Gallery in New York from January 19 to February 17, 2017. A more detailed discussion of *Over 21* can be found in Chapter 11.
3. For an overview of the many ways smell is playing a role in both art and design see Victoria Henshaw et al., *Designing with Smell: Practices, Techniques and Challenges* (New York: Routledge, 2018).
4. Rachel Herz, *The Scent of Desire: Discovering Our Enigmatic Sense of Smell* (New York: Harper, 2008); Avery Gilbert, *What the Nose Knows: The Science of Scent in Everyday Life* (New York: Crown Publishers, 2008); Michael Stoddart, *Adam's Nose, and the Making of Humankind* (London: Imperial College Press, 2015); Paolo Pelosi, *On the Scent: A Journey through the Science of Smell* (Oxford: Oxford University Press, 2016); Alexandra Horowitz, *Being a Dog: Following the Dog into a World of Smell* (New York: Scribner, 2016).
5. Carolyn Korsmeyer, *Making Sense of Taste: Food and Philosophy* (Ithaca: Cornell University Press, 1999); Barry C. Smith, ed., *Questions of Taste: The Philosophy of Wine* (Oxford: Oxford University Press, 2007); Raymond D. Boisvert and Lisa Heldke, *Philosophers at Table: On Food and Being Human* (London: Reaktion Books, 2016); Kevin W. Sweeney, *The Aesthetics of Food: The Philosophical Debate about What We Eat and Drink* (Lanham, MD: Rowman and Littlefield, 2017). In her recent book on historical preservation, Korsmeyer has explored important aspects of the third of the "lower" senses, touch. Carolyn Korsmeyer, *Things: In Touch with the Past*, (Oxford: Oxford University Press, 2019), 21–56.
6. Chantal Jaquet, *Philosophie de l'odorat* (Paris: Presses Universitaires de France, 2010). Useful articles on aesthetic issues in olfaction include Ann-Sophie Barwich, "Up the Nose of the Beholder? Aesthetic Perception in Olfaction as a Decision-Making Process," *New Ideas in Psychology* 47 (2017): 157–65; Emily Brady, "Sniffing and Savoring the Aesthetics of Smells and Tastes," in *The Aesthetics of Everyday Life*, ed.

Andrew Light and Jonathan M. Smith (New York: Columbia University Press, 2005), 177–93; Susan Feagin, "Olfaction and Space in the Theater," *British Journal of Aesthetics* 58, no. 2 (2018): 119–130; Frank Sibley, "Tastes, Smells, and Aesthetics," in *Approaches to Aesthetics: Collected Papers on Philosophical Aesthetics*, ed. J. Benson, B. Redfern, and J. Roxbee Cox (Oxford: Oxford University Press, 2001), 207–55; Marta Tafalla, "Anosmic Aesthetics," *Estetika: The Central European Journal of Aesthetics* 50:6, no. 1 (2013): 53–80.

PART I
WHAT CAN THE NOSE KNOW?

Overview

The Challenge of the Olfactory Arts

The scent consultant and artist Sissel Tolaas once collected sweat samples from men subject to anxiety attacks and encapsulated them in a special paint that she applied to the walls of an MIT gallery for her 2006 installation, *The Fear of Smell and the Smell of Fear*. As visitors moved around the room and touched the walls of the gallery, each odor was released in turn. Tolaas, who has a background in both chemistry and art and is on retainer to the International Fragrance and Flavor Corporation, has designed, among other things, a "Swedish" smell for IKEA stores and a "smell of death" for a German military museum. Her laboratory/studio in Berlin contains an archive of thousands of smell samples, ranging from dog poop and banana peels to old socks, exotic plants, and cigarette butts, along with many essential oils, synthetic fragrances, and the underarm pads she had her anxiety attack volunteers mail her. Tolaas's mission is to get people to appreciate their own body odors as well as the odors of their everyday environment, and her artworks are an expression of that campaign. For her there are no intrinsically "bad" smells, only a world full of thousands of complex and interesting scents waiting to be discovered. She is the John Cage of smell.

In 2012–2013 the New York Museum of Arts and Design's exhibition *The Art of Scent: 1889–2012* offered visitors an experience at the more pleasurable end of the odor spectrum. Twelve classic perfumes, each spritzed from a shallow indentation in the gallery walls, were accompanied by a title, date, and the creator's name, as if they were a series of paintings. The catalogue by the curator, Chandler Burr, was laced with tropes from art history, such as calling the perfume *L'Interdit* of 1957 "Abstract Expressionism" and *Eau d'Issy* of 1992 "Minimalism. By these presentational conceits, Burr hoped to convince visitors that perfumes are, as he told an interviewer, "actually works of art, beautiful and aesthetically important . . . equal . . . to painting, sculpture, music, architecture, and film."[1]

Works like *The Fear of Smell* and exhibitions like *The Art of Scent* raise two kinds of doubts in the minds of people who are given to thinking about art and aesthetics. First, are the works in these exhibitions "really" art, that is, are they *fine* art, or art with a capital "A," equal to painting, sculpture, and music?

Second, are things like learning to appreciate different kinds of odors through art installations or perfume exhibitions really worthy of being considered *aesthetic* experiences on a par with looking at paintings, listening to musical compositions, or standing in awe before a majestic waterfall?

Consider first the category of fine art, whose core of visual and aural arts would seem to exclude works derived from smells. From the late eighteenth century on, the fine arts were considered to be limited to works belonging to one of several canonical art forms such as poetry, painting, sculpture, architecture, classical music, and ballet. That list expanded by the mid-twentieth century to include certain kinds of photography, film, modern dance, and jazz. From the late 1950s on, the high-art list began to grow exponentially, as artists in all the fine arts experimented with new materials, new sounds, new movements, new technologies, new formats and contents. This opening of possibilities was particularly dramatic in the visual arts, as more and more artists turned to installation, performance, and participatory works that used any kind of found or constructed elements, from felt, fat, or teddy bears to sharks exhibited in formaldehyde. In this situation, it was inevitable that some artists would deliberately begin to use odors (although there had been isolated precedents in the early twentieth century). But once any gesture, activity, technique, or material, including odors, could be used to make works of "fine" or "major" art, the way was open for the elevation of numerous art forms once classified as "minor arts" to fine art status, whether those arts had been labeled design, decorative arts, crafts, entertainment, or recreation. The demand from the creators and appreciators of various "minor" arts to have them be recognized as "fine" arts has been broad, ranging from quilts to comics, cuisine to computer games, fashion to perfume. Not only have some of these efforts at elevation met with resistance by various critics, art theorists and philosophers, but most of the newly promoted art forms have still been discussed largely in terms of the traditional categories of form and content. But odors seem to present a problem when considered in these terms. How can we talk about form and content in works made of something that seems so formless, evanescent, and completely sensual as smells?

Similar questions can be raised about whether we can appropriately use the traditional concepts of aesthetic experience and judgment to interpret our response to artworks that focus on smells. Since the eighteenth century the aesthetic appreciation of art and the natural environment has often been guided by the conviction not only that the highest form of aesthetic experience and judgment, unlike mere sensory pleasure, must consider its objects "for themselves" or "disinterestedly," rather than as a means to something else, but also that our sensory response of pleasure or displeasure must have a cognitive or reflective component. Kant gave classic expression to this idea in his contrast between the reflective pleasures of taste, which involve a free play of the understanding and imagination, and the purely sensory pleasures he called the "agreeable." Judgments of beauty based on the pleasure of reflective taste could claim a form of universality, he argued, but disputes about the sensory pleasures of the agreeable would be pointless since there is no way to resolve disagreements over personal preferences.[2]

Although the concept of the aesthetic has gone through many permutations since Kant, and the concept of "disinterestedness" has fallen into disrepute, the demand that genuine aesthetic experience and judgment include a reflective or cognitive element has endured and would seem to be as difficult to apply to smell experiences as the traditional concept of fine art. Many people often have trouble identifying and naming smells, let alone describing and discussing them critically. The objects of vision or hearing, such as painting, sculpture, and music not only seem more objective and enduring, but easier to contemplate reflectively, "for themselves," than do objects of smell such as perfumes or artworks like *The Fear of Smell and the Smell of Fear*, which seem more likely to evoke purely sensory liking or disliking. In addition, the major fine arts have a long tradition of aesthetic criticism and theory behind them, whereas even the term "olfactory art" is of recent coinage. So it is only natural to ask: how can one exercise the mind critically in discussing something so vaporous, fleeting, and intangible as odors and putative "artworks" made of them? Contemporary olfactory artists like Tolaas or advocates of fine art status for perfumes like Burr are battling both a deeply engrained sense hierarchy and a tradition of aesthetic theory unfriendly to the so-called lower senses of taste and smell.

Chapter 1 will look more closely at some of the specific reasons often given for dismissing the use of odors and the sense of smell as vehicles of artistic creation and aesthetic appreciation. The first section, "The Historical Prejudice against Smell," briefly considers the history of prejudices against smell's aesthetic potential. The second and third sections, "The Case against Odors" and "The Case against the Sense of Smell," draw together in a more systematic way the most frequently made arguments against odors and the sense of smell and for each argument suggest counterarguments in favor of smell's cognitive capacities. The remaining chapters of Part I will then test those counterarguments against evidence from the contemporary neuroscience and psychology of olfaction.

But before launching into Chapter 1, let's look at some thoughts of a modern philosopher who unequivocally affirmed the philosophical importance of the senses and specifically of smell: Friedrich Nietzsche.

Notes

1. "Quoted in Barbara Pollack Scents & Sensibility," *ARTnews* 110, no. 3 (2011): 92. The catalog consisted of a short descriptive pamphlet and sample vials of each perfume. See Chandler Burr, *The Art of Scent: 1889–2012* (New York: Museum of Arts and Design, 2012).
2. Kant called both the taste of reflection and the taste of the agreeable "aesthetic" since both involve the senses, but subsequent philosophical usage has tended to treat the properly aesthetic as limited to reflective taste. Immanuel Kant, *Critique of Judgment*, trans. Werner S. Pluhar (Indianapolis: Hackett, 1987), 43–64.

Prelude
Nietzsche's Nose

> What magnificent instruments of observation we possess in our senses! This nose, for example, of which no philosopher has yet spoken of with reverence and gratitude.
>
> —Nietzsche, *Twilight of the Idols*

These words are part of Nietzsche's attack on the "concept-mummies"—philosophers like Kant, Hegel, and Schopenhauer—who claim that the body and the senses are deceptive.[1] Nietzsche turns the tables on them, arguing that even the sense of smell is a powerful instrument for "nosing out" the deceptions of those who ignore the body's role in thinking. Above all, for Nietzsche, the philosophical nose can detect anything that is likely to obstruct the path to strength and self-fulfillment. As he puts it in the *Genealogy of Morals,* every animal seeks to maximize its "feeling of power" and "abhors, just as instinctively and with a subtlety of sniffing out (*Witterung*) that is 'higher than all reason,' every kind of hindrance . . . on its path to power."[2] Among other things, the ability to "sniff out" that Nietzsche speaks of is an instinct for uncovering intellectual weakness and bad faith.

As Nietzsche wrote in a reappraisal of his first book, *The Birth of Tragedy*: "Whoever does not merely comprehend the word 'Dionysian' but comprehends *himself* in the word 'Dionysian' needs no refutation of Plato or Christianity or Schopenhauer—he *smells the decay*" (729). This thought leads Nietzsche to his most famous formulation of a personal ability to nose out deceit: "I was the first to *discover* the truth by being the first to experience lies as lies—*smelled*. . . . My genius lies in my nostrils" (782). One might think this only a rhetorical flourish, but it reflects Nietzsche's belief that smell is a model for the deep connection between thought and the body. Here, the "nose" stands for the body as a whole.[3] As he put it in an earlier passage of the *Genealogy of Morals*, "Whoever can smell not only with his nose but also with his eyes and ears, scents almost everywhere he goes today something like the air of madhouses and hospitals" (558).

For Nietzsche, philosophy was a way of life that required a complete integration of mind and body, thinking and sensing. Hence to go with him, one has to be willing to inhale the rarified air of the summits, "a *strong* air ... [P]hilosophy, as I have so far understood and lived it, means living voluntarily among ice and high mountains, seeking out everything strange and questionable in existence" (674). And what could be more strange and questionable for most Western intellectuals than the sense of smell? And so I place Nietzsche's reflections at the beginning of this exploration of the aesthetic powers of smell, the sense that many modern philosophers and intellectuals have either ignored or else denigrated as the lowest, most animalistic, and the least useful of the senses.

Notes

1. Friedrich Nietzsche, *The Portable Nietzsche*, trans. Walter Kaufmann (New York: Viking Press, 1954), 8. I have borrowed the title of Part I from Gilbert's *What the Nose Knows*.
2. Friedrich Nietzsche, *Werke in Drei Bänden* (Munich: Karl Hanser Verlag, 1960), 848. Walter Kaufmann, translates *Witterung* here with the Latinate term "discernment," which misses the fact that *Witterung* often refers to a dog or a person following a scent. See *Basic Writings of Nietzsche*, trans. Walter Kaufmann (New York: Modern Library, 1968), 543. The rest of the Nietzsche quotations are taken from Kaufmann's *Basic Writings of Nietzsche* with page numbers in parenthesis.
3. See Chantal Jaquet on the nose standing for the whole body in Nietzsche's thought. *Philosophie de l'odorat*, 410–25.

1
The Fear of Smell

If you have ever literally stopped to "smell the roses," as the old saying goes, you have probably discovered that most roses barely smell. That's because the breeding of commercial and show roses since the nineteenth century has been almost entirely focused on visual appearance and durability. Of course, it was not always so, as Shakespeare's Sonnet 54 reminds us,

> The rose looks fair, but fairer we it deem
> For that sweet odour which doth in it live.

Although there has recently been a revival of interest in planting roses with strong scents, the prevalence of the beautiful-to-look-at but almost odorless roses in our flower shops and many public gardens is just one index of the vision-centric bias that has dominated Western culture and especially philosophy with respect to art and aesthetics. As Carolyn Korsmeyer remarks: "In virtually all analyses of the senses in Western philosophy the distance between object and perceiver has been seen as a cognitive, moral and aesthetic advantage."[1] By contrast, the "proximal" senses of smell, taste, and touch have been held to be inferior precisely because they put us in physical contact with things. Let's briefly look at that tradition, which goes all the way back to Plato.

The Historical Prejudice against Smell

Although neither Plato nor Aristotle was as dismissive of the human sense of smell as some modern philosophers, they were hardly positive regarding smell's intellectual capacity or its aesthetic potential. In the *Timaeus*, for example, Plato embraced a view of the relation of language and smell that has bedeviled Western thought ever since: odors "lack names," he says, "because they do not consist of a definite number of simple types. The only clear distinction to be drawn here is twofold: the pleasant and the unpleasant."[2] As for the relation of smell to what we call aesthetic experience, in *Hippias Major* Plato claims that beauty comes only through the senses of hearing and sight, whereas "Everyone would laugh at us if we said . . . a pleasant smell is beautiful."[3]

Aristotle placed vision at the top of the sense hierarchy, but he did put smell in a middle position between the distance senses of vision and hearing and the proximal senses of touch and taste.[4] Like Plato he noted the lack of a special vocabulary for odors, claiming that smells are mostly named after tastes and then always accompanied by a pleasant/unpleasant designation. But Aristotle also claimed that humans actually have two kinds of smell, one of which he could have used to open the way to a positive aesthetics of smell. The first kind of smell is shared with other animals and is closely connected to taste, with the result that its pleasures and displeasures are "accidental," that is, they are tied to whether we are hungry or satiated, something modern researchers have also noted: when we are hungry a fruit or a steak will smell wonderful, but if we have eaten our fill, its smell no longer attracts, but may even repel. The second kind of smell for Aristotle is characteristic only of humans, and its objects are perceived as pleasant or unpleasant "in themselves," such as the smell of flowers.[5] As Chantal Jaquet observes, this second perspective could have opened the way toward some positive aesthetic reflections on smell, yet it did not in either Aristotle or his successors.[6] Moreover, Aristotle not only took a dim view of the olfactory art of perfume (see Chapter 13), but when it came to the arts of imitation in the *Poetics*, he discusses only the pleasures of vision and hearing, emphasizing that dramas and paintings please us by *how* the world is represented, something presumably smell and taste could not do.[7] The arts of imitation, in his view, are directed at the intellectual pleasures of vision and hearing and have a moral dimension, unlike the arts of perfumery or cooking, which are directed to the bodily senses.[8]

In medieval philosophy, Aquinas continued the tradition established by Plato, saying that we may "speak of beautiful sights and beautiful sounds, but not of beautiful tastes and smells."[9] And a low regard for smell and taste among the senses (touch fared a bit better) and their aesthetic potential continued on into early modern philosophy. (A notable exception was Spinoza, who wrote in the seventeenth century: "The wise man renews and refreshes himself with moderate food and drink, and also with scents, the beauty of plants in bloom, dress, music sports, theater.")[10] Most other philosophers had little appreciation for smell and taste. One idea that rationalists like Descartes and empiricists like Locke agreed on was that taste and smell give us only a subjective knowledge of "secondary qualities."[11] Even many of the Enlightenment philosophers of the eighteenth century who generally championed the senses as the basis of knowledge did not rank smell highly. Condillac's famous thought experiment, in which he imagined a statue awakening to life one sense at a time, made smell the first sense to awaken *because* it was generally regarded as the weakest of the senses, and in the end Condillac, although he did not disparage smell, still gave vision pride of place.[12] As for a possible role for smell in the fine arts, a typical view is that of Henry

Home, Lord Kames, who wrote: "The fine arts are designed to give pleasure to the eye and ear, disregarding the inferior senses."[13]

At the end of the eighteenth century, Kant, who so influentially articulated the new concept of the aesthetic as a reflective sentiment, in contrast to mere sensory satisfaction, said that smell and taste, unlike vision and hearing, make us primarily aware of our own bodily states rather than of their objects. As he put it in his *Lectures on Anthropology*, "If I see, then I attend to an object, but if I smell and taste, then I pay attention to . . . how my body is affected."[14] Accordingly, there is no place in Kantian reflective aesthetics for smell (or taste and touch), only for vision and hearing. Although the formal (visual) aspects of a flower constitute what Kant calls "free beauty," laying claim to universal assent, its pleasant smell "gives it no claim whatever: its smell delights one person, it makes another dizzy."[15] Hegel also excluded the "lower" senses from the genuine appreciation of art: "The sensuous aspect of art is related only to the theoretical senses of sight and hearing," whereas smell and taste are "excluded from the enjoyment of art."[16] Despite the higher value placed on smell by some materialist philosophers like Feuerbach or Nietzsche, the Kantian and Hegelian dismissal of smell dominated subsequent philosophy. Thus, the general contrast between the higher "theoretical" (or properly "aesthetic") senses of vision and hearing and the lower "bodily" senses of smell and taste can still be found from George Santayana and Edward Bullough at the beginning of the twentieth century to Roger Scruton, Dennis Dutton, Allen Carlson, Glenn Parsons, Jane Forsey, and other philosophers in the first decades of this century.[17] Moreover many of these and other traditional claims put forward to justify the depreciation of the artistic and aesthetic potential of odors and the sense of smell are still circulating in the culture at large. We need to turn now to the most important of these enduring negative claims as well as offer some contrary arguments in favor of the aesthetic and artistic potential of smell.

The Case against Odors

Let's begin by considering two classic objections to odors as a possible medium for artistic creation and potential objects of aesthetic interest, namely, that odors are *ephemeral* and lack *structure*, so that they cannot be ordered in a way that would make them bearers of meaning. Hegel, for example, dismissed odors as the product of whatever "is in the process of wasting away."[18] A little over a century later Monroe Beardsley claimed that tastes and smells cannot be arranged "in series, and so we cannot work out construction principles to make larger works out of them." Suppose, he goes on, you were trying to construct a scent-organ with keys by which perfume or brandy, or the aroma of new-mown hay or pumpkin pie would be wafted into the air. On what principle would you arrange the keys, as the keys of the piano are arranged by ascending pitch? How would you begin to look for systematic, repeatable, regular combinations that would be harmonious and enjoyable as complexes?[19]

For Beardsley there is simply not enough order within the sensory fields of taste and smell "to construct objects with balance, climax, development, or pattern," which explains why, he says, there have been no "taste-symphonies and smell-sonatas."[20]

Ironically, one of the more interesting olfactory artworks created early in this century is the *Olfactiano*, a scent piano of twenty-six keys on which its inventor, Peter de Cupere, played a *Scentsonata for Brussels* in 2004. Whether de Cupere's scent piano and sonata, or whether other more complex scent organs and musical works that have appeared since then, truly merit the appellation "fine art" is a topic for a later chapter. But at this point we can say that the existence of such works at least reopens the question of odors as a medium for art and of the sense of smell as a basis for aesthetic reflection in a way that Beardsley could not have anticipated. He was mostly writing before the postmedium turn of the art world began to move conceptual, installation, and performance works toward the important role they play today. Yet there are other philosophers who have more recently made claims similar to Beardsley's. Roger Scruton, for example, writes that smells "mingle, losing their character" and "remain free floating and unrelated, unable to generate expectation, tension, harmony, suspension or release."[21]

What can we say in reply to Beardsley and Scruton, apart from pointing out that many art museums have presented works of olfactory art and some of those works such as *Green Aria: A Scent Opera* arguably generate "expectation, tension, harmony"? In the first place, as Frank Sibley has pointed out, the volatility and evanescence objection to odors as an art medium runs counter to our aesthetic experience not only with other widely accepted aesthetic objects (storms at sea, birdsongs) but with many now well-established fine art forms, such as improvisational music and performance art.[22]

Second, the lack-of-structure objection, even without appealing to the de facto argument from the existence of works like *Green Aria*, overlooks the fact that serious students of perfumes and wines have always made considered aesthetic judgments based on taste sequences and odor structure (see Chapter 11). The lack-of-structure argument probably gets its initial plausibility from the fact that most people have little experience or training in analyzing complex odor constructions, whereas nearly everyone has some experience in appreciating the structure of works of visual and sound arts.

The Case against the Sense of Smell

When we turn from arguments against the artistic potential of odors to arguments against the aesthetic potential of the human sense of smell, the Western intellectual tradition has for the most part treated the sense of smell as *disreputable, deficient, deceptive,* and *dispensable*, hardly promising characteristics as a basis for developing an olfactory aesthetics.

Smell Is Disreputable

One argument (though perhaps more prejudice than argument) for the low status and neglect of the sense of smell in philosophy is smell's association with "animality" and the body on the one hand, and on the other, the closely related repugnance for the smell of "the Other," whether other races, ethnicities, or the lowest social classes. It is easy to understand the widespread association of smell with the less admirable qualities of the animal kingdom. Dogs obviously spend a great deal of their time sniffing the ground and each other's rear ends. And it is surely true that not only dogs, but many other mammals, including rats, as well as many unappealing insects crucially depend on their power of smell for survival. The most striking of these animal powers are related to pheromones, as anyone knows who has watched a trap-bag primed with Japanese beetle pheromone fill to the breaking point on a summer day. Such associations with dogs, rats, and insects certainly seem to lessen the dignity of smell. But as Aristotle noted, unlike other animals whose noses seem oriented entirely to food and sex, humans smell flowers just for pleasure.

Closely related to the association of smell with the animal side of our nature is repugnance at the smell of "the Other." As Hobbes observed long ago, there seems to be a general tendency to regard one's own bodily smell as pleasant, whereas the same kind of smell in others displeases.[23] Kant treated the sense of smell as disreputable because smell's connection with breathing means we are forced to inhale the odor of others—we literally ingest their smell.[24]

Because many East Asians have fewer apocrine glands and less body hair than Caucasians or Africans, they have much lighter body odor and have sometimes found the strong smell of Europeans off-putting. Of course, the fact that both Europeans and Africans have roughly the same number of apocrine glands has not impeded the racist belief among many whites that Africans have an inherently bad smell. Moreover, although there is a popular tendency to think of race primarily in visual terms (white, black, yellow, red), as the historian Mark M. Smith has shown in *How Race Is Made*, the tenacious myth that African Americans have an offensively strong body odor has been important in upholding whites' sense of superiority.[25] The offensive "smell" often attributed to recent immigrants to Europe and North America is often related to cultural differences in cuisine, since people who eat a lot of garlic, curry, or other strong spices will excrete these odors in their sweat. As for the class issue, almost every discussion of smell and class quotes George Orwell's comment that class distinction in the West can be summed up in four words, "*The lower classes smell.*"[26] But, since the time Orwell wrote this in 1937, improvements in housing and hygiene among the working classes in many developed countries have reached the point that it is only the poorest of the poor and the homeless who stink of heavy body odor.[27]

In a famous set of essays, *Dialectic of Enlightenment: Philosophical Fragments*, Max Horkheimer and Theodor Adorno joined together the Kantian themes of subjectivism and the fear of the other's odor invading one with the tropes of animality, race, and class:

> When we see we remain who we are, when we smell we are absorbed entirely. In civilization, therefore, smell is regarded as a disgrace, a sign of the lower social orders, lesser races, and baser animals.[28]

Horkheimer and Adorno say this in a late chapter titled "Elements of Anti-Semitism," and well they might, since a leading motif of Fascist propaganda during World War II was the "stinking Jew," a motif intended to persuade people it was necessary to eradicate such dangerous vermin. Since the irrational fear of the smell of others is largely immune to reasoning. we must pass on to three other negative characteristic of smell that are more amenable to philosophical discussion.

Smell Is Deficient

As William Lycan remarks, most modern philosophers who have written on perception have thought of vision "as perception itself, other sense modalities being conceived as vastly inferior."[29] And when philosophers have considered the cognitive powers of senses other than vision, they have acknowledged the informational acuity of hearing and the indispensability of touch, but taste and smell have often been considered so cognitively *deficient* that they could not contribute to informed aesthetic judgments. Yet even if we admit that smell and taste are inferior to vision, hearing, and touch as sources of general information, that does not diminish the importance of the kinds of information and the aesthetic pleasures they do provide, nor does it justify their past disparagement and neglect. Even so, many of those who regard smell as deficient also claim, as Aristotle did, that whatever information it offers us, the human sense of smell is extremely poor when compared to other animals. As we will see in the next chapter, recent scientific studies have given the lie to the supposedly total inferiority of human smell compared to other animals.

But the respect in which the sense of smell has been considered to be most cognitively deficient by many philosophers, as we have seen in the case of Kant, is that smell *deceives* us into thinking we are in contact with the real world when all we experience when we smell are our own sensations. Although the wider background of the subjectivity charge is the traditional depreciation the body in Western religion and philosophy that Nietzsche inveighed against, the controversial status of smell in the contemporary field of the philosophy of perception is sufficiently important for determining smell's place in art and aesthetics to deserve a separate discussion of smell's apparent deceptiveness.

Smell Is Deceptive

In contemporary philosophy of perception, the Kantian doctrine of the subjectivity of olfactory perception has often been justified by applying a vision-based model to smell. A vision-based model of perception stresses that we have a phenomenal experience of "seeing through" to individual objects located at a distance and at a particular place in a spatial field, and these individual objects are normally perceived as having edges and standing out as figures against a background. Since smell perception lacks most of these features, some philosophers like Scruton consider it uninformative about the external world and draw aesthetic consequences from this. As Scruton put it in an early work,

> Not every "sense" lends itself to aesthetic pleasure. . . . Visual experience is so essentially cognitive, so "opened out," as it were, on to the objective world, that our attention passes through and seizes on its object. . . . In tasting and smelling I contemplate not the object but the experience derived from it.[30]

The smell of a cushion, he says more recently, can exist without the cushion since smells "linger in the places where their causes have departed . . . I don't 'sniff through' the smell to the thing that smells, for the thing is not represented in its smell."[31]

During the last two decades, several philosophers of perception have challenged the Kant/Scruton position, arguing that the typical vision-based model of perception distorts both smell and taste. Louise Richardson, for example, stresses that our experience of an odor refers to something outside the body insofar as the simple act of breathing brings odor-carrying air "*into* the nostrils, *from without*" even though the air is not represented "as being at a distance," or as located at a specific point in space.³² Similarly, Clare Batty writes that "although olfactory experience does not 'pin' properties onto any particular thing," we experience odor properties as though "we are coming into contact with something external to us."³³ William Lycan has long criticized the claim that smell is only "a modification of our consciousness," and has recently proposed a two-level, "layering" account of smell that ties it to its sources. The first and basic level (similar to Richardson's and Batty's views) is that we experience odors as "clouds of molecules," reaching our olfactory bulb through the air; the second level connects us with odor sources "*by* representing" odors at the first level. "By smelling a certain familiar odor, I also smell—veridically or not—an actual rose or roasting lamb or my least favorite aunt." At this second stage, as Lycan points out, we could be objectively wrong in ascribing the rose odor that we smell to an actual rose if there were no rose nearby, or we could be wrong if there were an actual rose present but it was an odorless rose, or wrong if someone had sprayed an artificial rose smell in our vicinity. Thus, a smell experience can have two intentional objects at once, arranged hierarchically: in the case of "experiencing the rose smell in the absence of any rose—I am representing *both correctly and incorrectly*, the odor correctly and roses incorrectly."³⁴ Batty does not accept Lycan's idea that smell includes a second level that allows our smell experience also to refer (truthfully or mistakenly) to the odor's object-source. Instead, she insists that our sense of smell, *by itself*, can only perceive an odor as a kind of "smudge" on perception, vaguely located "here" around us, but not as emanating from a particular source, a position also embraced by Andreas Keller.³⁵ In order to locate a smell, she believes, we have to move our bodies in some way and/or use one of our other sense modalities like vision or touch if we are to establish whether the rose odor we smell is coming from an actual rose.³⁶ Yet even in Batty's account of smell perception as only able to locate odors vaguely "somewhere" around us, the experience of smell is not purely subjective, telling us nothing about the world outside, as Kant and Scruton claim.

The differences in perspective of Richardson, Lycan, Batty, and Keller on smell perception need not detain us further. What they agree on should be enough to show that the Kantian tradition according to which the sense of smell is deceptive because we *think* we are in contact with the world, but we are *really* just experiencing our own sensations, is hardly the last word on the cognitive powers of smell. I have taken us through these contemporary arguments against the Kantian tradition because the subjectivist understanding of smell is still alive in philosophical aesthetics, as we have seen in the case of Roger Scruton, who denies that smell and taste have sufficient cognitive resources to afford genuine aesthetic experience and sustain aesthetic judgments. If Lycan, Batty, and Richardson are right, Scruton's line of thinking about the aesthetic potential of odors and the sense of smell is based on a questionable philosophical claim.

Moreover, Scruton's specific application of that claim to aesthetics seems to me intuitively wrongheaded. I do not see why a person cannot objectively contemplate a lemon odor qua odor percept and compare it to the odor percept of strawberry, rather than simply contemplating their subjective experiences. Moreover, vision itself operates subjectively at times. One might say of lemon-yellowness while looking at a lemon in a Claesz still life ("it looks so real") rather than attending objectively to the color lemon-yellow within the painting as it relates to other colors—as an art critic or art historian would have us do. *Both* smell and vision seem capable of engaging the cognitive or "objective" component of aesthetic experience, just as both can be marshaled in a "subjective" or "interested" fashion, as the well-known fallibility of "eyewitness" testimony also shows in the case of vision.[37]

But Scruton adds another twist to his claim that smells tell us only about our subjective experience. He grants that we may sometimes *associate* meanings with odors, but he claims that in the case of visual or auditory objects like paintings or musical works we perceive the meanings as residing "*in* the object," whereas in the case of smell we are aware primarily of "what the object calls to mind." Although he grants that sounds, like smells, are also separate from the objects that emit them, he falls back on the claim that smells cannot be organized into structured works, as sounds can be in music. At most, he thinks, smells can have only the kind of "marginal aesthetic interest" we grant to "the sound of fountains where beauty is a matter of association" rather than an expression of meaning. But Scruton is wary of even granting odors this minimal aesthetic interest, preferring to "insist on the radical distinction" between objects of sight or hearing, "whose meaning can be directly seen and heard," and objects of smell that are really "objects of sensory enjoyment which acquire meaning only by the association of ideas."[38]

Here again, as in the case of perceiving the lemon-ness of the odor of a lemon, it seems to me that it may be possible to find meanings *in* certain perfumes and works of olfactory art, not just by associating them with ideas from other contexts. In the case of complex perfumes, for example, one may appreciate the way the various notes complement each other and unfold over time without necessarily associating them with any external theme or context. Naturally, in order to discover the meanings *in* the perfume or other work of olfactory art, we may need some knowledge of how such works are created, as well as some experience in appreciating them. Both Kevin Sweeney and Barry C. Smith have made similar arguments for the claim that taste qualities are *in* wine, not just in the perception of the taster.[39]

Smell Is Dispensable

Finally, smell has been considered dispensable, not only because of its supposed deficiency, deceptiveness, and disreputableness, but also because it is claimed to have only a few narrow uses and that humans have evolved beyond the animal condition of needing smell for foraging and reproduction. Kant claimed that smell is well-nigh worthless except for detecting rotten food or dangerous gases; it is the most dispensable of the senses, and it does not repay us to "cultivate and refine" it.[40] Similarly, Darwin believed that although the sense of smell is crucial for other animals to find food and warn them of predators, for humans, smell "is of extremely slight service, if any, even for savages in whom it is generally more highly developed than in the civilized races."[41] Freud also viewed smell as largely atrophied in humans and speculated that the decline in the importance of smell began as a result of our hominid ancestors adopting an upright position.[42] The general idea that the human sense of smell is of radically diminished power and utility lasted throughout most of the twentieth century. In the 1990s, the renowned developmental psychologist Howard Gardner was still claiming that smell is of "little special value across cultures."[43] Given this tradition of viewing smell as primitive and dispensable, it is no wonder that the AMA guidelines for impairment compensation by state courts and insurance companies consider the value of the total loss of smell to be worth 1%–5% compared to 85%–100% for vision.

Yet as those who have suddenly lost their sense of smell through an accident or illness know so well, the effect can be devastating. A world without scent is a world that has shrunken and lost its savor. Not only will flowers and woodlands have lost their fragrance but foods will have lost much of their flavor (which largely comes from smell), and one's spouse and children will have lost part of what unconsciously contributes to intimacy. It is no wonder that many people who have lost their sense of smell experience major bouts of depression.[44] A world without smell would lack much that gives everyday life some its deepest aesthetic satisfactions.

But a skeptic might object at this point that people *born* without a sense of smell seem to get on in life with little difficulty, and unless they happen to tell us they are anosmic, we might never know it. As for aesthetic experience, congenital anosmics would not miss the aesthetic satisfaction of smelling flowers and woodlands or other olfactory pleasures since they have never had such pleasures, whereas they can fully enjoy the vast majority of the human arts, which are primarily addressed to vision and hearing. Is the successful life of anosmics an argument against pursuing an olfactory aesthetics?

Fortunately, an anosmic philosopher, Marta Tafalla, has tackled this very question. Tafalla would agree with the skeptic that anosmics can for the most part lead a normal life, and that few who meet anosmics would guess they lack a sense of smell. Even the parents of many anosmic children seldom realize their child cannot smell until it reaches eight or ten years. Tafalla reports that in her own case every doctor who subsequently examined her told her that lacking a sense of smell would be of no importance to her life. It was only as a mature philosopher that she decided to confront the tradition going back to Kant that smell cannot be involved in genuine aesthetic experience and to examine whether anosmia results in an aesthetic deficit.[45]

In her essay "Anosmic Aesthetics" she argues that anosmics do experience serious restrictions on their aesthetic appreciation in many areas of life such as the experience of nature and not just the experience of olfactory artworks. Tafalla describes a walk with friends through a pine forest near Barcelona, during which not only could she not appreciate its scents as her companions could, but she was especially surprised when they became excited at noticing the smell of the sea long before it came into view. For her friends, she writes, what they could smell of the sea made the experience of the forest walk more aesthetically interesting, but Tafalla herself was aware of no difference. On another woodland walk she and her companions came across the carcass of a recently dead fox. Although Tafalla was fascinated at the fox's appearance, her companions experienced visceral revulsion at the stench and hurried away holding their noses. "For me the dead fox was a sad encounter, but not unpleasant, and it did not spoil the beauty of the place."[46] Although her friends could become excited by the smell of the distant sea or react instinctively to the smell of a putrefying carcass, she could neither "*perceive* the sea as near" nor "*feel*" revulsion at the smell of death. This shows, she writes, that anosmics' aesthetic appreciation of nature *is* diminished by their lack of smell since they "have no access to a certain kind of experience."[47] After exploring similar examples related to food, body odor, and gardens, Tafalla concludes that the absence of a sense of smell makes a profound difference to one's aesthetic experience.[48] Philosophers who dismiss the sense of smell as irrelevant to aesthetics, she suggests, are in effect adopting "voluntary anosmia."[49]

Although this chapter has challenged several of the main objections to the aesthetic and artistic importance of smell, none of my brief replies have been meant to suggest that smell has no aesthetic limitations in comparison to vision and hearing or that it equals them in cognitive power. Moreover, Beardsley's question about why there have been no taste symphonies or smell sonatas cannot be dismissed on the basis of a few examples when it comes to questions of aesthetic complexity and quality. Even Frank Sibley, one of the rare twentieth-century philosophers to develop an extended defense of including tastes and smells within the realm of the aesthetic, nevertheless had a low estimate of the aesthetic *value* of the artistic expressions of smells in perfume, wine, cuisine, and so on. His strategy in defense of including smell and taste within aesthetics against writers like Scruton was that on a series of aspects, such as intelligibility, contemplation, and qualitative description, the differences between the senses of taste and smell on the one hand, and vision and hearing on the other, are a matter of degree rather than kind. Yet despite Sibley's insistence that tastes and smells do belong within the general realm of the aesthetic, they remained for him of only minor aesthetic interest, and his summary of his defense of their inclusion within aesthetics sounds backhanded. "Even if their greatest values are still only slight and trivial," he writes, "it does not follow that they are never worth bothering with, or that they are not minimal *aesthetic* values on the lower end of a continuum."[50] Dennis Dutton is equally dismissive of the artistic and aesthetic possibilities of smell, claiming that smell's lack of the ability to give structured expression to the more serious emotions "seems to count decisively against smell as the medium for a self-subsistent art form that might someday stand with music, painting, drama, and literature."[51]

Parts III and IV of this book will show that the aesthetic value of many instances of the olfactory arts, including their expressive power, is far from trivial or minimal, but for now there is more work to do in securing the legitimacy of olfactory art and its aesthetic appreciation. Although I believe the arguments put forward in this chapter have shown that odors and the sense of smell are not as disreputable, deficient, deceptive, and ultimately dispensable, as some philosophers, scientists, and other intellectuals have claimed, we need to see if the kind of intuition-based and conceptual arguments I have offered can survive scrutiny in the light of contemporary neuroscience and psychology.

Before embarking on that task we should note that philosophers are divided on the issue of whether the natural and social sciences can illuminate conceptual issues in the field of aesthetics. Conversely, neuroscientists and psychologists are divided on the extent to which philosophical analysis can illuminate the empirical study of aesthetic experience. At the extremes, some neuroscientists have suggested that "neurosaesthetics" will simply replace philosophical aesthetics, whereas some philosophers have regarded empirical work as largely irrelevant to philosophy.[52] Most contemporary philosophers and neuroscientists adopt a more moderate approach. As the editors of a recent volume devoted to the issue summarize the middle ground, "Perhaps it is the responsibility of the philosophical theory builder to ensure that what is being claimed is consonant with or even supported by what the best current science tells us."[53] I agree. Not only do our reflections need to be consistent with the best current findings in the natural and social sciences, but research in the sciences, I believe, can greatly enrich aesthetic theory and analysis.

A recent example of such enrichment is Murray Smith's *Film, Art and the Third Culture*, in which he develops "a naturalized aesthetics of film."[54] Of particular interest for our purposes is Smith's method of "triangulation," which involves a transaction among three levels of evidence: the *phenomenological* or experiential (what mental acts "feel like"), the *psychological* (the capacities of the mind), and the *neuroscientific* (the brain networks and operations underlying mental capacities). No single level is adequate by itself, in Smith's view, since triangulation "occurs across different types of evidence and levels of inquiry."[55] Nor do the three levels form an epistemological hierarchy such that all reduce to neuroscience. Although brain and body are *ontologically* fundamental, in epistemological terms, Smith believes aesthetic theorists should seek a "reflective equilibrium" among the three levels of analysis.[56]

In general, the neuroscientific and psychological evidence on olfaction in the next two chapters not only will help us better understand the cognitive strengths and weakness of the sense of smell, but will also provide us with a baseline of insights useful for the remainder of the book. Yet before we plunge into brain anatomy and the experiments of psychology laboratories in Chapter 2, a brief literary interlude will better prepare us for the issues to come.

Notes

1. Korsmeyer, *Making Sense of Taste*, 12.
2. Plato, *Plato's Cosmology: The Timaeus of Plato Translated with a Running Commentary*, trans. Francis Macdonald Cornford (New York: Liberal Arts Press, 1957), 273. Yet in *Philebus* (51a–e) Plato suggests that the pleasures of singular odors, although "less sublime," are similar to the unmixed pleasures of abstract shapes, pure colors, and sounds. Plato, *The Collected Dialogues of Plato*, ed. Edith Hamilton and Huntington Cairns(Princeton: Princeton University Press, 1961), 1132–33.
3. Plato, *Collected Dialogues of Plato*, 1552.
4. Aristotle, *Nichomachean Ethics*, trans. David Ross, Reviewed by J. L. Ackrill and J. O. Urmson. (Oxford: Oxford University Press, 1998), 259.
5. For a discussion of these issues see David Ross's commentary on Aristotle, *Parva Naturalia* (Oxford: Clarendon Press, 1955), 209–17. In *De Anima* 421a.6–422a8 Aristotle complains that the human sense of smell is hard to define partly because it is so inferior to that of other animals. *Basic Works of Aristotle*, ed. Richard McKeon (New York: Random House, 1968), 573–74. See also Thomas K. Johansen, "Aristotle on the Sense of Smell," *Phronesis* 41 (1996): 6–7.
6. Jaquet, *Philosophie de l'odorat*, 44–45.
7. Kevin Sweeney makes this point in his chapter on Aristotle in *The Aesthetics of Food*, 49–50.
8. Aristotle, *Nichomachean Ethics*, 186. As David Summers points out, Aristotle even considered painting to be only a little higher than perfumery or cooking. *The Judgment of Sense: Renaissance Naturalism and the Rise of Aesthetics* (Cambridge: Cambridge University Press, 1987), 62. See also Korsmeyer, *Making Sense of Taste*, 24.

9. Donald McQueen, "Aquinas on the Aesthetic Relevance of Tastes and Smells," *British Journal of Aesthetics* 33, no. 4 (1993): 349. See also Umberto Eco, *The Aesthetics of Thomas Aquinas* (Cambridge: Harvard University Press, 1988), 57–8.
10. Spinoza, *Ethics*, trans. G. H. R. Parkinson (Oxford: Oxford University Press, 2000), 260. See Jaquet's on Spinoza in *Philosophie de l'odorat*, 76–8
11. The empiricist Thomas Reid at least gave the sense of smell a prominent place in his treatise on the senses. See Sweeney, *The Aesthetics of Food*, 108–10. See also Jake Quilty-Dunn, "Reid on Olfaction and Secondary Qualities," *Frontiers in Psychology* 4, no. 974 (2013), https:// doi.org/10.3389/fpsyg.2013.00974.
12. Étienne Bonnot Abbé de Condillac, *Philosophical Writings of Étienne Bonnot, Abbé de Condillac*, vol. I1, trans. Franklin Philips and Harlan Lane (Hillsdale NJ: Lawrence Erlbaum, 1982). There are, of course, differences throughout the tradition on how far the lower senses are from the higher, why they are lower, and on the ranking among the lower senses. For additional discussions of eighteenth-century thinkers see Jaquet, *Philosophie de l'odorat*, and Annick Le Guérer, *Scent: The Mysterious and Essential Powers of Smell*, trans. Richard Miller (New York: Random House, 1992), 164–78.
13. Henry Home, Lord Kames, *The Elements of Criticism*, vol. 1 (Hildesheim: Georg Olms Verlag, 1970), 6–7. I am grateful to Carolyn Korsmeyer for this reference.
14. Immanuel Kant, *Lectures on Anthropology*, trans. Robert R. Clewis, Robert B. Louden, G. Felicitas Munzel, and Allen W. Wood (Cambridge: Cambridge University Press, 2012), 67.
15. Kant, *Critique of Judgment*, 145.
16. G. W. F. Hegel, *Aesthetics: Lectures on Fine Art*, vol. 1, trans. T. M. Knox (Oxford: Clarendon Press, 1975), 38.
17. George Santayana, *The Sense of Beauty* (New York: Random House, 1955 [1896]), 68–69. Edward Bullough, "Psychical Distance as a Factor in Art and an Aesthetic Principle," in Stephen David Ross, *Art and Its Significance* (Albany: State University of New York Press, 1999), 465; Roger Scruton, *I Drink Therefore I Am* (London: Continuum, 2009), 122. Dennis Dutton, *The Art Instinct: Beauty, Pleasure and Human Evolution* (New York: Bloomsbury, 2009), 207–12. Glenn Parsons and Allen Carlson, *Functional Beauty* (Oxford: Oxford University Press, 2008), 167–95. Christopher Dowling, "The Aesthetics of Daily Life," *British Journal of Aesthetics* 50, no. 3 (2010): 226–38. Jane Forsey, *The Aesthetics of Design* (Oxford: Oxford University Press, 2013), 209–10.
18. Hegel, *Aesthetics*, 38.
19. Monroe Beardsley, *Aesthetics: Problems in the Philosophy of Criticism* (New York: Harcourt, Brace & World, 1958), 99
20. Beardsley, *Aesthetics*, 99.
21. Roger Scruton, *I Drink Therefore I Am*, 162. Dutton uses a similar argument in *The Art Instinct*, 207.
22. Frank Sibley, "Tastes, Smells and Aesthetics," in Sibley, *Approaches to Aesthetics: Collected Papers on Philosophical Aesthetics*, ed. John Benson, H. B. Redfern, and Jeremy Roxbee Cox (Oxford: Oxford University Press, 2006), 228.

23. Thomas Hobbes, *The Elements of Law Natural and Politic: Part I Human Nature* (Oxford: Oxford University Press, 1994), 46.
24. Kant, *Lectures on Anthropology*, 372, 433.
25. Mark M. Smith, *How Race Is Made: Slavery, Segregation and the Senses* (Chapel Hill: University of North Carolina Press, 2006).
26. George Orwell, *The Road to Wiggan Pier* (London: Victor Gollancz, 1937), 159.
27. Many libraries now exclude homeless people whose body odor is deemed offensive.
28. Max Horkheimer and Theodore W. Adorno, *Dialectic of Enlightenment*, trans. John Cumming (New York: Continuum, 1987), 184.
29. William G. Lycan, "The Slighting of Smell," in *Of Minds and Molecules: New Philosophical Perspective on Chemistry*, ed. N. Bhushan and S. Rosenfeld (Oxford: Oxford University Press, 2000), 274. Although Bence Nanay does not discuss smell in his recent *Aesthetics as Philosophy of Perception* (Oxford: Oxford University Press, 2016), he makes a strong case for drawing on contemporary theories of perception for doing aesthetics.
30. Roger Scruton, *The Aesthetics of Architecture* (Princeton: Princeton University Press, 1979), 114.
31. Scruton, *I Drink Therefore I Am*, 121.
32. Louise Richardson, "Sniffing and Smelling," *Philosophical Studies* 162 (2013): 401–19.
33. Clare Batty, "Smelling Lessons," *Philosophical Studies* 153 (2011): 169.
34. William Lycan, "The Intentionality of Smell," *Frontiers in Psychology* 5 (2014): 436–37, https://doi.org/10.3389/fpsyg.2014.00436.
35. Andreas Keller, *Philosophy of Olfactory Perception* (New York: Palgrave Macmillan, 2016). I discuss Keller's ideas in Chapter 3.
36. Clare Batty, "A Representational Account of Olfactory Experience," *Canadian Journal of Philosophy* 40 (2010): 511–38.
37. Frank Sibley makes a similar point in "Tastes, Smells, and Aesthetics," 224–25.
38. Scruton, *I Drink Therefore I Am,* 122.
39. Sweeney, *Aesthetics of Food*, 172–80; Barry C. Smith, "The Objectivity of Tastes and Tasting," in *Questions of Taste: The Philosophy of Wine*, ed. Barry C. Smith (Oxford: Oxford University Press, 2007.
40. Kant, *Lectures on Anthropology*, 67–70.
41. Charles Darwin, *The Descent of Man, and Selection in Relation to Sex* (Princeton: Princeton University Press, 1981), 24.
42. Sigmund Freud, *Civilization and its Discontents*, trans. James Strachey (New York: W. W. Norton, 1961), 46, 53. David Howes speculates that Freud's dismissal of smell's importance (even omitting the nose from his list of erotogenic zones of the body) may be a reaction to the fact one of Freud's patients nearly died from a botched nose operation by his one-time close associate Wilhelm Fliess, who thought the nose and sense of smell were a key to understanding neurosis. See David Howes, "Futures of Scents Past," in *Smell and History: A Reader*, ed. Mark M. Smith (Morgantown: University of West Virginia Press, 206.
43. Howard Gardner, *Frames of Mind: The Theory of Multiple Intelligences* (New York: Basic Books, 1993), 61.

44. Herz, *Scent of Desire*, 8–15.
45. Marta Tafalla, "A World without the Olfactory Dimension," *Anatomical Record* 2 (2013): 1287–96.
46. Marta Tafalla, "Anosmic Aesthetics," *Estetika: The Central European Journal of Aesthetics* 56, no. 1 (2013): 54.
47. Tafalla, "Anosmic Aesthetics," 55.
48. Tafalla, "A World without the Olfactory Dimension," 1291–94. See also Marta Tafalla, "Smell and Anosmia in the Aesthetic Appreciation of Gardens," *Contemporary Aesthetics* 12 (2014), http://hdl.handle.net/2027/spo.7523862.0012.019.
49. Tafalla, "Anosmic Aesthetics," 64. A psychologist who is aware of the deficits anosmics face, including its impact on aesthetic appreciation in the specifically Kantian "reflective" sense, is Richard J. Stevenson. See Stevenson's "An Initial Evaluation of the Functions of Human Olfaction." *Chemical Senses* 35 (2010): 14–15.
50. Sibley, "Tastes, Smells and Aesthetics," 254.
51. Dutton, *Art Instinct*, 212.
52. Greg Currie et al., eds., *Aesthetics and the Sciences of Mind* (Oxford: Oxford University Press, 2014), 10–12.
53. Currie et al., *Aesthetics and the Sciences of Mind*, 11.
54. Murray Smith, *Film, Art, and the Third Culture: A Naturalized Aesthetics of Film* (Oxford: Oxford University Press, 2017). Smith proposes a "non-reductive, cooperative" approach that seeks to bridge the gap between the "two cultures" that C. P. Snow wrote about (32–37).
55. Smith, *Film, Art*, 59–61.
56. Smith, *Film, Art*, 64–8.

Interlude
Calvino's "The Name, the Nose"

Although Italo Calvino's triptych of three interwoven tales, "The Name, the Nose," is focused primarily on smell's role in sexual attraction, I think it can help us appreciate more concretely what is at stake in the debates within both philosophy and the neurosciences about the nature of the sense of smell in general.[1] Calvino's little work exquisitely captures the mysterious power of individuals' distinctive scent. The first round of the interwoven tales begins with each of three males encountering the irresistible scent of a female. In the first, an eighteen-century sophisticate is dancing with woman at a masked ball whom he does not know, but, he says, "I felt I knew all in that perfume" (77). In the next tale, an early hominid is drawn to the smell of a particular female, mounting her in the crush of the running herd, and, in the third tale, a rock musician wakes up amid a tangle of drug-numbed, naked bodies on the floor of a murky London concert hall, and is drawn to the serene smell of one woman lying face down, whom he caresses and slips into as she rises slightly to welcome it.

None of the three males can see the face of the female whose scent holds him in thrall and each male subsequently, and obsessively, pursues the unknown female only by her smell. The man of the world goes to his friend, Madame Odile of Paris's leading perfumery, a shop he frequents, to see if he can describe the scent well enough for Madame Odile to give him the woman's address. When he gets to the shop, for all that he is a connoisseur of perfumes, he is suddenly at a loss for words to express the "streaked, rippling cloud" that had assailed his nostrils (71). As for the hominid, who is just beginning to learn to walk upright, he has discovered in the herd a female whose smell was not like that of the others, "not . . . for my nose" (72). But after he mounts her and the herd moves on, he cannot find her again; he only gets a faint trace of her odor mixed with the smell of another male, who in turn smells her odor on him. The two hominids fight until the other male is killed (79–80). The rock musician, having crawled out of the somnolent tangle of bodies in order to light the extinguished gas stove that he can smell, takes a break outside and then returns to search for the girl he knows "only by her smell" (76).

Yet even as Calvino emphasizes the power of smell in human attraction, he also suggests we may face inevitable frustration in its consummation. In the end, each female in the story dies before the male can even learn her name, whether killed by her consort, by a predator, or by leaking gas. The man of the world, having finally discovered the woman's address with the help of Mme Odile, arrives at the house only to find a wake in progress. The woman in the coffin is unrecognizable, her heavily bandaged face covered in a veil, yet he recognizes a hint of the perfume she had been wearing "merged with the odor of death" (81). The hominid, after vanquishing the female's furious mate, catches another faint emanation of her odor, but now it is coming from the pit into which the herd throws animals they've killed (82). The rock musician finally finds the girl's stiffened body in the darkness, but her face is covered by her hair as he pulls her out, and he struggles to distinguish her odor amid other smells in the ambulance, the emergency room, the morgue (83).

Calvino's three tales (if one can get past their masculinist assumptions) are a fascinating parable of the immediacy and emotional power of the human sense of smell and of our often futile attempts to give scents a name. Most importantly for our purposes, the stories exemplify many of the central issues about smell that are currently being investigated in both the sciences and philosophy: smell's cognitive powers and limits, its relation to the emotions, its role in social communication, the problem of naming and describing. In the next few chapters we will look at each of these issues in turn and have occasion to return to Calvino's anticipation of them forty years ago.

Note

1. Italo Calvino, *Under the Jaguar Sun*, trans. William Weaver (New York: Harcourt Brace, 1988), 65–83. "The Name, the Nose" is one of three tales he finished of a project to write a collection with a chapter on each of the senses. Subsequent references to "The Name, the Nose" in my text are to this edition.

2
The Neuroscience and Psychology of Smell I
What the Nose Can Do

Since the 1990s, when Linda Buck and Richard Axel, discovered the genetic code of olfactory receptors, and especially following their receipt of the Nobel Prize in 2004, there has been a spike of interest in the neuroscience of smell. Of course, compared to the long and intensive investigation of vision and hearing over the last two centuries, research on smell is still in its early stages, as is research on several other senses beyond the canonical five. Yet rapid advances are being made now that laboratory studies based on rodents can be supplemented by studying human olfaction using such devices as the electro-olfactogram, which can measure physiological changes in the nasal cavity, or fMRI imaging, which can reveal changes in brain activity. By now most of us have seen pictures of brain-imaging studies that allow researchers to identify which areas of the brain "light up" (are activated) under experimental conditions such as exposure to a particular odor and being asked to identify it. As the neuroscientist Jay Gottfried points out, until the last decade, fMRI techniques mostly gave us the "where," and the *amount* of neural activity in particular brain regions, but newer experimental techniques and mathematical models are beginning to help scientists understand some aspects of the "how," of the *patterns* of neural activity that give rise to perceptions.[1] Behavioral psychologists, in turn, have been able either to work with neurobiologists or to make their own brain-imaging studies to supplement experimental protocols.

Odors and the Human Olfactory System

Before exploring the workings of the sense of smell itself, we need to ask: what *is* an odor? Odors are volatile chemicals whose molecular weight is low enough to allow molecules to leave their source object, become airborne, and land on our nasal receptors. But what lands on our receptors is seldom a single molecule since the odors emitted by most sources are composed of dozens or even hundreds of molecules. A typical rose, for example may have over 250 molecules, coffee from 600 to 800. In general, we breathe in many kinds of odor molecules with each of our 23,000 breathes a day, although only about 10% of the air we take in is captured by the nasal receptor organ (epithelium).

But all of our odor experiences do not come from outside air entering through our two nostrils. Volatile odor molecules also reach our nasal receptors from food and drink via an opening at the back of our mouth (the nasopharynx) when we exhale. Experiencing odors through our nostrils is called "orthonasal olfaction," whereas experiencing odors via the nasopharynx as we exhale during eating or drinking is called "retronasal" olfaction. Retronasal smell, which began to be more intensively explored only in the 1980s, is crucial to our experience of flavor. In fact, most of what we experience as the flavor of our food and drink comes not from taste buds alone (which register only sweet, sour, salt, bitter, and umami), but from retronasal smell along with tactile "mouth feel" as the brain integrates information from taste, smell, and touch receptors. Thus when we lift a glass of wine and sniff, we orthonasally experience its aroma, but as we sip and swallow, the exhaled breath sends odor molecules back over the epithelium, and this retronasal smell helps complete our experience of the wine's flavor. Accordingly, the philosopher Barry C. Smith speaks of smell as a "dual sense," and neuroscientists are continuing to explore the respective roles of orthonasal and retronasal smell in the experience of flavors, a topic to which we will return in our final chapter, devoted to the role of smell in the artistry of contemporary cuisine.[2]

Another important aspect of many odor experiences is the trigeminal nerve, whose branches spread across the face and nose and have endings inside the nasal cavity. This is the nerve that makes certain odors not only *smell* a certain way but *feel* a certain way, cool (menthol), stimulating (eucalyptus), or burning (hot peppers). It also registers physical properties, like the tingling from a just-poured soda or the feel of the airflow when we sniff, and it causes tears when we chop onions.[3] An interesting artistic use of the trigeminal response is Peter de Cupere's *Tree Virus* (2008). It consisted of an igloo-like transparent structure, within which there was a dead tree sitting atop a huge ball impregnated with a combination of peppermint and black pepper. Many of those who entered soon started to tear up, and some even fled in pain.

Let's return to the process of orthonasal and retronasal smelling. What happens within the nose after odor molecules land on the receptors inside? Each odor receptor is broadly tuned to respond to a number of similar molecules. Although the exact mechanism whereby the receptors take up this information remains under debate, most olfactory scientists still accept a shape-based "lock and key" theory, although an alternative "vibration" theory has some supporters.[4] But for our purposes, how the brain processes the information from the olfactory epithelium onward to produce the experience or *perception* of smell is the more important issue for the purpose of understanding smell's capacity to support aesthetic reflection. In a first step, the tiny fibers of the receptors send a pattern of nerve impulses up to the *olfactory bulb* (actually two bulbs, one for each nostril). The olfactory bulbs in turn shape these impulses into mosaic patterns before sending them on to a variety of units in the brain that make up what is called the "primary olfactory cortex," among which the most important unit is the *piriform* (pear shaped) *cortex*.[5] As the neuroanatomist Donald Wilson sums up the latter's function, the piriform cortex is the key unit in configuring the information from the olfactory bulbs into distinct odor "objects" that stand out from the generalized odor background we are experiencing at any moment.[6]

What complicates the question of smell's cognitive capacity for supporting things like aesthetic judgments is that among the other brain units lying close to the piriform cortex are those responsible for emotion (amygdala) and memory (hippocampus), traditionally lumped together with the piriform cortex and other small units and called the "limbic system." Crucially, all these units are located in the lower area of the brain, often termed the "paleocortex" because it was the earliest part of the mammalian brain to develop in evolutionary terms. Over the course of millions of years, the later-developing "neocortex," or "upper" brain, in humans eventually became much larger and more complex, with six layers of cells to the paleocortex's three layers. The neocortex is the primary location not only of such higher-level activities as planning and reasoning, but also of the primary processing units for vision, hearing, and touch. Consequently, the association of smell with the paleocortex still suggests to some scholars that smell may be a kind of atavistic leftover. I remember telling an academic friend of mine that I was working on a book about smell and aesthetics, and he immediately quipped, "Oh, yes, the reptilian brain."

Yet the paleocortex also has innumerable reciprocal connections to the "higher" neocortex. In the case of smell, there are two links from the primary odor-processing unit, the piriform cortex (PC), to the frontal area of the neocortex called the *orbitofrontal cortex* (OFC). One link is direct and has many fibers; the other link is thinner and indirect and goes via the thalamus (vision, hearing, touch, and taste inputs all go through the thalamus to reach the frontal cortex). The importance of the existence of dual links between the primary olfactory cortex and the orbitofrontal cortex is that the OFC is the primary locus in the brain for such things as sensory integration, decision-making, and reward evaluation, all of which have a cognitive dimension even if the OFC is not the primary locus of abstract reasoning.[7]

Given these complex anatomical features of the brain's olfactory processing equipment, some of the disagreements over the cognitive potential of our sense of smell, and consequently of smell's availability for reflective aesthetic activity, derive from whether one focuses on the aspects of olfactory processing taking place in the paleocortex itself or whether one focuses on the neural connections up to the neocortex. At one end of the opinion spectrum, some neuroscientists, like Tim Jacobs, emphasize that odors must first access what he terms the "more primitive and subconscious regions of the brain where they influence mood and emotion," only later arriving at higher regions dealing with perception.[8] At the other end of the spectrum are neuroscientists such as Gordon Shepherd, who stress the direct connection from the piriform cortex to the frontal cortex above, where "volatile odor molecules are evaluated quickly at the highest level of the human brain."[9] Jay Gottfried also stresses that the flow of information from the odor receptors in the nose up to the OFC is "a short three-synapse arc," in contrast to other sensory modalities that are "elaborated over numerous synapses . . . prior to arrival in the OFC."[10]

But there is a third position, typified by Christine Merrick and colleagues, who suggest that it may be too early in the development of olfactory research to draw definitive conclusions about which particular neuroanatomical regions are the locus of olfactory consciousness. Instead, they propose that olfactory consciousness may arise through large-scale integrative neural networks oriented to voluntary action, a position similar to Bernard Baars's global network theory of general consciousness.[11] This third position is consistent not only with the widely accepted view among brain scientists that most sensory pathways exhibit "successive levels of convergence from specific sensory cortices to multisensory cortices, leading to maximal integration," but also with the broader psychological studies of cross-modal experiences typical of daily life such as those undertaken by Charles Spence and others who argue that *multisensory* perception is the norm.[12]

My reading of these exchanges on the brain pathways for smell is that they support the idea that although the olfactory system does not independently possess the extensive neocortical resources comparable to vision and hearing neither is it merely primitive and "reptilian." As Weiss, Secundo, and Sobel remark, when fMRI is applied to olfaction, "There remain a host of brain structures . . . routinely illuminated in imaging studies . . . [that] go well beyond the limbic system and include components of classic higher-order visual and auditory cortices."[13] Moreover, the biologist Robert Sapolsky points out that just because the human brain has developed different layers with distinct physiological characteristics over time does not mean that "evolution in effect slapped on each new layer without any changes occurring in the one(s) already there."[14] Even so, neuroanatomy alone is not likely to settle the question of whether the brain's olfactory system is capable of contributing to the kind of aesthetic reflection that goes beyond the mere expression of sensory pleasure or displeasure. To get a better idea of the cognitive potential of smell, we need to look more closely at the implications of ongoing neuroimaging and behavioral experiments.

The "Odor Object" Theory

As in the case of the neuroanatomy of smell, so in the case of the psychological studies of the peculiarities and powers of smell, there is general agreement on the broad outlines, but disagreements over the relative importance of particular features and their implications for cognition. In the rest of this chapter I will focus on those features of the sense of smell that have *positive* implications for smell's cognitively inflected aesthetic possibilities, whereas Chapter 3 will consider studies and theories that seem to support a negative view of smell's cognitive powers. Let's begin the positive case by considering smell's evolutionary function. Most psychologists are agreed that in evolutionary terms, the human sense of smell, like that of other mammals, has from early on served as an attraction/repulsion system in relation to food sources, predators, and mating. Moreover, because our physical and social environment is filled with thousands of odors, some of them potentially dangerous, others potentially useful or simply pleasing, an underlying evolutionary function of the human smell system is to quickly and accurately draw from the mass of molecular information passing through our nasal passages what is most crucial to survival and flourishing at any given moment.

Currently, the leading neuropsychological hypothesis on how this happens is called the "odor object" theory. The main point of the "odor object" theory is that our sense of smell tends to weave the enormous amount of molecular information coming our way into singular perceptual "objects" that the brain can compare to other smell unities stored in memory. In this way, a unique odor "signature" can be recognized by the brain as an entity against a background of other odors, for example, in Calvino's story, the odor of "the" female amid all the odors emanating from the hominid herd and its surroundings. As we noted earlier, the innumerable outputs from the nasal receptors seem to be gathered by the olfactory bulb into a mosaic-like pattern before being sent on to the piriform cortex, where this pattern is further formatted as an "odor object" and matched with already known odor unities, for example, Calvino's "man of the world" in the perfume shop trying to match the scent of the woman with whom he had danced. According to one of the most influential versions of the odor object theory, developed by Donald Wilson and Richard Stevenson in *Learning to Smell*, if there is no match between a novel pattern from the olfactory bulb and those already on store in the piriform cortex, the novel pattern will be stored in turn as a new odor object.[15]

Although there are various versions of the odor object theory, what is common to all is the idea that our primary odor-processing mechanism is one that constructs singular percepts as a blend of volatile molecules that are "separated from the surrounding clutter of volatiles to stand out as an entity reflecting a putatively unidentified specific source (e.g. a melon's odor in the market)."[16] I remember a mundane personal experience of this phenomenon. I was asked to bring home a bunch of cilantro from the grocery, but when I got to the produce section, the cilantro was mixed in with several kinds of parsley and other greens, whose leaf forms my untrained eye could not identify, and for some reason there were no labels to follow. Fortunately, by bringing bunches with differently shaped leaves up to my nose, I could easily separate out "cilantro" by its smell. Of course, in this situation as in many others, further confirmation of an odor's actual source, may need to come from the cooperation of the other senses, especially vision and touch.

One of the principal supports of the odor object theory has been a series of studies over the years showing that when human subjects are given a mixture of odors, they cannot with any accuracy distinguish more than three. These experiments have been repeated using mixtures of both unknown chemicals and familiar everyday odors with similar results, and perhaps most tellingly, the subjects in several of these experiments have included not only ordinary individuals, but also a few experts from the flavor and perfume industries, who only do marginally better than the untrained.[17] Thus, the processing of odor information seems to be configural from its earliest shaping at the olfactory bulb level to its processing in both the piriform cortex and later stages involving the orbitofrontal cortex. An analogy is often drawn between such configural processing of odors and the way we process the visual recognition of faces and some other objects.[18] Moreover, a number of fMRI studies have offered confirmation that certain patterns of response occur in the posterior area of the piriform cortex that correspond to subjects' perception of broad odor categories such as minty, woody, or citrus. Other studies have shown that the olfactory system, like other sensory modalities, engages in what is called "predictive coding," an anticipatory response of great adaptive advantage.[19] As Gottfried remarks in his review of recent studies related to odor objet perception, in the brain the "codes of odor object categories are arranged in much the same way that visual object categories (for example, houses, cows, and chairs) are organized."[20]

The philosophical arguments by Batty, Lycan, and Richardson that we examined earlier to the effect that the sense of smell is able to represent and that what it represents initially is something external to us seem in the main supported by empirical research done in connection with the "odor object" theory. But, as in all things philosophical, there is considerable disagreement among specialists in the philosophy of perception on whether this "something external" should be called an *object*. Andreas Keller, whose recent *Philosophy of Olfactory Perception* offers a careful and comprehensive discussion of smell from the perspective of the philosophy of perception, agrees with Batty in rejecting the idea that what we perceive when we detect an odor should be called an "object." Like Batty, he argues that what we experience as smells lacks the kind of spatial location, dimensionality, and figure-ground characteristics typical of visual objects.[21] Yet as Barry Smith points out, Batty's version of the antiobject position is based on a highly artificial view of olfactory experience in which one is not allowed to consider inputs from the other senses. Fortunately, we do not need to choose sides in this quarrel since even Batty and Keller accept the fact that humans are remarkably good at detecting and discriminating among odors, whatever category name we put on "what" is perceived. But given the unsettled nature of the argument in the philosophy of perception, I will continue to put the term "odor object" in quotation marks.

Detection, Discrimination, Learning

The "odor object" theory itself implies an impressive general cognitive power for smell. After all, it is no small feat to be able to constitute singular odor entities and keep track of them out of the confusing mass of odor molecules that whirl through our noses every day. The fact that we perceive these "objects" may be one of the reasons humans are as good at odor detection and discrimination as they are. There is also evidence of "blind smell," that is, the ability to physiologically register and respond to odors of which we are not conscious.[22] Humans are able detect very slight concentrations of odorants, for example, one part per billion of the odorant ethyl mercapan, which is added to propane as a warning against gas leaks. That is equivalent to three drops of odorant in an Olympic-size swimming pool.[23] Many dog owners can identify their dogs by smell alone, and humans wearing kneepads have been able to track a scent though the grass with considerable success.[24] Thus, although many animals have larger numbers of odor receptors than we do and make far more concerted use of them, the great superiority of their smelling apparatus is sometimes outweighed in our case by the way we use other brain areas to analyze the information gathered by our smell receptors. As the olfactory psychologist Avery Gilbert puts it, for many animals "a smell is a call to action, a trigger for a biologically hard-wired survival response," whereas "human cognitive abilities turn smells into symbols.[25]

Naturally, both detection and discrimination are affected by age, with a gradual decline in ability until some people over eighty may become effectively anosmic, which may explain the higher proportion of deaths from asphyxiation among the elderly. Gender also plays a role in detection and discrimination, with many studies of children and adults showing that in general women have greater ability than men. Finally, there is the issue of just how many odors humans can discriminate. A common number bandied about in lay writing for some time has been ten thousand, but in 2014 a team of researchers using a mathematical model based on current data estimated the number at one trillion, much greater than the half million sounds and several million colors we can potentially discriminate. Although some of the mathematical assumptions behind the one trillion figure have been challenged, many researchers would agree that the number of smells we can potentially discriminate is high indeed and probably on a par with visual and auditory discrimination.[26]

In addition to powers of detection and discrimination, an equally important positive indictor of the cognitive potential of smell are the many studies that show how quickly humans can learn to identify new smells. Although there is some evidence that certain responses, such as the disgust or fear elicited by the smell of putrescence from rotting flesh, may be innate, most smell responses seemed to be learned.[27] In fact, research over recent decades has confirmed that learning smells and tastes, like learning sounds and rhythms, already begins in the womb. Several studies have demonstrated that newborns will show a more positive response to the odor of things their mothers have eaten such as garlic or anise than the newborns of mothers who have not eaten such things. Moreover, neonates quickly learn to distinguish the mother's breast odor from that of other lactating females.[28] The Rush University Medical Center premature unit has taken advantage of prenatal and early postnatal olfactory learning by giving parents of infants confined to the unit six-inch cloth pads to wear against their skin and leave with their infants, thus assuring the parents that this deep olfactory connection is being sustained.[29] Even young children are able to accurately distinguish among odors by general criteria such as edibility. One such experiment not only showed a high classificatory agreement among children ages four to eleven, but more importantly, their classifications did not show a necessary "overlap with hedonic categorization, suggesting that young children can separate emotional from more cognitive operations while processing odors."[30]

As for adults' ability to learn new smells, not only are adults able to track a scent as fast as they can crawl, but in a quite different experiment people who apparently had a specific anosmia for the steroid androstenone were trained to detect it after repeated exposure.[31] In another learning experiment, a single 3.5-minute exposure to a minty odorant turned the subjects involved into temporary "mint experts" whose perceptual learning was accompanied by increases in activity in both the posterior piriform cortex and the OFC that suggested the "neural representation of odor quality can be rapidly updated through mere perceptual experience."[32] Indeed, Gottfried and Wu have even been able through aversive training to teach subjects to distinguish between ordinarily indistinguishable enantiomers (odor molecules identical in chemical structure except for right- or left-hand orientation). After putting the results of their successful experiments with similar evidence from other laboratories, they remark on the "tremendous perceptual and neural plasticity of the olfactory system," which enables the sense of smell "to promote adaptive responses to foods, friends, foes, and mates."[33] Given the olfactory system's rapid learning abilities, it is no surprise that some AI researchers concerned with machine learning in situations where training samples are limited have begun to explore using olfactory rather than vision circuits as a model.[34]

Social Communication

The sense of smell's powers of detection, discrimination, and learning not only allow it to play a crucial role in detecting dangerous odors or appealing foods, but also, as Gottfried and Wu suggest, to play a role in social communication. In nonhuman species, the most obvious aspect of communication using odors is sexual signaling between male and female insects or mammals via pheromones. Yet talk of a "human pheromone" is surrounded by such a tangle of confusion and controversy that many scientists prefer the term "chemosignal" for the more general communicative role of smell. Reserving the human sex pheromone issue for the following interlude, let's consider some leading examples of proven olfactory communication, or "chemosignaling."

Take the idea that there is a detectable "smell of fear." In the first chapter, we discussed Sissel Tolaas's artwork *The Fear of Smell and the Smell of Fear*, in which she covered gallery walls with a paint encapsulating the odor of sweat from men subject to anxiety attacks. She used men since male axillary secretions (underarm sweat) are generally more potent than those of females. The idea that the human "smell of fear" is detectable, at least to other animals such as dogs and horses, is old and widespread. But is there any scientific evidence that humans can actually detect the smell of fear in others and discriminate it from similar smells? There is some, although as with many scientific issues, the evidence turns out to be more complex and interesting than common beliefs.

Over the last two decades there have been a score of studies that have attempted to determine whether people can detect a fear-like or anxiety odor in others. A favorite setup has been to expose one group of subjects to a comedy movie and another to a horror film while they wear underarm pads that collect their sweat. Then a third group sniffs the pads and either responds to a questionnaire or engages in some cognitive or emotional task. Among the experimental effects of smelling fear sweat versus exercise sweat are an increase in startle reflex, an increased likelihood of interpreting ambiguous facial expressions as fearful ones, and even exhibiting a fearful expression oneself.[35]

In a summary of various fear studies up through 2015, Weiss, Secundo, and Sobel conclude: "Taken together, this body of research strongly suggests that humans discriminate the scent of fear from other body odors and that this chemosignaling influences behavior."[36] As interesting as this conclusion is, we need to keep in mind that in nonexperimental settings chemosignaling is seldom consciously processed and that in everyday encounters where people have bathed and put on deodorants and cologne or perfume, even subliminal detection might be difficult. To provide deeper support for the cognitive capacity of smell in communicating, we need to find evidence of a more direct involvement of smell messaging in everyday situations.

In one sense smell communication actually begins at birth, as we have seen, based on the neonate's experience of the particular odor mix of the mother's amniotic fluid, which is immediately updated through the sensory-cognitive mobilization of delivery and the search for a nipple.[37] As children pass from infancy into childhood, they can recognize siblings by smell, and this continues into adulthood. Yet when we move beyond the capacity of smell to discriminate and recognize kin and close friends with whom we are in regular contact to the question of whether chemosignals can influence emotion and behavior with respect to sexual attraction outside kinship, the issue becomes more controversial. One kind of evidence of the existence of gender-related or "sexual" chemosignaling is studies like the one in which sweat from the armpits of men watching erotic videos was presented via a nebulizer to female subjects who were lying in a brain scanner; the women showed increased activation in the OFC, hypothalamus, and fusiform cortex.[38] Conversely, exposing male subjects to emotional tears from females who had been watching sad movies not only reduced the level of sexual attractiveness of pictures of women's faces that the men were subsequently shown, but also reduced activity in their hypothalamus and fusiform cortex.[39]

Some communication theorists and philosophers of language might balk at considering many of these examples of "chemosignals" to be genuine acts of communication. After all, the basic definition of a communication, whether by word or gesture, normally involves a sender, a message, a medium, and a receiver. And in many models of communication senders are not really senders unless they *intentionally* send a message. Simply giving off a body odor is not an intentional act of communication in most situations. Nor does it seem right to speak of odors resulting from natural processes like exercise or anxiety as "messages." They are signs of a sort and they carry information, even information that may lead a perceiver to take a more or less voluntary and intentional action. But in themselves they are mute. They are situational facts, not actions.

Of course, there *can* be intentional olfactory signals. Probably the best-known natural olfactory signal is the lowly fart. Although it is normally involuntary, as kids well know, in the right circumstances they can let one go—"pull my finger." Adults have even used flatus as a political weapon. In an incident connected with the Democratic Party convention of 2016, one group of disgruntled Bernie Sanders supporters planned a "fart-in" to protest the nomination of Hillary Clinton, a tactic that goes all the way back to the 1960s, when the community organizer Saul Alinsky threatened a fart-in at a symphony concert to disrupt the cultural life of the Rochester, New York, socialeconomic establishment.[40] By now there is even a body of scientific evidence on the physiological mechanisms of farting, including the chemical composition (hydrogen sulfide and methyl mercapan), gender inflection (women's are smellier, men's bigger), and cultural status (in some cultures it is no more offensive than coughing).[41] There are, of course, more serious intentional means of direct smell communication, such as the traditional use of incense or perfume, topics I will discuss in Parts III and IV.

Our initial question was whether there is scientific evidence to support the idea that the human sense of smell, contrary to the view of the Kant-to-Scruton tradition, has enough intelligence to fund cognitively informed aesthetic discussion and judgment. We have not only canvassed smell's basic cognitive equipment and general mode of operation, but also seen that smell clearly has considerable powers of detection and discrimination and that people can quickly learn to identify new "odor objects," even if the term "odor object" is philosophically controversial. And we have also seen that smell functions as a mode for signaling fear, recognition, offense, or attraction. With respect to the latter, its time to pause and consider the "human pheromone" issue before going on to look at empirical evidence of smell's cognitive *limitations* that could be interpreted as seriously threatening the philosophical prospects of an olfactory aesthetics.

Notes

1. Jay A. Gottfried, "Structural and Functional Imaging," in *Handbook of Olfaction*, 3rd ed., ed. Richard L. Doty (Hoboken. NJ: Wiley Blackwell, 2015), 279.
2. Smith, "The Chemical Senses," 325–30.
3. As Rachel Herz points out, in many cases it is not the actual "smell" coming from some activity like chopping onions or eating habaneros that might make us avoid it, but the painful trigeminal reaction. Herz, *Scent of Desire*, 47–48.
4. The major alternative to the lock-and-key theory, which is based on the fit between the shapes of odor molecules and receptors, is Luca Turin's vibration theory, but it would take us too far afield to enter into those debates. See Turin's *The Secret of Scent: Adventures in Perfume and the Science of Smell* (New York: Ecco, 2006). For a discussion of the clash between the two theories and its issues and its implications for the philosophy of science see Ann-Sophie Barwich, "Making Sense of Smell: Classifications and Model Thinking in Olfaction Theory," PhD diss. (University of Exeter, 2013).

5. For an extensive discussion of processing in the olfactory bulb see Gordon M. Shepherd, *Neurogastronomy: How the Brain Creates Flavor and Why It Matters* (New York: Columbia University Press, 2012), Chapters 5–10.
6. Donald A. Wilson et al., "Cortical Olfactory Anatomy and Physiology," in Doty, *Handbook of Olfaction*, 209–23.
7. Donald A. Wilson et al., "Cortical Odor Processing in Health and Disease," *Odor Memory and Perception: Progress in Brain Research* 208 (2014): 277.
8. Tim Jacob, "The Science of Taste and Smell," in *Art and the Senses*, ed. Francesca Bacci and David Melcher (Oxford: Oxford University Press, 2011), 183.
9. Shepherd, *Neurogastronomy*, 110.
10. Jay A. Gottfried and Christina Zelano, "The Value of Identity: Olfactory Notes on Orbitofrontal Cortex Function," *Annals of the New York Academy of Sciences* 1239 (2011): 139, https://doi: 10.1111/j.1749-6632.2011.06268.x. One can draw both positive and negative consequences from the direct connection to the OFC. On the one hand, it shows that odor processing is not confined to some "primitive" part of the brain; on the other hand, it also means that the initial olfactory information receives less elaborated processing than the visual and auditory information that passes through the thalamus and other areas.
11. Christina Merrick et al., "The Olfactory System as the Gateway to the Neural Correlates of Consciousness," *Frontiers in Psychology* 4, no. 1011 (2014): 1–15, https://doi.org/10.3389/fpsyg.2013.01011. Bernard J. Baars, "Multiple Sources of Conscious Odor Integration and Propagation in Olfactory Cortex," *Frontiers in Psychology* 4, no. 930 (2013): 1–4, https://doi.103389/fpsyg.2013.00930.
12. Kingston Man et al. "Neural Convergence and Divergence in the Mammalian Cerebral Cortex: From Experimental Neuroanatomy to Functional Neuroimaging," *Journal of Comparative Neurology* 521, no. 18 (2013): 4097, https://doi: 10.1002/cne.23408. See also Tim Bayne and Charles Spence, "Multisensory Perception," in *The Oxford Handbook of Philosophy of Perception*, ed. Mohan Matten (Oxford: Oxford University Press, 2015), 603–20.
13. Tali Weiss, Lavi Secundo, and Noam Sobel, "Human Olfaction: A Typical Yet Special Mammalian Olfactory System," in *The Olfactory System: From Odor Molecules to Motivational Behaviors*, ed. Kensaku Mori (Tokyo: Springer, 2014), 190.
14. Robert M. Sapolsky, *Behave: The Biology of Humans at Our Best and Worst* (Harmondsworth, UK: Penguin, 2017), 23.
15. Donald A. Wilson and Richard J. Stevenson, *Learning to Smell: Olfactory Perception from Neurobiology to Behavior* (Baltimore: Johns Hopkins University Press, 2010).
16. Thierry Thomas-Danguin et al. "The Perception of Odor Objects in Everyday Life: A Review on the Processing of Odor Mixtures," *Frontiers in Psychology* 5, no. 504 (2014): 6, https://doi: 10.3389/fpsyg.2014.00504.
17. Thomas-Danguin, "Perception of Odor Objects," 5–6.
18. Shepherd, *Neurogastronomy*, 81–84.
19. Gottfried and Zelano, "The Value of Identity, 138–48.
20. Gottfried, "Structural and Functional Imaging," 279.
21. Keller, *Philosophy of Olfactory Perception*, 71–84.

22. Lee Sela and Noam Sobel, "Human Olfaction: A Constant State of Change-Blindness," *Experimental Brain Research* 205, no. 1 (2010): 13–29, https://doi.org/10.1007/s00221-010-2348-6.
23. In another case, ordinary consumers detected an off-odor in a batch of bottled water that the producer's finely calibrated detection machinery missed. Yaara Yeshurun and Noam Sobel. "An Odor Is Not Worth a Thousand Words: From Multidimensional Odors to Unidimensional Odor Objects," *Annual Review of Psychology* 61 (2010): 223, https://doi: 10.1146/annurev.psych.60.110707.163639.
24. Jess Porter et al., "Mechanisms of Scent-Tracking in Humans," *Nature Neuroscience* 10, no. 1 (2007): 27–29, https://doi: 10.1038/nn1819.
25. Gilbert, *What the Nose Knows*, 65–66.
26. C. Bushdid, Marcelo O. Magnasco, et al., "Humans Can Discriminate More Than One Trillion Olfactory Stimuli," *Science* 343, no. 6177 (2014): 1370–72, https://doi.10.1126/science.1249168.
27. Arnaud Wisman and Ilan Shrira, "The Smell of Death: Evidence That Putrescine Elicits Threat Management Mechanisms," *Frontiers in Psychology* 6, no. 1274 (August 2015), https://doi.org/10.3389/fpsyg.2015.01274. See also Richard J. Stevenson, "An Initial Evaluation of the Functions of Human Olfaction," *Chemical Senses* 35 (2010): 9–10. Some researchers find the innate vs. acquired polarity "hollow." See Benoist Schaal, "Emerging Chemosensory Preferences: Another Playground for the Innate-Acquired Dichotomy in Human Cognition," in Zucco, *Olfactory Cognition*, 261.
28. Benoist Schaal, "Prenatal and Postnatal Human Olfactory Development: Influences on Cognition and Behavior," in Doty, *Handbook of Olfaction*, 314–20.
29. Armando L. Sanchez, "Aromatherapy for Preemies: Heart-Shaped Cloths Carrying Scent of Moms and Dads Help the NICU Babies Promote Bonding," *Chicago Tribune*, December 2, 2018, 1.9. I am grateful to Judy Shereikis for calling this article to my attention.
30. Schaal, "Prenatal and Postnatal," 310.
31. Weiss et al. "Human Olfaction," 177–202.
32. Gottfried, "Structural and Functional Imaging," 297.
33. Jay A. Gottfried and Keng Nei Wu, "Perceptual and Neural Pliability of Odor Objects," *Annals of the New York Academy of Science* 1170, no. 1 (2009): 328–30, doi: 10.1111/j.1749-6632.2009.03917.x.
34. Some AI researchers, using the moth's olfactory system as a model, found that if two machines were given less than twenty samples, a machine designed on the model of a moth's olfactory circuits was better at recognizing handwritten digits than a machine based on a vision circuit model. Thus, in a situation where data are sparse and must be quickly processed (self-driving cars), the more rapid and distributed approach based on an olfactory model may be superior to one based on vision. Jordana Cepelewicz, "New AI Strategy Mimics How Brains Learn to Smell," *Quanta Magazine*, September 18, 2018, https://www.quantamagazine.org/new-ai-strategy-mimics-how-brains-learn-to-smell-20180918/.
35. Gottfried, "Structural and Functional Imaging," 298.

36. Weiss et al., "Human Olfaction,"180.
37. Schaal, "Prenatal and Postnatal," 321.
38. Gottfried, "Structural and Functional Imaging," 298.
39. S. Gelstein et al. "Human Tears Contain a Chemosignal," *Science* 331 (2011): 226–30, https://doi: 10.1126/science.1198331. Another research group claimed they were unable to replicate the Gelstein findings, but Noam Sobel offered a convincing reply. "Revisiting the Revisit: Added Evidence for a Social Chemosignal in Human Emotional Tears," *Cognition and Emotion* 31, no. 1 (2017): 151–57, https://doi.10.1080/02699931.2016.1177488
40. Kristen Salyer, "Bernie Sanders Supporters Say They'll Fart in Protest of Hillary Clinton," *Time*, July 19, 2016, http://time.com › Ideas › politic. To protest the Kodak's discriminatory labor practices, Alinsky threatened to buy one hundred tickets to the Rochester Symphony for one hundred African Americans who would be fed beans the day of the concert.
41. Gilbert, *What the Nose Knows*, 29–30.

Interlude
The Pheromone Myth

Although the internet is rife with ads for "pheromone fragrances" with names like "Alfa Mashio" or "Holy Grail" for men or "Alfa Donna" and "MAX Attract Silk," for women, all guaranteed, as one ad puts it, to "turn you into a sexual magnet," its mostly baloney. Many people who use the term "pheromone" in relation to human sexual attraction are engaging in wishful thinking and misleading hyperbole—a true pheromone does not attract, it compels. A genuine pheromone is an odor emitted by a female insect or mammal that automatically triggers a mating response in any male close enough to scent it—which could be up to two miles in the case of a male dog, as you are likely to know if you have had a female dog in heat. Or, as Edward O. Wilson remarks,

> Consider a female moth calling for a mate. Her sex attractant must be unique to her species. Small quantities must travel far—up to kilometers out in some cases—and must be read and trigger a response in the right kind of mate, not another kind, or worse, in a spider or a moth-hunting wasp.[1]

There are, of course, other kinds of true pheromones among insects and mammals besides sexual ones, for example warning, navigational, and tracking pheromones, the latter used by ants. As the chemist Paolo Pelosi points out, social insects like ants and bees maintain their hive organization by means of smells that strictly regulate individual roles within the colony, and these odors are true pheromones.[2]

There is evidence that in the distant evolutionary past millions of years ago, our human ancestors had a sexual pheromone system, which was, as it is in many mammals today, an odor system *separate* from the main olfactory system, and this human chemosignaling system centered on what is called the vomeronasal organ. In fact, a vestige of the human vomeronasal organ may emerge during human fetal development but disappears by birth.[3] Bottom line: humans no longer have the most basic physiological equipment for true sexual pheromone interaction. Consequently, the idea that someone might eventually discover the human sexual pheromone is a myth. Moreover, even if there were an irresistible attractant, would you really want every horny member of the opposite sex camping on your doorstep? As for our central question concerning smell's cognitive capacity to support aesthetic reflection and discussion, the fact that we are no longer susceptible to the operation of pheromones argues in favor of smell's ability to make fine aesthetic discriminations.

In fact, Michael Stoddart's *Adam's Nose, and the Making of Humankind* goes so far as to claim that the loss of the vomeronasal organ and thus the loss of an operational human pheromone system was *the* turning point in human evolution. That is because its loss "freed our ancestors from slavish response to sexual smells and so, in freeing them, made them human."[4] Of course, Stoddart is well aware that there are many other claims in evolutionary theory as to the crucial factor that made us human, but he thinks these other developments *followed* on the loss of the vomeronasal system among our distant forebears. Stoddart's basic argument is that human sociality can exist only where the compulsory link between smell and sexual response has been broken. Whatever the relative importance of other factors in making us human, Stoddart is surely right to emphasize that the transformation of sexual smells from a physiological compulsion into a form of social communication is part of what distinguishes us as humans from many other species. Calvino captured this idea in "The Name, the Nose," whose humanoid is human-like precisely because he is attracted by the particular scent of a particular female that differs from that of all other females in the herd. And that kind of particularity reflects another characteristic of smell that is related to sexual selection; most of us do have a unique scent, a scent signature as it were.

Part of what is responsible for our unique odor imprint is our HLA/MHC (human leukocyte antigen / major histocompatibility complex) profile. In humans HLA genes and MHC molecules work together to help our immune system recognize foreign substances such as viruses. Some studies have shown that women who are not on hormones prefer men who have an HLA/MHC profile that is different from their own. The reason for this, it is suggested, is that if two people with very different HLA/MHC complexes were to mate, it would likely create more diversity in the genes of any offspring. This is sometimes discussed as a pheromone-like phenomenon, but obviously, neither HLA/MHC differences nor other such reproduction-related olfactory attraction effects, such as men preferring the smell of women who are ovulating, are the equivalent of the compulsive sex pheromones that drive behavior in other species.[5]

Of course, there are innocent-enough uses of the term "pheromone," uses that simply suggest the idea of a signature body odor and how it might be attractive to others either naturally or as modified by a perfume or cologne. The olfactory artist Clara Ursitti, for example, created a participatory artwork called *Pheromone Link™ Scent Library* for an art gallery in Toronto in 2001. It consisted of a room with conventional romantic symbolism, a circular red couch and a cluster of cardboard tubes arranged in a heart shape (the "Scent Library") hanging on the wall. Participants deposited into one of the tubes a numbered T-shirt they had worn long enough for it to be impregnated with their scent, and other participants chose the one that smelled best to them, dropping their vote in a box. The winning T-shirt got a bottle of whiskey and a package of chocolate-flavored condoms. In addition, if anyone wished, Ursitti would arrange for the two people to meet at a local bar the next night.[6]

Notes

1. Edward O. Wilson, *The Origins of Creativity* (New York: Norton, 2017), 62–63.
2. Pelosi, *On the Scent*, Chapter 5.
3. Pelosi, *On the Scent*, 229–34.
4. Stoddart, *Adam's Nose*, 20.
5. Accessible general discussions of the function of HLA/MHC in avoiding inbreeding can be found in Herz, *Scent of Desire*, 126–32, and Stevenson, "Initial Evaluation," 11–12.
6. See Ursitti's website, https://www.claraursitti.com.

3
The Neuroscience and Psychology of Smell II
What the Nose Can't Do

We have seen some of the ways that the human sense of smell is cognitively good at detection, discrimination, and learning as well as being able to play a role in social communication. This evidence lends support to the analytic arguments of Chapter 1 that smell is not wholly subjective, with little cognitive capacity and practical use. Yet there is another body of scientific evidence that shows aspects of the sense of smell that are not so cognitively keen, evidence that could be used to support the Kant-Scruton depreciation of aesthetic potential of smell. Such cognitive limitations might raise doubts about the smell system's capacity to support the kind of differentiation and reflection essential to genuine aesthetic discussion and judgment. Although I will treat each of smell's putative cognitive deficiencies separately, it will quickly become apparent that they overlap and intertwine.

Four Counts against Smell's Cognitive Adequacy

A first count against smell's cognitive adequacy for aesthetics can be derived from our very ability to learn odors and odor associations rapidly. As we have seen, because there is little formal education of the sense of smell, odor associations tend to arise idiosyncratically. One could argue that many people have developed individual aversions or attractions that could easily skew any attempt at moving toward reasoned aesthetic discussion and evaluation. Moreover, there is even some variability in people's basic olfactory equipment since, unlike vision, where there are only three receptor types, there are over three hundred for smell, and the number of active genes varies slightly by individual. For example, some people have what are called "specific anosmias," the inability to smell certain substances that other people can easily detect. Among the most studied is the steroid androstenone, a chemical in boar pheromone. Based on projections from various experiments, half the population can't smell it at all and the other half is split down the middle between those who find it pleasurably musky and those who find it unpleasantly urinous.[1]

A second and more serious count against smell's cognitive potential is its much stronger connection with the emotions than vision or hearing. Calvino's three males in pursuit of a particular female by smell alone certainly seem emotionally driven. In the typical psychology experiment concerned with smell and emotion, people are exposed to pictures or sounds of some object, along with the odors of the same kind of object, and, when asked which feels more emotional, a significantly greater percentage say the odors provoked more emotion. Rachel Herz has performed numerous experiments along this line, leading her to conclude that smell more than any other sense is the sense of emotion.[2] Another kind of evidence for the greater emotionality of smell than vision and hearing comes from experiments that show the primacy of hedonic (pleasant/unpleasant) terms used in ratings over terms for specific qualities. Richard J. Stevenson observes that "the hedonics for smell feels more direct and visceral than the hedonics associated with vision and audition," especially in the experience of disgust. For example, the sight of fake dog feces may or may not be distasteful, but the artificial smell of dog shit nearly always repels as much as the real thing.[3]

A third characteristic of smell that implies a lack of cognitive potential is the fact that, as laboratory studies have shown, most people to have a poor ability to name or describe smells. Remember the great difficulty Calvino's "man of the world" has in describing the elusive odor of the woman he is desperate to find despite the fact that he is a connoisseur of perfumes. Various studies have shown that most people are unable to name by smell alone the odors of between 20% and 50% of the household items they use regularly. One well-known psychologist decided to supplement his controlled experiments with an informal test at home and held an open jar of peanut butter under the nose of a blindfolded family member who eats peanut butter almost every day, and the person could not name it![4] One reason sometimes given for these failures is what is often referred to as the "poverty of language" when it comes to odors. Whereas most languages have extensive and nuanced vocabularies for colors and sounds, smell vocabularies are highly limited, and most people end up just referring to odor sources or simple variations on the pleasant/unpleasant axis. But there also seem to be some underlying physiological reasons for naming failure. According to research by the neuroscientists Jonas Olafsson and Jay Gottfried, the neurological pathways for smell processing in the brain have fewer connections to the brain's language areas than do the neural pathways for vision and hearing. Whereas vision has "multiple entry points into the lexical-semantic network," smell relays are only thinly connected to cortical areas that could enhance "odor-object representations with lexical-semantic content."[5]

As if the naming difficulty weren't bad enough, there is often a "top-down" influence on odor processing, or "cognitive penetration," that can also make many people not only uncertain but also quite fickle when it comes to identifying and naming odors. As far back as the 1890s psychologists have been able to convince people that a particular odor was present when the putative source was emitting nothing but odorless air. (The professor uncaps a jar and asks, "How many of you smell something?" Hands go up.) Moreover, a number of more recent experiments have shown that specific verbal labels can easily influence what people think they smell. When Rachel Herz exposed experimental subjects to two identical mixtures of isovaleric acid and butyric acid under the name "Parmesan cheese" first and under the name "vomit" later, most people reacted positively when the odor was called Parmesan cheese but expressed disgust when it was called "vomit." Some wanted to leave the room! After the experiment was over, many of the students according to Herz "would *not* believe that it was the same odor that they were smelling on both occasions."[6]

A final characteristic of smell that bodes ill for a cognitive aesthetics of smell is that for most people, smell seems to operate largely unconsciously. Most people pay very little attention to odors, and when they attempt to summon up the memory or image of an odor, they have great difficulty. Part of the problem is the phenomenon of *habituation*. William James gave a famous auditory example of habituation: we cease to notice a ticking clock in the room when deeply engaged in an activity, but begin to hear it again when the activity ceases. Although a similar phenomenon occurs with smell, habituation seems to be both more rapid with smell and much harder to overcome by voluntary recall. As for the attempt to imagine odors, many people simply declare they cannot form a mental image of an odor in the way they can imagine a sight or a sound. Yet there is evidence from fMRI studies that people who are asked to make a conscious effort at reattending to or imagining an odor actually register activity in the same areas of the brain used for real smelling.[7] Given the importance of imagination in aesthetic experience, this difficulty in consciously imagining odors also bodes ill for an olfactory aesthetics.

What Is Smell's Dominant Characteristic?

There are many other special characteristics of smell we might examine, but those we have outlined are enough to make sense of three leading olfactory psychologists' overall characterization of smell that suggest its cognitive weakness. Each psychologist makes one or more of the characteristics we have discussed the defining characteristics of smell. Thus Rachel Herz is convinced that emotionality is the primary attribute of the sense of smell, and emphasizes not only the results of behavioral research but the proximity of the primary smell-processing brain area (piriform cortex) to the primary emotion-processing brain area (amygdala), concluding that "there is a privileged and unique anatomical relationship between the neural substrates of emotion and olfaction, and as such it seems that odors are *inherently more emotional and less cognitively analyzed* than other stimuli."[8] The implication is that if smell were to represent anything or to be involved in aesthetic judgment, it would only be through a cloud of uncontrolled emotion, hardly a reliable basis for aesthetic discussion.

Noam Sobel explains people's difficulty identifying and naming odors by reducing all smell judgments to a single defining characteristic, the hedonic or pleasant/unpleasant axis, a phenomenon that also reinforces the idea of smell's emotionality and linguistic failures. In a lively article, "An Odor Is Not Worth a Thousand Words," Sobel and Yaara Yeshurun not only argue that our initial reaction to an odor is always hedonic, but also that no one, not even experts, makes a genuine qualitative judgment about smell, only a hedonic one. "The one thing humans can and do invariably say about an odor is whether it is pleasant or not. We argue that this hedonic determination is *the key function* of olfaction."[9] This means that the "boundaries of an odor object are determined by its pleasantness, which . . . like an emotion—remains poorly delineated with words."[10] We seem to be all the way back to Plato's idea that the most we can say about any smell is whether it is pleasant or unpleasant.

Although the psychologist Peter Köster would agree with Herz that emotion is a highly important characteristic of our sense of smell and with Sobel that most smell judgments include a strong hedonic component and are only weakly articulated, he insists that smell's true defining feature is its *unconscious* nature. His "misfit" theory of olfaction holds that, in evolutionary terms, smell's warning function is primary, and that we are unconscious of the smells around us for the very good reason that we only become conscious of one of them when something turns up that doesn't *fit* our learned expectations. Köster suggests that the unconscious nature of smell also partly explains naming difficulties. "In everyday life . . . we almost never name odors and the odors that are most important to us (the odors of our surroundings and the people we know) are usually non-nameable."[11] He sums up his misfit theory this way: "Odors are probably not meant to be identified. They are the silent emotional reminders of the surroundings and situations with which they are linked by unconscious association and . . . should . . . be recognized as the ephemeral and unnoticed providers of feelings of safety and comfort."[12]

Finally, Richard J. Stevenson and Tuki Attuquayefio have offered a theory to explain all of the limitations of human smell: its emotionality and linguistic failures, the predominance of hedonics, and the lack of conscious awareness. Whereas the other psychologists we discussed tend to see these characteristics as arising from their adaptive evolutionary function, Stevenson and his coauthor argue that the most parsimonious explanation of smell's unique features is that smell's dominant characteristic is its cognitive weakness: "Olfaction's unusual features may be attributed to its *limited . . . neocortical resources.*" In their view, as the neocortex expanded in the course of evolution, smell remained restricted to the paleocortex, where it lacks the resources to form "ideas to communicate within the brain and between people."[13]

The philosopher Andreas Keller, who has carefully reviewed the scientific literature on olfaction in *Philosophy of Olfactory Perception*, seems at one point to embrace a similar view of smell as so emotional, hedonic, and tongue-tied that it is impervious to cognition. He opens his chapter on olfaction and cognition this way:

> Olfaction is often considered the most animalistic and primitive of our senses. Odor stimuli induce desires, emotions and physiological responses that make us respond to certain smells in automatic ways. Reason is powerless to intervene. In contrast, it is difficult to talk about smells, or even to name them.[14]

If we put Herz's, Sobel's, and Köster's characterizations of smell together with Stevenson's claim about a lack of neocortical resources, and Keller's remark about smell's automatic responses that reason is "powerless" to affect, the outlook for a cognitively informed olfactory aesthetic experience looks bleak. Whereas Kant and Darwin suggested that the human sense of smell is not *worth* cultivating, the theories we have just been considering cast doubt on whether it would even be *possible* to cultivate it to a significant degree, and Köster's view of smell's evolutionary function could be taken to imply that it would be wrong to even try. Does this mean that the negative tradition concerning odors and smell in art and aesthetics that reaches from Kant and Hegel to Scruton, Dutton, and several other contemporary philosophers is right after all?

Measuring the Abilities of Olfactory Experts

But before we give up on the cognitive possibilities of smell, we need to remember, as Sibley pointed out, that we are not asking if smell (or taste) can *equal* the cognitive powers of vision and hearing. We are asking whether smell has enough cognitive resources to rise above purely emotional and hedonic preferences when necessary and provide evidence for the kind of critical discussion of olfactory artworks that Hume, for example, envisioned for more traditional arts like literature, painting, and music. That suggests a useful next step, namely, to inquire whether or not the best critics of olfactory art and design works are, in fact, able to adequately articulate and rationally justify their appreciations. Surely, an understanding of the aesthetic potential of any sensory mode should pay particular attention to the abilities and perspectives of those who, as Hume argued, show a combination of sensitivity, knowledge, comparison, and extensive practice, and whose "joint verdict" could be used as the criterion of correct aesthetic judgments. Indeed, in Hume's famous essay "The Standard of Taste," a prime illustration of this principle is the story of Sancho's dispute with his friends over whether the wine in the keg they were drinking from tasted of iron or leather. When the keg was drained, an iron key with a leather thong was found at the bottom, showing, Hume suggested, that there was, indeed, a cognitive element to the taste dispute between Sancho and his friends. Following Hume's lead, then, I believe we should turn to the best critics of those arts involving smell.

Naturally, the emergence over recent decades of a number of works of olfactory art, such as those of Tolaas, Ursitti, or de Cupere that we have already mentioned, has produced art criticism and theoretical reflection to go with it, most notably, the work of Jim Drobnick, but I will save a discussion of such art criticism for a later chapter in which I discuss the kind of contemporary scent artworks that are typically shown in galleries and museums. Here I will focus on trained perfumers since, like trained wine experts, perfumers' cognitive abilities with respect to smell have been studied by neuroscientists. Of course, many people enjoy making fun of wine criticism, and perfume criticism hardly exists outside a few books by specialists and some blogs on the internet. Moreover, there is nothing many of us like more than seeing haughty, self-anointed experts exposed as no better than the rest of us (although I'm not sure that pretentious critics are any more prevalent in the realms of wine and perfume than in the realms of painting, music, and film). In the case of actual perfume and wine expertise, there is, in fact, evidence from recent studies of both perfumers and wine specialists that the human sense of smell may have more cognitive potential than suggested by a negative interpretation of neuroscience and behavioral evidence that is based on randomly selected populations.

The particular experts who were the subjects of the studies I have in mind are professional perfumers who typically have undergone two to three years of rigorous training, followed by several years of apprenticeship. Of course, olfactory experts of any kind are small in number. It is estimated that there are some five hundred professional perfumers worldwide and about 150,000 trained professionals in the wine industry. To date there have been some fifty empirical studies of wine professionals and a half-dozen of perfumers. Even so, many of these studies are worth considering, particularly since their results are consistent with the results of studies of experts in other areas such as professional athletes and professional musicians.[15]

Before looking at these recent olfactory studies, however, we need to mention the widely cited work of Laing and Livermore that shows perfumers run into a discrimination limit in identifying components in an odor mixture that is similar to the average untrained person.[16] But this limit does not by itself demonstrate an across-the-board failure of olfactory expertise. First of all, even within these limits experts were generally better at identifying components than nonexperts. Moreover, as the neurobiologist André Holley has observed, perfumers are trained in a particular tradition of "notes" and "accords," so that "when confronted with a mixture prepared by experimenters ignoring professional habits and traditions, perfumers must rely solely on their general detection and recognition capacities, and their performances are more like those of naïve subjects."[17] Holley's point is also a reminder of the built-in limitations of laboratory experiments that attempt to isolate smell from the other senses. Our daily sensory experiences are almost always multimodal, so that the other senses normally play a role in any exercise of smell, and smell is often a real, if unnoticed, partner in experiences we might think of as solely visual or auditory. My point is not to question the validity of the typical odor mixture tests, but to suggest that by themselves they do not show that olfactory experts are not really experts or that smell cannot be cognitively cultivated. After all, the ability to discriminate a large number of components in a single mixture is hardly the whole of olfactory cognition, and in several other areas, such as description, imagery, and brain plasticity, there is now clear evidence that expert perfumers and flavorists perform at a relatively high level compared to novices or even trainees. In what follows I want to look at three studies of such olfactory expertise.

The language and hedonics issue was addressed in a study done in 2014 by Caroline Sezille and colleagues at the University of Lyon's Neuroscience Center that compared the differences in linguistic descriptors used by professional perfumers and flavorists with those of trainee cooks and untrained individuals. Each group was exposed to twenty odorants pretested to ensure a wide hedonic range.[18] Each individual in the experiment was given two tasks, first to rate the pleasantness of the odorant and then to describe it as "precisely as possible." Although the pleasantness ratings for all groups were similar, there were striking differences in the kinds of descriptors used. As in similar previous studies, the experts processed the odors "more deeply on a lexico-semantic level, with few hedonic references." Specifically, their descriptions were longer, more precise, and more consistent, as well as being semantically richer and more expressive than those of either the trainees or the untrained.[19] These results are not surprising since we know that professional flavorists and perfumers undergo extensive training in a common vocabulary and are expected to be able to offer qualitative, not merely hedonic, responses. Crucially, Seville and her colleagues' study of perfumers and flavorists confirms several studies done with wine experts as well as the experience of specialists in brandy, beer, cheese, fish, and other fields, where experts tend to use analytic terms and novices tend toward holistic and hedonic terms.

The Seville study is also in line with neuroscience imaging studies that have questioned Sobel and Yeshurun's reduction of all human smell experience to the hedonic axis. Although hedonic evaluation are often all that many people can articulate of their initial response to an odor, Jonas Olafsson and colleagues have performed experiments on the temporal dynamics of olfactory response that question the primacy of the hedonic response. In their studies, most people first identify an odor quality, "It's strawberry," then fractions of a second later make their hedonic judgment, "It's pleasant."[20] And recall Benoist Schall's comment on a study of odor ranking by children five to eleven years old: their classifications did not show a necessary "overlap with hedonic categorization, suggesting that young children can *separate emotional from more cognitive operations* while processing odors."[21] Although these few studies showing that genuine expertise is possible in the realm of smell may be challenged or qualified by further research, they seem consistent with the intuition that most educated people are able to distinguish their qualitative judgments in general from their purely hedonic ones. Barry C. Smith has forcefully argued a similar point about the informed judgments of wine experts. Even if inclinations of liking/disliking inevitably enter into experts' judgments, experts are able to move deeper into understanding a particular wine's quality whether they personally like the wine or not.[22]

The next two studies of olfactory experts that I want to consider offer evidence of smell's capacity for cognitive cultivation based on neuroscience imaging of brain plasticity. The first plasticity study concerns "structural brain plasticity," that is, changes at the anatomical level in the *amount* of gray matter present in relevant brain areas. Just as brain studies of higher-performing musicians and athletes have indicated increased gray matter in relevant motor areas of the brain, so the study by Delon-Martin et al. in 2013 detected a larger gray-matter volume in perfumers' olfactory processing areas such as the piriform cortex and the orbitofrontal cortex than in the brains of novices. Moreover, the greater amount of gray matter was positively correlated with age and experience in the professional perfumers but negatively correlated with age in control subjects.[23] In addition to confirming the old adage of "Use it or lose it" for normally functioning subjects, these results are also consistent with the studies of people suffering from late onset anosmia or hyposmia, whose brains show gray-matter atrophy in olfactory-related areas. The studies also agree with the results of the 2010 study by Frasnelli et al. showing a correlation of olfactory bulb volume with higher identification scores and larger orbitofrontal cortex volume with better discrimination performances.[24] Taken together with these other studies, the

Delon-Martin demonstration of gray-matter increase correlating with experience in perfume experts provides strong evidence that the sense of smell can indeed be cultivated to a high degree and such abilities maintained.

Equally impressive results regarding the cultivation of olfactory expertise come from a 2012 study of "functional brain plasticity," that is, changes in the *activity levels* of relevant brain areas with respect to odor imagery. Specifically, Jane Plailly and colleagues' study of experts' functional brain plasticity offers strong confirmation that olfactory mental imagery can reactivate memory traces within the piriform cortex, and do so much better among experts than nonexperts. The study involved fourteen beginning perfumery students who had completed two years of training and fourteen experienced perfumers, most of them well known in the profession, who had between five and thirty-five years of experience. The experiment had two parts: in the first part each subject was exposed to twenty chemicals from the list of three hundred that student perfumers are expected to learn; in the second phase, a series of twenty chemical names were flashed in random order on a screen and the subjects were asked to form a mental image of the smell if they could. Roughly 92% of both students and professionals claimed to have formed images.[25]

During the passive perception phase, both students and professionals showed similar activation of the relevant areas of the piriform cortex. But during the imagery sessions, there were striking differences between the two groups. The experienced professionals could "quickly—and even instantaneously—imagine most odors, whereas students had difficulty with this task and could only imagine odors by deliberately focusing their attention."[26] Even more striking was the fact that the fMRI recordings showed markedly lower levels of activity in the olfactory processing areas of the professionals' brains compared to the students. At first glance, one might think that it should have been the opposite, but, in fact, a lower level of activity means that the professionals had to expend less effort than the students.

Moreover, differences among the professionals' *level* of enhanced efficiency / reduced effort were correlated with the number of years of experience, lending support to the idea that "mental imaging of odors develops from daily practice and is not an innate skill."[27] This result from experienced perfumers is consistent with brain studies of professional musicians and golfers that have shown similar kinds of functional brain activity decreases associated with performance gains. Researchers who have done brain studies on musicians, such as Martin Lotze et al.'s study of violinists, have concluded that over time, professionals learn to control their movements more or less automatically, resulting in less cortical activity in motor areas, thereby freeing up additional brain resources for enhanced performance.[28] Similarly, Plailly and colleagues conclude that professional perfumers "progressively develop more efficient strategies in their field of expertise, allowing them to liberate additional resources for other aspects of artistic performance such as the creation of new fragrances."[29]

If we put together the results of these recent studies of olfactory expertise—the experts' superior naming and descriptive ability, the structural/anatomical increase of gray matter in olfactory areas, and the functional brain activity decrease in those same areas—*and* keep in mind that these results are consistent with the results of similar studies in other domains such as music or athletics, we must grant that there is now beginning to be neurological evidence that the cognitive powers of the human sense of smell are capable of being cultivated to a high degree. Even neuroscientists like Olafsson and Gottfried, who showed that the brain circuitry for smell has more limited connections with language areas than it does for vision, accept that studies like that of Plailly et al. show that the general neural processing limitations on smell can be overcome through training and practice, leading to expertise that far exceeds the performance of the typical subjects of most psychological experiments.[30] Moreover, R. L. Stevenson, who expressed such a dim view of smell's cognitive resources, has also accepted the possibility that "extensive practice can produce increases in neocortical processing power for smell sufficient to propel what may be unconscious processes in naïve participants into conscious ones for experts."[31]

Interpreting Olfactory Experts' Abilities

Of course, our appeal to recent studies of smell experts is unlikely to convince everyone who supports the negative tradition regarding smell's lack of cognitive power. The skeptics will emphasize the small number of professional perfumers in the world and the small number of studies of them compared to the critical mass of studies of nonexperts showing that most people are unconscious of the odors around them, are heavily influenced by emotion, tend to make simplistic hedonic judgments, and are unable to identify and name most odors. Given Andreas Keller's forceful assertion of smell's strong connection to the emotions and its weak connection to language that I cited earlier, one might think that he would support the skeptics, but that is not the case.

In fact, in *Philosophy of Olfactory Perception*, Keller embraces an understanding of the brain as a complex of overlapping networks that opens the way to recognizing a modest role for cognition in the sense of smell. He notes that "olfaction is bidirectionally connected to other [sensory] modalities and cognitive processes" so that "massive feed back from higher brain areas" provides the neural correlates for such things as the "cognitive penetration" of olfaction (as when we react differently to isovaleric acid depending on whether it is labeled "cheese" or "vomit.")[32] Keller then uses the phenomenon of attention to explain how various levels of cognitive awareness affect what and how we smell. Thus, if we assume that all our sensory modalities have evolved in order to guide behavior, it makes sense that most of the time we do not consciously perceive the odors around us. In most situations, according to Keller, the fact that we register odors in the brain without consciously noticing them guides us toward immediate actions: inhale / hold the breath, swallow / spit out, approach/avoid, stay/ go.[33] These are the situations, I assume, that Keller had in mind when he wrote the passage on the emotional hedonics of smell that I quoted earlier, in which he speaks of automatic responses where "reason is powerless to intervene." But in addition to such situations prompting immediate binary responses, Keller suggests there are other situations in which we may have time to choose among several relevant behaviors, such as when we are writing a wine review or trying to locate a gas leak in a school building. Then, rather than an immediate emotional reaction, there can be conscious attending to olfactory qualities.[34]

As Keller points out, the fact that most perceptual attention in humans is primarily attached to vision and hearing does not mean "the complete absence of attention" in the case of smell.[35] And I believe this point about attention also holds for complaints about smell's weak connection to language when compared with vision and audition; "comparatively weak" in relations to vision or hearing does not mean the complete absence of linguistic capacity to express smells. Thus, on the issue of whether the human sense of smell has enough cognitive and linguistic resources to support intelligent aesthetic creation and discussion, even on Keller's account the answer should be a cautious yes.

But the case for smell's cognitive adequacy to underwrite reflective aesthetic experience and judgment is actually stronger than Keller's analysis implies. As we have noted, despite the small number of studies of olfactory experts, the results of studies like those of Sezille, Delon-Martin, and Plailly are supported by similar studies of expertise in wine tasting, music performance, and athletics. And when we add to those outcomes, the earlier evidence from Chapter 2 that even untrained people are excellent at detection, discrimination, and learning, and evidence of what Gottfried and Wu refer to as the "tremendous perceptual and neural plasticity" of our sense of smell, I believe we have strong scientific support for the belief that smell can be cultivated to a degree sufficient for sophisticated artistic creation and critical aesthetic reflection.

Moreover, we should not forget the multisensory character of our everyday perceptual experience. Smell profits mightily from the other senses' abilities and offers something of its own in return. In fact isolationism regarding the senses partly underlies philosophical disagreements about the cognitive capacities of smell. Treating smell in isolation is part of what lies behind the differences between the philosophers Lycan and Batty concerning whether smell represents indirectly by representing odors (Lycan) or represents a vague "something" in the immediate surroundings (Batty). Batty is explicit that she is talking about what smell can do completely on its own so long as we do not draw on any of the other senses and hold our bodies immobile, not even moving our head.[36] That is an interesting thought experiment, but, as she herself notes, it is obviously not the way we normally operate. When I went to the store to find cilantro and my vision failed to tell me by the shape of leaf which bunches among the various green herbs were cilantro, I was fortunately not immobilized, but could bring different bunches of herbs to my nose, letting my sense of smell tell me which was cilantro. And the odor was not a vague something in the surrounding air but was localized since I felt the leaves against my nostrils. Equally important, there were many contextual aspects of the situation that narrowed my options. By contrast, it is likely that if I had volunteered for a laboratory experiment and was exposed to a puff of cilantro odor while blindfolded or lying in an fMRI scanner and asked to name it, I would fail. But when it comes to the smells in a food market or to those that are part of olfactory artworks in a gallery or museum, they are not only experienced multimodally, but are often accompanied by a variety of contextual clues and in the case of scent artworks in a gallery or museum, may even include an artist's statement.

My overall conclusion—that the sense of smell is cognitively much stronger than the mainstream Western intellectual tradition has been willing to admit— may seem like a modest result for such an extensive discussion of the empirical and theoretical work on the neuroscience and behavioral psychology of smell. But given the long negative intellectual tradition in the West of either ignoring or disparaging the sense of smell and given the philosophical tradition of denying the artistic and aesthetic potential of smell, it has seemed important to consider the strongest evidence against the cognitive potential of smell before examining other evidence from the neurosciences and the philosophy of perception that suggests smell can in fact be cultivated sufficiently for artistic creation and aesthetic reflection.

But the arguments I have just given from neuroscience studies of experts, even when corroborated by similar studies of experts in other fields and supplemented by arguments from the philosophy of perception, still do not add up to an overwhelming case. For one thing, there are so few olfactory experts; for another, we can't all spend two or more years in perfumery school learning to identify two hundred to three hundred odors. We need evidence that ordinary people can learn to attend to odors and to appreciate the olfactory arts in ways similar to those they have learned for looking and listening with attention as they appreciate music, painting, and literature. Fortunately, in addition to the neuroscience studies of experts, there is evidence from an array of social science and humanities disciplines to support the idea that nonprofessionals can learn to cultivate their sense of smell. Part II will draw on that evidence to answer the specific claims that the sense of smell is of little use to humans, and that we are inevitably tongue-tied when it comes to expressing smell experiences due to an essential poverty of language for olfaction. At this point, as a transition to that discussion, I will close Part I by drawing further on neuroscience and psychology as well as philosophical analysis to counter the charge that the sense of smell is too emotional to sustain reflective aesthetic experiences and judgments.

Notes

1. Herz, *Scent of Desire*, 28–29.
2. Herz, *Scent of Desire*, 11–18, 114–16.
3. Richard J. Stevenson and Tuki Attuquayefio, "Human Olfactory Consciousness and Cognition: Its Unusual Features May Not Result from Unusual Functions but from Limited Neocortical Processing Resources," *Frontiers in Psychology* 4, no. 819 (2013): 4, https://doi.org/10.3389/fpsyg.2013.00819.
4. Yeshurun and Sobel, "An Odor Is Not Worth a Thousand Words," 326.
5. Jonas K. Olofsson and Jay A. Gottfried, "The Muted Sense: Neurocognitive Limitations of Olfactory Language," *Trends in Cognitive Sciences* 19, no. 6 (2015): 314–21, https://doi: 10.1016/j.tics.2015.04.007.
6. Herz, *Scent of Desire*, 57.
7. Sela and Sobel, "Human Olfaction," 13–29.
8. Rachel Herz, "Odor Memory and the Special Role of Associative Learning," in *Olfactory Cognition: From Perception and Memory to Environmental Odours and Neuroscience*, ed. Gesualdo M. Zucco, Rachel S. Herz, and Benoist Schaal (Amsterdam: John Benjamins, 2012), 101. Italics mine.
9. Yeshurun, and Sobel, "An Odor Is Not Worth a Thousand Words," 219. Italics mine.
10. Yeshurun, and Sobel, "An Odor Is Not Worth a Thousand Words," 230.
11. Egon P. Köster, Per Møller, and Jozina Mojet, "A 'Misfit' Theory of Spontaneous Conscious Odor Perception (MITSCOP): Reflections on the Role and Function of Odor Memory in Everyday Life," *Frontiers in Psychology* 5, no. 64 (2014): 6, https://doi: 10.3389/fpsyg.2014.00064.
12. Köster, "A 'Misfit' Theory," 7.
13. Stevenson and Attuquayefio, "Human Olfactory Consciousness," 9.
14. Keller, *Philosophy of Olfactory Perception*, 117.
15. Jean-Pierre Royet, Jane Plailly, Anne-Lise Salve, Alexandra Veyrac, and Chantal Delon-Martin, "The Impact of Expertise in Olfaction," *Frontiers in Psychology* 4, no. 928 (2013): 8, https://doi: 3389/fpsyg.2-13.00928.
16. Andrew Livermore and David G. Laing, "Influence of Training and Experience on the Perception of Multicomponent Odor Mixtures," *Journal of Experimental Psychology: Human Perception and Performance* 22, no. 2 (1996): 267–77, https://doi.org/10.1037/0096-1523.22.2.267. For a recent review of the topic see Thomas-Danguin et al. "The Perception of Odor Objects in Everyday Life."

17. André Holley, "Cognitive Aspects of Olfaction in Perfumer Practice," in *Olfaction, Taste and Cognition*, ed. Catherine Rouby et al. (Cambridge: Cambridge University Press, 2002), 21.
18. Caroline Sezille et al., "Hedonic Appreciation and Verbal Description of Pleasant and Unpleasant Odors in Untrained, Trainee Cooks, Flavorists, and Perfumers," *Frontiers in Psychology* 5, no. 12 (2014): 1–8, https://doi: 3389/fpsyg.2014.00012.
19. Sezille et al., "Hedonic Appreciation," 7.
20. Jonas K. Olofsson et al., "A Time-Based Account of the Perception of Odor Objects and Valences," *Psychological Science* 23, no. 10 (2012): 1224–32, https://doi.10 .1177/ 0956797612441951
21. Schaal, "Prenatal and Postnatal Development," 310. Italics mine.
22. Smith, "The Objectivity of Tastes and Tasting," 55–56.
23. Chantal Delon-Martin, Jane Plailly, et al., "Perfumers' Expertise Induces Structural Reorganization in Olfactory Brain Regions," *NeuroImage* 68 (2013): 55–62, https:// doi: 10.1016/j.neuroimage.2012.11.044.
24. J. Frasnelli, J. N. Lundstrom, et al. "Neuroanatomical Correlates of Olfactory Performance," *Experimental Brain Research* 201 (2010): 1–11.
25. Jane Plailly, Chantal Delon-Martin, and Jean-Pierre Royet, "Experience Induces Functional Reorganization in Brain Regions Involved in Odor Imagery in Perfumers," *Human Brain Mapping* 33, no. 1 (2012): 224–34.
26. Plailly, "Experience Induces Functional Reorganization," 232.
27. Plailly, "Experience Induces Functional Reorganization," 231.
28. Martin Lotze et al. "The Musician's Brain: Functional Imaging of Amateurs and Professionals during Performance and Imagery," *NeuroImage* 20 (2003): 1817–29, https://doi: 10.1016/j.neuroimage.2003.07.018.
29. Plailly, "Experience Induces Functional Reorganization," 233
30. Olofsson, and Gottfried, "The Muted Sense," 318–19.
31. Stevenson, "Human Olfactory Consciousness," 9.
32. Keller, *Philosophy of Olfactory Perception*, 117
33. Keller, *Philosophy of Olfactory Perception*, 117–24.
34. Keller, *Philosophy of Olfactory Perception*, 177–78.
35. Keller, *Philosophy of Olfactory Perception*, 160.
36. Batty, "Representational Account of Olfactory Experience," 511–38.

4
Smell, Emotion, and Aesthetics

> Smell is the archetypical sense of emotion.
> Intellect is for the eyes and ears.
> —Michael Stoddart, *Adam's Nose*

Stoddart's claim that smell is to emotion as sight and hearing are to intellect echoes the views of those psychologists who treat smell as almost purely a matter of emotion with little or no cognitive resources.[1] But such a stark opposition between emotion and intellect is deeply misleading in the light of the psychology and philosophy of emotion. In saying this I am not questioning the abundant evidence that odor experiences bear a higher affective charge than those of vision or hearing. Rather, the first half of this chapter will show that the emotions themselves often have a cognitive component and that a robust cognition often requires an emotional aspect to operate effectively. The second half of the chapter will show that aesthetic experience and judgment also have an indispensable affective as well as cognitive aspect, so that smell's strong emotional charge is not per se an impediment to its participation in reflective aesthetic experiences and judgments.

The Intelligence of the Emotions

The conventional opposition between reason and emotion is far too simplistic despite the fact that one can trace aspects of it back as far as Plato. In Plato's tripartite division of the soul, reason must control the will, which must control the passions and appetites. One of Plato's objections to the art of tragedy was that it stirred up the passions rather than calming them. Of course, this was not Plato's last word on the passions; in the *Symposium* he celebrated the power of Eros to lead us by stages from the love of beautiful bodies up to the love of Beauty and the Good for their own sake. Yet the best corrective in ancient philosophy to a radical conflict of reason and emotion can be found in Aristotle's argument that the ethical person is one whose appetites, senses, and emotions are well trained to follow a middle path. Although many religious and philosophical thinkers in the Western tradition have warned of the ways in which emotion

Art Scents. Larry Shiner, Oxford University Press (2020) © Oxford University Press.
DOI: 10.1093/oso/9780190089818.003.0009

can undermine and distort morality and thought, many of those same thinkers have, like Aristotle, viewed the emotions as not only educable, but, like Spinoza, have seen affectivity itself as a critical support for intellectual insight. Indeed, in contemporary philosophical reflection on the emotions, a number of writers, such as Michael S. Brady in *Emotional Insight: The Epistemic Role of Emotional Experience*, have not only explored the more specifically intellectual feelings of curiosity, wonder, excitement, and intellectual courage, but even argued for the cognitive value of such "paradigm emotions" as fear, anger, joy, and pride.[2] John Deigh was not exaggerating when he remarked in the *Oxford Handbook of Philosophy of Emotion* of 2010 that the idea of "emotion as essentially a cognitive state ... now prevails among philosophers and psychologists."[3]

One thing common to both the philosophy and the psychology of emotion is that the emotions are often viewed as "intentional"; that is, they are typically directed *at* something; for example I am angry *at* or *with* someone who has offended me. Although research has shown that some of the basic emotions do have a distinctive body state and may sometimes occur with little cognitive input, the cognitive aspect of many emotions has been attested by other kinds of research, such as the demonstration that emotions involve appraisals of a situation. Some of these studies stress "patterns of salience," suggesting that emotions are cognitively significant in helping us focus our attention and identify what is important amid the welter of information that constantly assails us.[4]

Naturally, positions on the nature of cognitive-emotional integration vary widely among both neuroscientists and psychologists, but it is safe to say that the kind of sharp dichotomy between emotion and cognition implied by statements like Stoddart's has been replaced by a more complex and nuanced view of cognitive-emotional interaction, including evidence that cognitive resources are often required if the amygdala is to recognize and respond to threats. As the editors of the 2013 *Handbook of Cognition and Emotion* remark, "It has become increasingly apparent that cognition and emotion often interact and are perhaps not isolated entities."[5]

One well-known avenue of research supporting an essential connection between emotion and cognition has come from studies of brain damage such as Antonio Damasio's widely read *Descartes' Error: Emotion, Reason, and the Human Brain*. Descartes's error, according to Damasio, lay in separating mind and brain, reason and emotion. Drawing on studies of people with brain damage, Damasio showed that those who had a poor capacity to experience emotion were seriously limited in their ability to make intelligent decisions. In one case, a patient spent a half hour deliberating over which of two dates to select for his next appointment, rationally calculating every possible obligation and circumstance that could come up, and might have gone on indefinitely unless Damasio had not gently suggested the man take the second date. From cases like this, Damasio concluded that there is a "crucial role" for feelings in navigating the many daily decisions whose multiple conditions would tie us in knots if it were not for the focusing power of the emotions. The emotions do not make the decision for us but allow us to deliberate and choose among fewer options.[6]

Among philosophers, as one might expect, there are several competing theories as to what constitutes the core of the cognitive aspect of emotion as well as emotion's specific contribution to cognition. For our purposes, we do not need to choose among them, but I will briefly mention two main types in order to give something of the flavor of the cognitive mainstream in contemporary philosophy of emotion. First, and oldest, is what is actually called "the cognitive theory," typified by the approaches of Robert Solomon and Martha Nussbaum, according to which emotions are sometimes likened to judgments involving "propositional attitudes." Second, there are "perceptual" theories that range from those of Ronald de Sousa or Amélie Rorty, which liken emotions to perceptions of the external world that provide us frameworks such as "paradigm scenarios" for orienting cognition, to that of Jesse Prinz, who emphasizes bodily response as part of situational appraisals.[7]

Carolyn Price's recent version of a perceptual theory suggests that we think of the cognitive aspect of emotions as involving "grounds" (a global response based on past experience that is adequate for an initial, rapid response) rather than "reasons" (a carefully worked out case based on explicit evidence). Accordingly, like Damasio, she suggests that the wise course in life is to follow neither "head" nor "heart" alone, since we need both for sound understanding; attending to both of them prompts us to "think again," a suggestion that takes in both the warnings of Plato against the dangers of emotion and the counsel of Aristotle to cultivate the emotions toward moderation.[8]

Price's view also fits well with Jenifer Robinson's suggestion that we think of emotion not as a state but as a *process*. Although many situations may provoke an initial physiological reaction or affective appraisal with little higher cognitive input (similar to Ekman's universal "basic emotions"), this is rapidly followed, often in milliseconds, by a developing assessment in the higher cortex. And even our initial "gut reaction" contains important information, although we may not be explicitly conscious of it. But what is particularly interesting about Robinson's process view for our concern with the alleged uniqueness of smell's emotionality compared to the "intellectuality" of vision and hearing is that most of the neuroscience research she reviewed in developing her philosophical understanding of the emotions was based on studies of vision and hearing, which turn out to be subject to the same general processes as smell: an immediate physiological reaction followed fractions of a second later by input from higher cortical regions.[9] The cognitive differences between smell, on the one hand, and vision or hearing, on the other, then, are a matter of degree and do not justify blunt contrasts that imply that smell is overwhelmingly emotional and somatic whereas vision and hearing are primarily intellectual and cerebral.

This brief survey of some current psychological and philosophical studies of the cognitive aspect of emotions should be enough to show that it would be a gross error to assume that the greater affectivity or "emotionality" of the sense of smell as compared to vision or hearing means that smell lacks the capacity to support reflective activities. Price's description of what she calls cognitive "grounds," a rapid global response based on past experience, could be seen as a cognitive version of the sort of immediate "stay/flee" reactions of the smell system that Keller described as impervious to reason. But as Robinson's notion of emotion as process suggests, it is not so much that smell is "impervious" to reason in such stay/flee situations as that there is often no time for deliberation, just as there is often no time for deliberation when the stimulus is visual or auditory. Price's and Robinson's views suggest that it may not be just on rare occasions when one needs to write a wine review or locate a gas leak that smell involves a cognitive dimension, but that even in some situations of apparently "automatic" reaction, our olfactory response may be "grounded" in past cognitively informed experiences. If we accept the current philosophical arguments and scientific evidence that the emotions themselves contain a cognitive element that is educable, and that feelings play a positive (as well as sometimes disruptive) role in cognition, then the fact that people's sense of smell is strongly marked by emotional and hedonic reactions, does not imply that smell is incapable of providing a basis for critical aesthetic reflection and discussion.

Aesthetics and Emotion

A second major argument against the claim that the strongly affective character of smell is fatal to aesthetic reflection and communication concerns the concept of the aesthetic itself. Just how one might define "the aesthetic" has been a highly controversial matter within philosophy ever since the term was coined by Alexander Baumgarten in the mid-eighteenth century and gradually displaced the terms "taste" and "beauty" over the course of the nineteenth century. As an adjective, it has been used by philosophers and critics to speak of aesthetic experience, aesthetic appreciation, aesthetic judgment, and something called "the aesthetic attitude," as well as aesthetic concepts and properties, even aesthetic emotions. As Elisabeth Schellekens notes, the general idea of the aesthetic has been taken to include several kinds of mental states, including the pleasurable appreciation of form, detachment from practical concerns, perception of qualities such as elegance or balance, attention to the ways artworks embody formal and expressive qualities, awareness of emotional or representational content in art, awareness of formal qualities in nature, and so on.[10] But Paul Guyer has shown in his *History of Modern Aesthetics* that despite all this variety, most of modern philosophical aesthetics since Kant has been preoccupied with the cognitive aspect of aesthetic judgments, whether it be Kant's own theory of the play of our cognitive powers of imagination and understanding or Hegel's conviction that a proper experience of art reveals higher truths of the Spirit. Yet Guyer also notes that there has been a minority tradition that has treated emotion as a crucial part of aesthetic experience, and he suggests that the historically most interesting thinkers have been those who have tried to integrate aspects of all three approaches: revealing truth, the play of our mental powers, and emotional impact.[11] Clearly, theories of aesthetic experience and judgment that embrace both its cognitive and its emotional dimensions, unlike purely cognitive theories, should have ample room for exploring how all the senses, including touch, taste, and smell, can be involved in aesthetic communication. This outcome, as we saw earlier, was one of the aims of Frank Sibley's pioneering exploration of the aesthetic possibilities of taste and smell. In the remainder of this chapter, I want to flesh out the claim that a fully adequate idea of aesthetic experience and judgment must include an emotional dimension by examining a few historical and contemporary examples.

A key figure from ancient philosophy is again Aristotle, whose *Poetics*, although concerned with ancient Greek drama, points the way toward understanding the role of emotion in all arts. For Aristotle, the best tragic poets are able to rouse the emotions of pity and fear and at the same time provide a catharsis of those emotions. Probably as much ink has been spilled over just what Aristotle meant by catharsis as any concept in the history of aesthetics. Yet despite intense disagreements on details, I think most interpreters today would agree that the popular notion that catharsis signifies "purging" the emotions in the sense of getting rid of them is wrong. Rather, many interpreters see catharsis as having at least three dimensions. One of those dimensions does indeed involve a kind of "living through" the aroused emotions in a way that brings a certain calm. But this calming is neither a purging in the sense of elimination, nor is it the sole aspect of catharsis, but is bound up with two others. For Aristotle, catharsis involves the mind as well as the feelings, something we could call a "deepening of understanding." Catharsis means achieving the calm that accompanies gaining insight into the nature of what has happened and why and how it relates to life as a whole. The power of drama and of many other art forms changes not just "how we see the world," as a popular cliché would have it, but at the same time changes how we *feel* the world. A third aspect of catharsis, often overlooked, concerns the way the emotional and cognitive insight of catharsis also represents a spiritual cleansing.

Aristotle's joining of the emotional and cognitive in the experience of tragedy was echoed down through the centuries in the famous phrase of the Roman poet and critic Horace: the aim of poetry is to "please and instruct." This joining of pleasure with a cognitive and moral view of the experience of art, as Guyer argues, "was the core of Western thinking about art for centuries before the formation of a specialized discipline of aesthetics in the eighteenth-century."[12] Indeed, in the early eighteenth-century debates over the objectivity of taste that led up to the Kantian version of aesthetic cognitivism, there were influential figures in both France (Abbé Du Bos) and Britain (Edmund Burke) who made the "passions" central to what they called "taste" and what we now call aesthetic judgment.

By the time Kant published his *Critique of the Power of Judgment* in 1790, he had at his disposal the new term "aesthetic," and Kant used it in its generic meaning of something involving the senses to embrace both what he called the merely "agreeable" (matters of subjective pleasure and preference) and the "taste of reflection" (the free play of our cognitive powers of imagination and understanding). Kant's focus on the free play of our mental powers explicitly excluded both "charm and emotion" (*Reiz* and *Rürung*) from the realm of genuine aesthetic judgment, and both Hegel and Schopenhauer focused on the aesthetic as the revelation of truth to the exclusion of emotion.[13] By the twentieth century, many philosophers had begun using "aesthetic" to refer only to the kind of thing Kant had called the taste of reflection, although they explained it in a variety of ways. It was not until the late nineteenth century that thinkers such as Dilthey and Nietzsche in Germany and Santayana in America began to integrate emotion into their concepts of aesthetic experience and judgment. In the first half of the twentieth century, two very different and influential philosophers, John Dewey and R. G. Collingwood, also joined emotion and cognition in their theories of aesthetic experience.[14]

In contemporary analytic aesthetics, views on the emotional component in aesthetic responses to the arts vary widely, with Jesse Prinz perhaps most firmly insisting that our aesthetic response to art is fundamentally emotional, specifically the emotion of wonder.[15] There have been especially lively philosophical discussions of the role of emotion in the aesthetic appreciation of music and literature, such as Jenifer Robinson's *Deeper Than Reason: Emotion and Its Role in Literature, Music and Art*. Although G. Gabrielle Starr is a literary scholar rather than a philosopher, as she remarks in *Feeling Beauty: The Neuroscience of Aesthetic Experience*, "Aesthetic value is both thought and felt; it is something 'cognitive,' 'sensory,' and 'emotional' " all at once.[16] There is still a lot of theoretical work to be done in showing how each of these aspects of aesthetic value interacts with the others, but it suffices for the point I am making that the role of emotion in appreciating art and nature is no longer on the periphery of mainstream aesthetics but at its center.

Finally, and most directly related to the place of emotion in the aesthetic exercise of the sense of smell, there is the extensive work done on the aesthetic appreciation of wine by analytic philosophers such as Barry C. Smith, Kevin Sweeney, and Cain Todd, each of whom gives an important place to the emotional aspect of appreciation. As Barry C. Smith has said about the aesthetics of wine: "Wines, like music, can give rise to aesthetic emotions; wonder, surprise, delight, disappointment, and fascination," emotions that "take us beyond mere liking and suggest an appreciative engagement with a wine."[17] As I will argue in a later chapter, the best perfumes can also give rise to "wonder, surprise, delight, disappointment and fascination," and take us far beyond mere liking or disliking. And, of course, the same is true of the odors that accompany theatrical productions, or that are incorporated into installation or performance works like those of Tolaas, Ursitti, or de Cupere. The many kinds of contemporary olfactory arts appeal via our sense of smell to both our reason and our emotions, often leading us to aesthetic reflection and discussion in a way not unlike what occurs in our engagement with music, painting, or literature.

This chapter's survey of the place of cognition in emotion, of emotion in cognition, and of emotion in aesthetic experience and judgment has given us good reason to dismiss the idea that because the sense of smell has such a strong emotional and hedonic component, it cannot be the basis of an olfactory aesthetics. First, we have seen that the emotions themselves not only have a cognitive dimension but that cognition has an emotional dimension that is indispensable to its fullest exercise. Second, we have just shown that a truly robust account of aesthetic experience and judgment will necessarily include not only a cognitive but also an indispensable emotional element. Moreover, as we will discover in Part III, artists and designers who use odors in their work often cite the strong affective dimension of smell experiences as one of the advantages of odors as a medium for art and design.

I want to close this chapter by revisiting a remarkable olfactory experience reported by the psychiatrist Oliver Sacks in his book *The Man Who Mistook His Wife for a Hat*. A twenty-two-year-old medical student came to talk with Sacks about a vivid dream he had had after taking psychoactive drugs. He dreamed he was a dog, and when he awoke, his sense of smell had become so heightened that

> all other sensations, enhanced as they were, paled before smell.
> I went into the clinic, I sniffed like a dog, and that sniff recognized, before seeing them, the twenty patients who were there. Each had his own olfactory physiognomy, a smell-face, far more vivid and evocative, more redolent, that any sight face.[18]

So acute was the student's sense of smell, Sacks says, that this student could "smell their emotions—fear, contentment, sexuality—like a dog. He could recognize every street, every shop, by smell—he could find his way around New York infallibly, by smell."

Does this sound too good to be true? I thought so when I first read it thirty years ago; surely this student was exaggerating, if not confabulating. Why did Sacks just take him at his word and report it as if it were true? Imagine my surprise, then, when I discovered in Sacks's 2015 autobiography that Sacks was *himself* the student he describes in that essay from 1985.[19] Given our earlier questioning of the suggestion that smell is almost purely emotional and hedonic, I find the following lines describing the younger Oliver Sacks's drug-induced experience to be particularly revealing for our purposes: "Smell pleasure was intense—smell displeasure, too—but it seemed to him less a world of mere pleasure and displeasure than a whole aesthetic, a whole judgment, a whole new significance, which surrounded him."[20]

Here, in a nutshell, is the main point I have been making in this chapter: as important as the emotional and hedonic aspects of our smell experiences may be, they are capable of taking us beyond mere liking and disliking. The exercise of our sense of smell contains within itself a "whole aesthetic," capable of critical judgments and articulations of significance, and I don't think it takes drugs to put it in motion.

Notes

1. Stoddart, *Adam's Nose*, 236. The emphasis on emotion as opposed to cognition is typical of the work of Rachel Herz, but aspects of that view are widely shared among psychologists.
2. Michael S. Brady, *Emotional Insight: The Epistemic Role of Emotional Experience* (Oxford: Oxford University Press, 2013), 9–11.
3. John Deigh, "Concept of Emotion in Modern Philosophy and Psychology," in *The Oxford Handbook of Philosophy of Emotion*, ed. Peter Goldie (Oxford: Oxford University Press, 2010), 26.
4. Andrea Scarantino and Ronald de Sousa, "Emotion," *The Stanford Encyclopedia of Philosophy* (Winter 2018 Edition), Edward N. Zalta (ed.), https://plato.stanford.edu/archives/win2018/entries/emotion/.
5. Michael D. Robinson, Edward R. Watkins, and Eddie-Harmon-Jones, eds., *Handbook of Cognition and Emotion* (New York: Guilford Publications, 2013), 4.
6. Antonio R. Damasio, *Descartes Error: Emotion, Reason and the Human Brain* (New York: G. P. Putnam's Sons, 1994), 173.

7. Scarantino and De Sousa, "Emotion," *Stanford Encyclopedia of Philosophy* offers a brief summary of several perceptual views. See also Jesse J. Prinz, *Gut Reactions: A Perceptual Theory of Emotion* (Oxford: Oxford University Press, 2006).
8. Carolyn Price, *Emotion* (Cambridge: Polity Press, 2015).
9. Jenifer Robinson, *Deeper Than Reason: Emotion and Its Role in Literature, Music and Art* (Oxford: Oxford University Press, 2005), 57–99.
10. Elisabeth Schellekens, *Aesthetics and Morality* (London: Continuum, 2007), 23.
11. Paul Guyer, *A History of Modern Aesthetics*, vol. 1: *The Eighteenth Century* (Cambridge: Cambridge University Press, 2014), 9–29.
12. Guyer, *History of Modern Aesthetics*, 20–21.
13. Kant, *Critique of Judgment*, 68–72. For Hegel art is the sensuous embodiment of the Idea; for Schopenhauer art frees us momentarily from both the will and the emotions.
14. John Dewey, *Art as Experience* (New York: G. P. Putnam's Sons, 1934); R. G. Collingwood, *The Principles of Art* (Oxford: Oxford University Press, 1938).
15. Jesse Prinz, "Emotion and Aesthetic Value," in *The Aesthetic Mind: Philosophy and Psychology*, ed. Elisabeth Schellekens and Peter Goldie (Oxford: Oxford University Press, 2011), 71–88.
16. G. Gabrielle Starr, *Feeling Beauty: The Neuroscience of Aesthetic Experience* (Cambridge, MA: MIT Press, 2013), 16.
17. Barry C. Smith, "The Aesthetics of Wine," in *The Encyclopedia of Aesthetics*, 2nd ed., ed. Michael Kelly (Oxford: Oxford University Press, 2014), 286.
18. Oliver Sacks, *The Man Who Mistook His Wife for a Hat* (New York: Summit Books, 1985), 150.
19. Oliver Sacks, *On the Move: A Life* (New York: Alfred A. Knopf, 2015), 133, 254.
20. Sacks, *Man Who Mistook His Wife for a Hat*, 150.

PART II
SMELL REDEEMED
Language, Culture, and Memory

Overview
A Biocultural Approach

Now that we have answered the charge that smell is too subjective and emotional to sustain aesthetic reflection and judgment, we need to refute two other claims that, if true, would make cultivating the sense of smell or developing an olfactory aesthetics seem hardly worth the trouble. The first claim, going back at least to Kant, is that the sense of smell is of little use to humans, an idea Darwin also held, viewing smell as little more than an evolutionary vestige. The second claim is that smell's connection to language is so weak, and the potential of human languages for expressing smell is so poor, that serious aesthetic discussion involving smell would be extremely difficult if not impossible. To answer these charges, we will need to turn from a focus on neuroscience and psychology to several other disciplines—history, anthropology, linguistics, and literature—if we are to demonstrate that the sense of smell has in fact proved its usefulness in many cultures, including those of the West, and that peoples in many cultures are able to cogently articulate olfactory experiences, something even true of many Western poets and novelists.

First, I should offer a methodological caution. As in the case of psychological studies showing that smell responses are highly emotional, I am not challenging the validity of psychological studies that seem to show smell's linguistic weakness. I am simply claiming that the conclusions of many of these studies may apply primarily to the average contemporary Westernized experimental subject, and may be partly artifacts of Western cultural history, not universal characteristics of human beings as such. Thus, psychologists like Köster are not wrong to emphasize the unconscious nature of smell for the average contemporary Westerner. Nor is Sobel wrong to claim that in odor identification and naming, most urbanized Westerners do a miserable job, and have great difficulty getting beyond hedonic labels. But these are not necessarily universal, biologically determined human limitations. Although there do seem to be some underlying physiological bases for smell's linguistic limitations, such as those Olafsson and Gottfried have suggested are partly responsible for naming difficulties, neurological constraints on the sense of smell do not express themselves identically in all cultures nor in the same way at all times in the same culture. Determining the respective roles of biology and culture in the historical and cultural variability of the exercise of the sense of smell is a matter that must be established through both empirical research and informed debate on the implications of such research.

The position I am taking on the relation of culture and biology is similar to that of the neuroscientist Aniruddh Patel concerning the respective roles of evolution and culture in shaping the brain's musical capacities. As Patel points out, some evolutionary theorists have believed that the human capacity for music making emerged because it had survival (sexual selection) value for our ancestors, whereas cultural theorists have tended to claim that the brain's capacity for music making is a purely historical and cultural development. As Patel argues, this is a false dichotomy since it suggests that if something like music (or in our case art) is merely a cultural product, we could easily do without it.[1] Rather, Patel argues for what many neuroscientists call "gene-culture coevolution," the way some cultural developments can lead to genetic modifications.[2] Although it would take far more evidence than now exists to corroborate such a gene-culture coevolution of our ability to create and enjoy olfactory artworks, the more general biocultural approach suggested by Patel's work has obvious parallels to Murray Smith's "third culture" approach to film that involves triangulating evidence from experience, psychology, and neuroscience.[3] Smith's method served him well for the visual and auditory aesthetics of film, but as we concluded at the end of Chapter 3, answering the charges that smell is a near useless evolutionary vestige or that humans and human languages cannot adequately express smell requires we supplement neuroscience and psychology with evidence from the social sciences and humanities.[4] Thus, the approach I will be taking in Part II is closer to Dominic Lopes's proposal for a "liberal naturalism," in which "history, anthropology, and sociology stand to aesthetics on a par with psychology, neuroscience, or biology."[5]

Yet the general implication for human olfaction of any form of bioculturalism is the same: it would be as wrong to claim that the olfactory characteristics shown by modern Westernized individuals are universal human limitations determined solely by evolutionary selection, as it would be to claim that smell's fundamental characteristics are determined solely by historical and cultural factors. Neither biological reductionism nor cultural autonomism can give us a just measure of the strengths and limitations of the sense of smell for creating works of art or appreciating the world aesthetically. Thus, although the chapters that follow will focus on the ways culture shapes the sense of smell, they are meant to complement, not compete with, the neurological and psychological evidence already presented.

Part II will draw evidence from sensory history (Chapter 5), from anthropology and linguistics (Chapter 6), and from literature and social psychology (Chapters 7 and 8). Taken together, these chapters, along with the interlude "Fragrant Asia," will demonstrate the possibility of a vigorous olfactory aesthetics and complete the groundwork for the more direct exploration of the contemporary olfactory arts in Parts III and IV. Part II begins with a prelude on some recent evolutionary theories that suggest the human sense of smell is not a useless vestige. as Darwin claimed, but may have played its part in our becoming fully human.

Notes

1. Aniruddh D. Patel, *Music, Language, and the Brain* (Oxford: Oxford University Press, 2008), 400–401.
2. On gene-culture coevolution in general see Robert Boyd and Peter J. Richeson's *Not by Genes Alone* (Chicago: University of Chicago Press, 2004). On the link between the arts and evolution see Stephen Davies, *The Artful Species* (Oxford: Oxford University Press, 2012), Chapters 8–12.
3. Dominic McIver Lopes, *Aesthetics on the Edge: Where Philosophy Meets the Human Sciences* (Oxford: Oxford University Press, 2018), 16.
4. Thus, my methodology differs from Smith's insofar as I will make the social sciences and humanities an important fourth level of analysis without denying the ontological priority of brain and body.
5. Lopes, *Aesthetics on the Edge*, 9. See especially the chapter "Aesthetic Theory and Aesthetic Science: Prospects for Integration," 77–97.

Prelude
Darwin, Smell, and Evolution

In September 1838, Darwin paid a visit to Jenny, an orangutan in the London Zoo, bringing with him several objects to test her sensory interests. She liked the taste of peppermint and "listened with great attention" to a harmonica; she also liked the feel of a silk handkerchief and seemed "to relish the smell of Verbena."[1] Some thirty years later, in the *Descent of Man* (1871), Darwin wrote several influential pages on the evolutionary roots of human musicality.[2] Unfortunately, Jenny's relish of the smell of verbena seems to have inspired no similar investigation of the possible evolutionary roots of human olfactory arts. On the contrary, by the time of the *Decent of Man*, Darwin had concluded that the human sense of smell is little more than a vestige from some prehuman ancestor and asserted, as we saw earlier, that it is "of extremely slight service even to savages."

Darwin offered no specific arguments for this opinion in the *Descent of Man*, but recently John P. McGann, a neuroscientist at Rutgers University, has uncovered a probable source of some later scientists' arguments for what McGann calls the "myth" that humans have a poor olfactory system. He suggests that a major source of the myth is a claim concerning the olfactory bulbs made by the renowned neuroanatomist Paul Broca in 1879. Because the human olfactory bulbs are small and represent a smaller proportion of the brain in humans than in other mammals, Broca concluded that the olfactory bulb had atrophied in the course of evolution, reflecting a decrease in the importance of smell. Broca's views were picked up and repeated by many psychologists, including Freud, well beyond the mid-twentieth century. A second argument based on mammalian differences has since been added to the bulb size claim to support the idea that in the course of evolution smell became of little significance to humans. This is the fact that although humans have around one thousand genes for coding the receptors in the nose, over the course of evolution the number that actually get expressed has declined to less than four hundred, whereas some of our relatives, like Old World monkeys, still have seven hundred active genes and the New World squirrel monkey has a full mammalian set of almost one thousand. Moreover, some researchers argue that the drop in the number of active genes occurred at roughly the same time primates acquired full color vision, further confirming the idea that the human sense of smell sharply declined in power and importance millions of years ago as humans came to rely increasingly on vision. As McGann shows, however, all three of these arguments—the appeals to differences in olfactory bulb size, to the smaller number of active genes in humans than in some primates, and to a supposed sharp drop in the number of genes expressed occurring at the same time as the emergence of full color vision—have been questioned in research done since 2000.[3] Research and debate over these kinds of issues is ongoing, but as McGann's essay makes clear, the traditional Darwinian view that our sense of smell is an evolutionary vestige of little importance should no longer be treated as an unchallengeable fact.

Moreover, there are at least two other, more general evolutionary theories that offer positive alternatives to the idea that our sense of smell is an unimportant vestige. First, recall Michael Stoddart's claim that a crucial turning point in the evolutionary emergence of our humanness was the loss of the vomeronasal system, the kind of system whose triggering in other animals by a pheromone results in automatic sexual pursuit. Even if Stoddart overemphasizes the general importance of the loss of the vomeronasal system, that loss did enable our *orthonasal* smell system to become more fully human by making the role of smell in mating a matter of communication rather than impulsion. From that time on, humans, like Calvino's fictional humanoid, could now be attracted to the particular smell of a particular individual.

A second evolutionary theory supporting the importance of smell implicates *retronasal* smell in making us human. Richard Wrangham argues in *Catching Fire: How Cooking Made Us Human* that part of what led to an enlarged human brain was the social interaction surrounding the introduction of fire and cooking.[4] Wrangham argues that cooking food not only made food easier to chew and increased its flavor, but also necessitated complex social arrangements for acquiring, storing, protecting, preparing, and consuming food. Critics of Wrangham's claim point out that the big jump in brain size occurred well before the development of fire and cooking. But even if cooking emerged later, it was still early in human evolution and decisively shaped the future of our species. The neuroscientist Gordon Shepherd has gone on to draw a specific connection between cooked food, sociality, and the sense of smell. Shepherd argues that given the key role retronasal smell plays in creating food flavors, our sense of smell was a significant evolutionary adaptation.[5]

Shepherd's evolutionary argument begins by contrasting the human nose and its role in flavor perception to the snouts of other mammals, such as dogs, with which humans have been unfavorably compared concerning smell since Aristotle. Certainly dogs far outdo us in *orthonasal* smelling (sniffing), not only because they have many more smell receptors, but also because dogs' relatively long snouts and their specially shaped nasal openings are designed for maximum input while sniffing the ground. But the opening of a dog's *retronasal* passage (nasopharynx) is much farther back in its mouth than in humans, and the passage to the nasal receptors is much narrower than in humans. As a result, our nasal receptors seem to be able to rapidly take in a larger volume of retronasally experienced scent molecules from chewing our food than dogs can. When we chew and swallow and exhale, the odor-laden air is immediately pushed up into the nasal passage and across the epithelium, and our perception of flavor results from the union of the scent signals coming from the epithelium in the nose with the taste, tactile, and sound signals coming from the mouth. The greater importance of retronasal smell for humans than for dogs may be why dogs tend to scarf their food and we tend to savor it and especially enjoying eating and talking at leisure with others.[6]

Shepherd's theory also addresses one of the other main arguments for the view that smell is an evolutionary vestige of little importance, namely, the supposed drop in receptor numbers that accompanied our distant ancestors developing forward-facing eyes and full color vision. Shepherd argues that the important role of retronasal smell in the brain's perception of flavor means that the decline in the number of receptors was probably offset by our larger brain size and more complex mechanisms for *integrating* smell, taste, touch, and sound information. Thus, "Despite the declining numbers of receptor genes, the brain processing mechanisms of the smell pathway, culminating in the neocortex, *bestow a richer world of smell and flavor on humans than on other animals.*"[7] Shepherd's overall conclusion is that "current studies are already revealing capabilities of human smell that go far beyond the traditional view. Rather than being weak and vestigial, human smell appears to be quite powerful."[8]

Of course, there are several competing theories about what it is that makes us distinctively human, and all such theories are bound to have a large speculative component. But what is important about Wrangham's theory as parsed by Shepherd in relation to smell is not whether the emergence of cooking was *the* turning point that "made us human" (I would place my bet on toolmaking and language). What we should take from this debate on the evolutionary role of cooking and smell is that the very early development of retronasal smell along with cooking and human sociality, even if they were not the turning point that "made us human," nevertheless upsets the confident claim that the human olfactory system is a prehuman vestige of little consequence.

But if Darwin was wrong in *The Descent of Man* about smell being a useless vestige, he may have been right about the importance of *aesthetics* in natural selection, and this also indirectly favors a more positive role for smell in human evolution, especially given Stoddart's point about the loss of the vomeronasal organ. For a long time the received wisdom among theorists of evolution was that aesthetics played only an ancillary role in natural selection. According to this standard view, if female birds, for example, seem to chose the male with the most beautiful plumage, song, or courtship dance, it had nothing to do with aesthetics but was really because what looks beautiful to us was actually a marker of reproductive health and vigor for them. But the Yale ornithologist Richard Prum has argued in *The Evolution of Beauty* (2017) that Darwin was right and that sexual choice among birds does have a strong aesthetic component.[9] Even more encouraging for the importance of smell in evolution is the fact that the zoologist Michael Ryan makes a similar case for the role of aesthetics, including smell, in sexual selection among mammals. Whereas Plum considered primarily the visual and auditory courtship displays among birds, Ryan, who is a leading expert on frogs, also devotes a separate chapter of *A Taste for the Beautiful* (2018) to the place of smell in sexual selection among both humans and animals. As he points out, sexual selection means that traits of attractiveness that enhance an animal's mating success will evolve even if the traits hinder survival to some degree—think of the clumsy peacock with its enormous tail or the fact that certain frogs who attract mates with their loud croaking or moths who attract mates by strong odors also expose themselves to their most feared predator, bats, who are also likely to hear or to smell these mating invitations. Naturally there is a cost-benefit balance at work here or peacocks along with certain species of frogs and moths would have disappeared long ago.[10] Although we can easily appreciate a peacock's tail or the songs of certain birds, it is harder for us to appreciate or even imagine the delight caused by the odors emitted by a fruit fly, a moth, or even a deer. One is reminded of Thomas Nagel's famous essay "What Is It Like to Be a Bat?" dealing in part with the difficulty of knowing what another creature experiences. Ryan writes, "Extrapolating on Nagel, when we catch a whiff of a buck's musk during his rut, we might not share the same ecstasy as a doe, but if we probe her olfactory system, we can at least understand why she is in ecstasy."[11] Ryan goes on to discuss the role of odors in human sexual communication, especially the HLA/MHC complex that we discussed earlier, which provides additional evidence of smell's roots in human evolution.

Prum's and Ryan's reanimation of Darwin's idea of sexual selection interestingly intersects with Stoddart's point about the impact of the loss of the vomeronasal nasal organ, since that loss meant that human mating was no longer triggered by pheromones but has come to depend in part on aesthetic factors, among which are smell and touch, as well as sight and hearing. And humans, thanks to their technological prowess, are able to enhance their natural odor with natural or artificial scents to make themselves even more aesthetically appealing. Of course, the various evolutionary theories we have examined, which project hypotheses covering millions of years, all have a large speculative element. Even so, it seems to me that there is now beginning to be evidence regarding both orthonasal and retronasal smell that supports the idea that the human sense of smell has positive roots in evolution and is not a vestige in the course of disappearing, but in fact, may have had a role in perpetuating the human species. But there is a less speculative kind of evidence that can offer at least a partial answer to the question of the usefulness of smell and whether all humans are as unaware of odors or respond to them in a purely hedonic manner as do most urbanized Westerners: the history of human cultures. Although there is an enormous amount of work still to be done in historical and anthropological sensory studies, some fascinating discoveries have already been made that give the lie to Darwin's idea that the sense of smell is of little use to either "primitive" or "civilized" peoples.

Notes

1. Charles Darwin, *Charles Darwin's Notebooks, 1836–1844: Geology, Transmutation of Species, Metaphysical Inquiries* (Cambridge: Cambridge University Press, 1987), 554.
2. Charles Darwin, *The Descent of Man, and Selection in Relation to Sex* (Princeton: Princeton University Press, 1981).
3. John P. McGann, "Poor Human Olfaction Is a 19th Century Myth," *Science* 356, no. 597 (2017): 1–6, https://DOI: 10.1126/science.aam7263.
4. Richard W. Wrangham, *Catching Fire: How Cooking Made Us Human* (New York: Basic Books, 2010).
5. Shepherd, *Neurogastronomy*, 224–32. Of course neither Darwin nor Broca could have considered the central role of retronasal smell in human nourishment since it only began to be more intensely explored in the 1980s after an influential article by Paul Rozin on smell as a dual sense. Shepherd, *Neurogastronomy*, 17.
6. Shepherd, *Neurogastronomy*, 19–27. My italics.
7. Shepherd, *Neurogastronomy*, 111.
8. Shepherd, *Neurogastronomy*, 13–14.
9. Richard O. Prum, *The Evolution of Beauty: How Darwin's Forgotten Theory of Mate Choice Shapes the Animal World—and Us* (New York: Doubleday, 2017).
10. Michael J. Ryan, *A Taste for the Beautiful: The Evolution of Attraction* (Princeton: Princeton University Press, 2018), 8.
11. Ryan, *Taste for the Beautiful*, 35.

5
The Dialectic of Deodorization
Smell in Western History

One kind of indirect support for whether some human activity is rooted in evolution, as Stephen Davies points out, is whether a certain kind of behavior is both universal and ancient.[1] Certainly, there are signs of the use of incense and perfumes in almost all civilizations going back many thousands of years. In the West, for example, perfume-burners have been found in the Mediterranean region dating from around 5500 B.C. in the form of "Mother Goddess" figures whose heads are flattened on top and have a small hole with signs of burning.[2] Although this chapter's historical evidence is interesting in itself, I present it in order to set up three rather simple points. The first is that since odors and the sense of smell once played much more complex medical and social roles in Western culture that necessitated a greater attention to smells in general and to certain olfactory arts, smell could in principle play a more important and more conscious role again, something already happening with both the emergence of sensory studies in several academic disciplines along with the increasing prominence of various olfactory arts. The second point is that although the function of odors has become primarily aesthetic rather than utilitarian over the past two hundred years, this in no way diminishes the importance of the sense of smell unless we regard aesthetic experience itself as of little value. The third point is that because the rich olfactory history of the West has been obscured and largely forgotten in the process of "deodorization" it is possible that contemporary Westerners' unconsciousness of odors and their difficulty in naming them in psychological experiments may not be completely the result of biological limitations, but may have been amplified by historical changes.

The first section of this chapter will briefly present evidence for the important role incense and perfume have played in religion, medicine, and social relations throughout Western history from ancient Egypt through the eighteenth century. The second section will investigate the process by which these social, medical, and spiritual roles narrowed beginning in the early modern period, leading to a process that is often referred to as the "deodorization" of Western cities.[3] Of course, as, Mark M. Smith, a leading proponent of sensory history, points out, we should not exaggerate changes to the general smellscape of the West as if the distant past lived amid unbearable stenches compared to our relatively fragrant present.[4] But, as I will show in this chapter, there is abundant evidence of a profound shift in the role and image of both perfume and incense over the course of Western civilization from its roots in ancient Egypt, Greece, and Rome to the present. By the early twentieth century, for example, perfumes, which formerly served a wide range of functions, had been reduced in the West to little more than an optional adornment. And incense, which had also had wide medical and household as well as social and religious functions, had been reduced to a few rituals in Orthodox, Catholic, and some Anglican churches or relegated to an atmospheric enhancement in some boutiques and homes. These changes have no doubt contributed their part to Western intellectuals and philosophers ignoring and depreciating the sense of smell and to the general public's largely forgetting the formerly pervasive uses of scents and no longer paying much attention to them in daily life.

Given how young the field of sensory history is and how vast the time span that needs to be explored here, it would be foolhardy to offer more than a sketch focusing on a few well-established examples.[5] As I do so, I will occasionally use the combined term "perfume/incense" to remind us that for most of Western history, the term "perfume" had a far larger and more respectable scope than it does today and often included what we call incense within it (after all the Latin root of "perfume" is *per* (through) + *fumare* (to smoke or burn). Among the other differences between ancient and modern perfumes is that until the technique of distilling alcohol was perfected by Arab scientists in the ninth century and began to be used as a vehicle for perfumes in Europe in the late fourteenth century, most of what we call perfumes today came in the form of powders, pastes, or oils.[6]

Our Forgotten Olfactory Past

The Egyptians can truly be said to have led a scented life—and afterlife. The best-known aspect of this is the Egyptian practice of embalming that involved removing the vital organs and refilling the major cavities of the body with scented resins, after which it would be anointed with perfumed oils. At the end of the process, a priest wearing a jackal mask would say: "Thou hast received the perfume which shall make thy members perfect."[7] But the Egyptians not only scented their dead, they scented almost everything: statues of gods, civil ceremonies, banquets, houses, clothes, and their bodies.[8]

After the Hebrews escaped from Egypt, one of the first things God commanded them to do was to build an altar from which blood sacrifice would send up "a pleasing odor," and a second altar upon which to "burn fragrant incense every morning" (Exodus 29:18, 30:7) Moreover, the priests themselves were to be anointed with a fragrant oil containing "liquid myrrh . . . sweet smelling cinnamon . . . aromatic cane . . . and olive" (31:23–25). Another important reference to anointing with perfumed oil came from the idea of the Messiah (literally, "the anointed one"), who, as Deborah Green interprets Isaiah 11:3, "will be able to smell the truth rather than depend on what his eyes see or his ears hear."[9] (When I first read this, I couldn't help thinking of Nietzsche's idea of nosing out falsehood and bad faith.) The Hebrews also adorned their bodies with perfumes, as we know from the Song of Songs, which features two lovers extoling each other's perfumed scent, a topic to which we will return in Chapter 13.[10]

The ancient Greeks developed rich spiritual, medical, and social uses of perfume/incense. A notable civil use of scent was to open the Athenian Assembly by summoning the gods with incense, or as one text has it, by summoning "like to like." The classicist Ashley Clements points out that the term "like to like" (*omoia*) implies that sacrificial aromas were considered to be "equivalent to the divinities they were intended to attract."[11] Similarly, Jean-Pierre Vernant writes: "The gods smell fragrant; their presence is made manifest not only by intensely bright beams of light but also by a marvelous smell."[12] At the end of Euripides's *Hippolytus*, for example, Artemis manifests herself to the dying Hippolytus not visually but as a fragrance. "Oh breath of fragrance divine!" Hippolytus exclaims, "the goddess Artemis is in this place!"[13] If odors, by their invisibility and mobility, offered an ideal medium for expressing aspects of the divine in early Greek history, later thinkers from Plato on tended to view human-divine relations in increasingly rationalized ways; by the third century C.E., the Neoplatonist Porphyry could write that the important thing in sacrifice is not the incense, but "the disposition and manner of those who sacrifice."[14]

Although some ancient philosophers may have had reservations about the spiritual efficacy of incense, the medical uses of perfume/incense unquestionably thrived in ancient Greece. As the medical historian Laurence Totelin remarks, although smell played a smaller role in diagnosis than touch and sight, in therapeutics "smell substances were omnipresent, to the point where it is sometimes difficult to draw a boundary between ancient perfumery and ancient pharmacology."[15] For example, there was a scent cure for a "displaced" uterus (hysteria) based on the belief that the uterus "delights in pleasant smells and goes towards them; and . . . flees fetid smells."[16]

The Romans of the imperial era were even more given to using perfume/incense in a multitude of spiritual, practical, and social ways.[17] Not only was perfume/incense present in religious rites, in the scenting of theaters, and in medicine, but both men and women wore perfumes, including soldiers, who often perfumed their hair. Christians in the Roman Empire naturally shared many of the basic Roman attitudes toward odors, such as the general association of good smells with good things—divinity, flowers, healthy bodies, morality—and bad smells and stenches with demons, decay, illness, immorality. Christians also participated in most of the daily Roman uses of perfume/incense, for example, household fumigation, massages at the baths, medical interventions, and even personal adornment. Given the importance of incense in Roman and other Near Eastern religious and ceremonial practice and its cost, it is no accident that in the Gospel of Matthew two of the three gifts of the Magi to the infant Jesus are frankincense and myrrh.

Yet for the first three hundred years after the crucifixion, Christians did not use incense in worship due to the prevalence of incense in Roman civic cults and the martyring of many Christians who refused to participate by offering incense to the emperor.[18] But once Christianity became the official state religion around 390, incense and perfumed oils entered Christian ceremonies as part of the Eucharist and baptism. In *Scenting Salvation: Ancient Christianity and the Olfactory Imagination*, Susan Harvey suggests that an important theological assumption undergirding this newly enriched olfactory piety was that Christians, like Jews, and unlike Manichaeans or some Neoplatonist philosophers, regarded matter and the human body positively since it was God's creation.[19]

Yet there were also powerful antimaterial and anticorporeal tendencies in early Christianity, as reflected in the ascetic and monastic movements, for example, the famous "pillar saints" like Simeon Stylites, whose deliberate mortifications of the flesh included self-inflicted wounds that led to unbearable stenches. In many cases, however, the stench of self-mortification might be overcome at a saint's death when a pleasing "odor of sanctity" might miraculously appear.[20] The belief in the odor of sanctity would continue to play a role in Christianity all the way through the early modern era, as reflected in the use of saints' relics such as the miracle-working fragrant hand from the corpse of Saint Theresa of Avila, or as suggested in Dostoyevsky's *The Brothers Karamazov*, when Alyosha's faith is deeply shaken because the corpse of the saintly Father Zossima instead of emitting an odor of sanctity begins to stink.

Christian theology also developed a doctrine of the "spiritual senses."[21] This belief had obvious parallels with the Platonic and Neoplatonic idea of the soul's gradual ascent to the vision of the Good, leaving bodily loves and sensory perceptions behind. Of course, most of the laity and parish clergy did not wrestle with the ontological conundrum of how one should interpret Paul's "For we are the pleasing aroma of Christ to God" (2 Cor. 2:15) or theological phrases like "divine perfume" or "the fragrance of Paradise," but simply used these images in hymns and homilies.[22]

Many of these uses of perfumes and incense in early Christianity continued throughout Western Europe from the Middle Ages into the nineteenth century. Yet some of them gradually narrowed or disappeared beginning with the Reformation. Thus, we now need to consider how the narrowing of the many uses of perfumes, incense, and aromatic plants came about and what its implications are for the philosophical status of the sense of smell today, and accordingly for its aesthetic and artistic potential.

The "Deodorization" of Western Societies

I will examine two historical moments in the narrowing of the uses of perfume/incense and the corresponding reduction in the importance of the sense of smell in Western culture. The first moment was the diminishing of the spiritual uses of incense as a result of the Protestant Reformation, a process that coincided with a revival of the use of perfumes for adornment. The second moment begins in the mid-eighteenth century and culminated in the early twentieth century with the elimination of the medical uses of perfumes and the gradual "deodorization" of cities as part of sanitary reform. To help us more fully appreciate the impact of these changes, I will preface my discussion of each historical moment with a brief indication of the broad spiritual and practical functions of perfume/incense that still existed and were about to be challenged.

As a background to the Reformation's rejection of the spiritual use of incense and the belief in the odor of sanctity, we need to keep in mind the rich olfactory piety that still existed in the early Renaissance. The philosopher Marsilio Ficino, translator of Plato and Plotinus, suggested a parallel of odor and spirit to help his readers understand the power of scents. "Since each of them—that is, odor and spirit—is a certain vapor, and like is nourished by like, no doubt the spirit and the person with a lot of spirit receives great nourishment from odors."[23] But it was Montaigne, in the late sixteenth century, who most cogently and beautifully expressed the joint spiritual and aesthetic function of scent within Christian experience: "The use of incense and perfumes in churches, so ancient and widespread in all nations and religions, was intended to delight us and arouse and purify our senses to make us more fit for contemplation."[24]

Although moderate reformers like Erasmus might argue that pure prayers are "a perfume to God more pleasant than any incense," more radical reformers like Calvin sought to base all decisions on scripture and eliminate any accretions to worship made since the New Testament times, including the idea of the odor of sanctity.[25] Luther was more moderate on incense than Calvin, writing in 1523, "We neither prohibit nor prescribe candles or incense. Let these things be free."[26] The Anglican Church, as one might expect, also took a middle way. Although some puritan-leaning Anglican clergy vehemently opposed incense, it was still used here and there in seventeenth-century England (both Herrick and Milton viewed it positively), and, although it was never officially banned, incense use largely disappeared in the eighteenth century, only to be revived by the High Church movement of the nineteenth.[27]

These incense controversies might seem like minor parochial quarrels, yet they are an indication of an important narrowing of the spiritual use of scents that takes on added significance when considered in the context of a revival of the personal use of perfume among the elite that had begun even before the Reformation.[28] In *The Ephemeral History of Perfume: Scent and Sense in Early Modern England* Holly Dugan brings the evidence of an increased availability of perfume ingredients in English market towns of the Reformation period together with the religious controversy over incense to argue that the sixteenth century saw the beginning of "a large-scale cultural shift from censing to sensing" in Protestant areas of Europe.[29]

As important as this first moment of narrowing the uses of incense was in reducing the scope of smell experience in the West, it was the second moment, involving the collapse of the medical uses of perfume/incense along with the urban sanitary campaigns of the nineteenth century, that most radically undermined the importance of the sense of smell in the West. If we are to appreciate the extent of these changes, we also need to keep in mind just how important the therapeutic uses of perfume/incense still were down into the nineteenth century, a period during which "healing smells were a regular feature of the sick room."[30]

The most dramatic medical use of perfume/incense in the early modern period was as a plague preventive. Homes were fumigated by burning aromatic woods or perfume powders, by spreading aromatic herbs such as rosemary, and by perfuming both bedding and clothing. In addition, the well-off could afford to wear pomanders, a circular case, sometimes artfully made of silver, filled with strong perfume ingredients such as amber, musk, or civet along with various herbs.[31]

Mercifully, outbreaks of the plague were intermittent, but chronic stenches such as the smell of excrement or decaying carcasses were also viewed as threatening to health. Louis-Sébastien Mercier wrote of Paris in 1782,

> If I am asked how anyone can stay in this filthy haunt . . . amid an air poisoned by a thousand putrid vapors, among butchers' shops, cemeteries, hospitals, drains, streams of urine, heaps of excrement . . . I would reply that familiarity accustoms the Parisians to humid fogs, maleficent vapors, and foul-smelling ooze.[32]

Although people had complained of such stenches before Mercier, they were often regarded as inevitable and unavoidable, especially since excrement was a valuable fertilizer. Moreover, as we have seen in earlier chapters, olfactory habituation allows humans to adapt to all but the most astringent odors.

Alain Corbin has argued that by the mid-eighteenth century in France there was beginning to emerge among the bourgeoisie and nobility a greater sensitivity to strong odors. "Our nerves have become more delicate," is the way one author in the 1765 edition of the *Encyclopédie* put it.[33] At the same time, some chemists such as Lavoisier were beginning to deny the efficacy of either strong perfumes or incense for getting rid of obnoxious smells, claiming that perfumes only mask dangerous odors. By 1818, the official French pharmaceutical code had embraced the chemists' demonstration of the ineffectiveness of perfume and incense against stenches, sounding the death knell for their use in one major medical context.[34] In addition, officials across Europe now pursued a number of other methods for eliminating stenches such as draining, paving, and ventilating.

It was Britain, however, that led the way in sanitary reform. In early nineteenth-century London, as in Paris, human excrement was still deposited in cesspools, then emptied into carts and hauled to fertilizer-processing plants outside the city. Edwin Chadwick, who still believed that smell can cause disease, argued in his famous 1842 report *The Sanitary Condition of the Labouring Population* that the answer to the smell and threat of human excrement was not to use chemicals, but to trap and remove it through dedicated sewer pipes. By the 1850s, many London communities had pipes emptying human waste into the Thames. Eventually, after the "Great Stink" of 1858, when lack of rainfall left the Thames unusually low and a heat wave sent up a terrible stench from the river, Parliament finally voted a huge project to gather all the main pipes and run them underground alongside the river, taking human waste far downstream.[35] This was the now famous Thames Embankment, and should you ever be in London strolling along the Embankment Gardens amid the statues of poets and the fragrance of spring flowers, you might pay tribute to the excrement that runs under your feet.

As these kinds of sanitary and deodorizing measures were being carried out in England and eventually across the continent and in the Americas, a final step in the elimination of almost any medical role for smells took place. In the 1860s and 1870s, Louis Pasteur's and Robert Koch's independent demonstration of the germ theory of disease proved that smells were not even the cause of diseases, let alone a cure. Some smells might still be a symptom, but by the 1880s in both Europe and the United States many physicians gradually began to give up on them as a serious source of either diagnosis or treatment, although laypeople took longer to abandon the belief that malodors were a threat.[36] By the early twentieth century a three-thousand-year tradition of smell's crucial role in medicine was dead.

As that tradition died, the process of deodorizing cities and towns continued apace, targeting one smelly industry and practice after another, but now increasingly propelled by aesthetic rather than medical urgency. Moreover, as people's tolerance of one kind of smell led to its elimination, a *dialectic of deodorization* developed: the elimination of one offensive odor made people less tolerant of other strong smells, which people demanded be eliminated in turn. In the late twentieth and twenty-first centuries, even things as benign smelling as coffee-roasting plants have been forced out of some cities, and a few places have even banned the wearing of perfume in public buildings. In 2000 Halifax, Nova Scotia, banned wearing scented products anywhere in public. For some people the ideal city now seems to be a largely odorless one.

Philosophical Implications

What are the implications for the possibility of an aesthetics of olfactory art of these episodes from the rich olfactory past of the West, leading to the reduction of perfume to purely aesthetic uses and the narrowing of incense use? The first implication is that the pervasive uses of perfumes and incense in so many aspects of life from the ancient world until a century and a half ago, surely contradicts those who regard the human sense of smell as primitive and useless and as incapable of involvement in matters of mind and spirit. On the contrary, for almost three thousand years scents and the sense of smell not only played a vital role in medicine and civic ceremonies, but in many aspects of social, religious, and personal life. Moreover, whereas modern thinkers from Kant to Darwin and Freud denigrated smell as animalistic and primitive, from the ancient Greeks down through the Renaissance and beyond smell was considered by many people a spiritual sense and odors a medium of communication with the divine.

A second implication of the West's olfactory history arises from the fact that as many of the medical and other practical and spiritual uses declined, expanded aesthetic uses emerged. Thus, aesthetic theory needs to take account of what one could call a kind of "reodorization" that has taken place in the West from the mid-twentieth century on.[37] This suggests that far from being a useless vestige, as the negative intellectual tradition has assumed, our sense of smell continues to play a vital, if philosophically unacknowledged and unexplored, role in our lives. Or, to put it another way, what people have once done, they can do and are doing again, by finding new or renewed aesthetic uses for odors and the exercise of the sense of smell. Part III will explore some of those contemporary aesthetic uses of odors in theater, film, music, and installation art, as well as in perfume creation. And Part IV will examine the crucial role of smell in creating a positive olfactory environment in cities and buildings and of giving our food and drink enticing aromas and flavors.

A third implication of the historical episodes we have examined is that, although there are indeed underlying biological reasons for certain aspects of the olfactory deficits that contemporary Western subjects exhibit in psychological experiments, *part* of the reason contemporary middle-class, urbanized Westerners don't notice most odors or find it easy to talk about them may be historical. Given the history we have just traced, along with the neglect, misrepresentation, or outright denigration of smell by many intellectuals, it is not surprising that most people have let their olfactory capacities remain undeveloped.

The next chapter will complement this chapter's evidence of historical changes in the role of smell in Western culture, with evidence of a wide-ranging ability to articulate smell experiences among non-Western cultures across the globe. Specifically, I will present evidence to suggest that both Darwin's so-called savages and many non-Western "civilized" peoples not only continue to give odors and the sense of smell a more important social role than Westerners do, but are able to articulate their experiences of smell in ways that run counter to the idea of a universal "poverty of language" for expressing smell.

As a transition to that discussion, I present a brief interlude on the historical and contemporary uses of perfume/incense in India, China, and Japan, an interlude that, besides its intrinsic interest, will reinforce the lesson that odors and the olfactory arts of perfume and incense have played and still play important social, religious, and aesthetic roles globally.

Notes

1. Stephen Davies, *The Artful Species* (Oxford: Oxford University Press, 2012), 45–49.
2. Philippe Marinval, "Fragrances from Prehistoric Times to Ancient Gaul," in *Perfume: A Global History*, ed. Marie-Christine Grasse (Paris: Somology Art Publisher, 2008), 31–32.
3. Constance Classen, David Howes, and Anthony Synnott even speak of the deodorization of Western cities as an "olfactory revolution." *Aroma: The Cultural History of Smell* (London: Routledge, 1994), 78.
4. Mark M. Smith, "Smelling the Past," in Smith, *Smell and History*, ix–xxv. Mark S. R. Jenner also warns against overdrawing the differences between past and present in "Civilization and Deodorization? Smell in Early Modern English Culture," in *Civil Histories: Essays Presented to Sir Keith Thomas*, ed. Peter Burke, Brian Harrison, and Paul Slack (Oxford: Oxford University Press, 2000), 127–44.
5. In addition to Mark M. Smith's anthology, the general reader can consult Jonathan Reinarz's thematic synthesis, *Past Scents: Historical Perspectives on Smell* (Urbana: University of Illinois, 2014).
6. It makes sense in most contexts to translate the Latin *unquenta* as "perfume" rather than "unguents" or "oil" since ancient Roman writers "almost always use [the term] with primary reference to scent." Shane Butler, "Making Scents of Poetry," in *Smell and the Ancient Senses*, ed. Mark Bradley (London: Routledge, 2015), 75.

7. Bob Brier and Hoyt Hobbs, *Ancient Egypt: Everyday Life in the Land of the Nile* (New York: Sterling, 2009), 63.
8. Lise Manniche, *Sacred Luxuries: Fragrance, Aromatherapy and Cosmetics in Ancient Egypt* (Ithaca: Cornell University Press), 1999.
9. Deborah A. Green, "Fragrance in the Rabbinic World," in Bradley, *Ancient Senses*, 148.
10. But within the Hebrew Bible, the Song of Songs jostles uneasily with prophetic works that treat the smell of sacrifice and incense as secondary to the command of justice (Isaiah 11–17). For additional discussion see Deborah A. Green, *The Aroma of Righteousness: Scent and Seduction in Rabbinic Life and Literature* (University Park: Pennsylvania State University Press, 2011).
11. Ashley Clements, "Divine Scents and Presence," in Bradley, *Ancient Senses*, 46.
12. Marcel Detienne, *The Gardens of Adonis: Spices in Greek Mythology*, trans. Janet Lloyd (Princeton: Princeton University Press, 1994), xxvii.
13. Clements, "Divine Scents and Presence," 57.
14. Susan Harvey, *Scenting Salvation: Ancient Christianity and the Olfactory Imagination* (Berkeley: University of California Press, 2006), 22–23.
15. Laurence Totelin, "Smell as Sign and Cure in Ancient Medicine," in Bradley, *The Ancient Senses*, 29.
16. Totelin, "Smell as Sign and Cure," 28.
17. For a general discussion of the smellscape of ancient Rome see Neville Morley, "Urban Smells and Roman Noses," in Smith, *Smell and History*, 33–49.
18. The Jews had earlier obtained legal exception on the grounds that theirs was an ancient religion.
19. Harvey, *Scenting Salvation*, 105–14.
20. Harvey, *Scenting Salvation*, 156–200.
21. Harvey, *Scenting Salvation*, 169–80. For a discussion of the various meaning of "spiritual sense" in Christian theology see Paul L. Gavrilyuk and Sarah Coakley's "Introduction" in *The Spiritual Senses: Perceiving God in Western Christianity*, ed. Paul L. Gavrilyuk and Sarah Coakley (Cambridge: Cambridge University Press, 2012).
22. Harvey, *Scenting Salvation*, 235–37. After the Romans destroyed the Second Temple in 70 C.E., Rabbinic Judaism confronted the issue of whether incense should play a role in Jewish religious ritual, but did not include it. Green, "Fragrance in the Rabbinic World," 157.
23. Marsilio Ficino, *Three Books on Life*, trans. Carol V. Kaske and John R. Clark (Binghamton, NY: Renaissance Society of America, 1989), 221–22. The quotation is from Chapter 18, called "On Nourishing the Spirit and Conserving the Soul through Odors." Note the parallel to the Greek notion of incense and the divine as "like unto like" mentioned above.
24. Michel de Montaigne, *The Complete Essay of Montaigne*, trans. Donald M. Frame (Stanford, CA: Stanford University Press, 1957), 221.
25. The Erasmus quotation is from Holly Dugan, *The Ephemeral History of Perfume: Scent and Sense in Early Modern England* (Baltimore: Johns Hopkins Press, 2011), 29. On Calvin see David Robertson, "Incensed over Incense: Incense and Community in Seventeenth Century Literature," in *Writing and Religion in England*,

1558–1689: Studies in Community Making and Cultural Memory, ed. Anthony W. Johnson (Farnham, UK: Ashgate, 2009), 392.
26. Martin Luther, *Luther's Works*, vol. 53 trans. Ulrich S. Leupold (Philadelphia: Fortress Press, 1955), 25.
27. Robertson, "Incensed over Incense," 393–409. Robertson gives a vivid picture of the debate that raged all through most of the seventeenth century.
28. The widespread use of perfumes for adornment had virtually disappeared in many areas of Western Europe during the early Middle Ages, but began to revive in the courts of southern Europe after the crusaders brought perfumes back from the Middle East, and by the early Renaissance, perfume use had expanded beyond royal courts and the nobility as ingredients became available in larger market towns.
29. Dugan, *Ephemeral History*, 17. She finds a particularly piquant example in the plot of a popular saint's play, the *Digby Mary Magdalene*, in which perfumes play a key role in Mary's sexual fall under the guidance of the Lady Lechery and King Lust. The play's focus on the conflict between worldly perfumes and holy incense, Dugan argues, reflects a fascination with the increasing use of perfumes for aesthetic and seductive purposes at the time (40–47).
30. Richard Palmer, "In Bad Odour: Smell and Its Significance in Medicine from Antiquity to the Seventeenth Century," in *Medicine and the Five Senses*, ed. W. F. Bynum and Roy Porter (Cambridge: Cambridge University Press, 1993), 63.
31. Annick Le Guérer, *Scent: The Mysterious and Essential Powers of Smell*, trans. Richard Miller (New York: Random House, 1997), 39–108.
32. Quoted in Alain Corbin, *The Foul and the Fragrant: Odor and the French Social Imagination*, trans. Miriam L Kochan (Cambridge, MA: Harvard University Press, 1985), 54.
33. Corbin, *The Foul and the Fragrant*, 73.
34. Corbin, *The Foul and the Fragrant*, 70.
35. David S. Barnes, "Confronting Sensory Crisis in the Great Stinks of London and Paris," in *Filth: Dirt, Disgust, and Modern Life*, ed. William A. Cohen and Ryan Johnson (Minneapolis: University of Minnesota Press, 2005), 103–29.
36. Melanie A. Kiechle emphasizes the gradualness of the process in the United States and the increasing separation of the approach of health professionals from that of ordinary people. *Smell Detectives: An Olfactory History of Nineteenth Century Urban America* (Seattle: University of Washington Press, 2018).
37. Other scholars have also noted the return of the use of odors in a variety of aesthetic contexts over the past century. See Robert Jütte, "Reodorizing the Modern Age," in Smith, *Smell and History*, 170–86.

Interlude
Fragrant Asia

Early in the great Indian epic the Mahabharata, scent plays a key role in the birth and marriage of Satayavati, great-grandmother to both the Pandava and Kaurava princes, whose quarrel motivates the events of the epic, including its best-known segment, the Bhagavad Gita. As King Vasa was out hunting amid the "delightful perfume of spring flowers," he began lustfully thinking of his wife Girikā and spontaneously ejaculated. A fish swallowed his semen and Satyavati was born from the fish. Unfortunately, although she possessed exceptional beauty and goodness, she smelled like a fish. One day she encountered a wandering sage who offered her a boon, and, of course, she asked for a fragrant body. "Therefore, her name, 'Fragranced,' was renowned on the earth," and people could smell her wondrous scent from nine miles away. A neighboring king caught a whiff of it on the wind, followed it, and ended up marrying her, thus launching the great dynasties of the Mahabharata.

James McHugh, who recounts this story in *Sandalwood and Carrion: Smell in Indian Religion and Culture*, suggests that the role of smell in this tale is typical of Sanskrit literature in that smells are not primarily triggers of memory, as in much Western literature, but impulses to action: people notice them from afar and seek out their source.[1] McHugh argues that between 500 and 1500 C.E. the South Asian elites developed an olfactory culture of a breadth and sophistication that is "difficult for us to imagine in our relatively deodorized world."[2]

McHugh has found three Sanskrit treatises devoted to the creation and use of perfumes and incense. One of them, the *Girdle of Hara* (Siva), is addressed to the "Cultivated Man" who is "solely intent on perfecting a life of righteousness, wealth, pleasure and fame."[3] The reader familiar with classical Hinduism will recognized in the first three characteristics, the *trivarga*, the three aims of worldly life: *dharma* (righteousness), *artha* (wealth and power), and *kama* (pleasure), with *moksha* (liberation) as the ultimate spiritual aim. The goals of the *trivarga* are even more clearly at the center of another text, called the *Essence of Perfume* (*Gandhasara*), which begins: "This treatise . . . provides for a rite of worship of the gods with incense and auspicious perfumes; makes men thrive; provides the results of the *trivarga* . . . pleases kings, and delights the mind of the cultivated lady."[4]

A striking thing about these writings, McHugh points out, is that they demand the reader not only recognize the names of unusual plants, understand complex processes, and employ some mathematics, but also appreciate the metaphors and metrical rhythm of many formulae. The names of the various formulas ranged from "Arouser of Kama," (after Kamadeva, the god of love), through flowers, gods, and kings to the playful "Fracas" and "Who Goes There?"[5] But my favorite is "Yaksa Mud." Yaksas were spiritual beings whose scent was so wonderful that by comparison "our greatest perfume is equal to mud."[6] "Mud" might also refer to the dark ruddy color of some versions of Yaksa Mud since most ancient Indian perfumes were pastes rubbed on the body and so could be seen as well as smelled. Even today, some images of the Hindu god Venkateśvara have a white camphor mark on the face.[7]

Down through the centuries, perfume/incense has remained omnipresent in India not only for personal adornment by those who could afford it, but for use in religious and social ceremonies, for fumigating and scenting dwellings, and for many medical conditions, and it remains an important part of Indian life today. Given the similar role of perfume/incense in Arab and Persian cultures of the past, the establishment of the Mughal empires simply enriched the span of ingredients and techniques. Of course, the effects of British colonial control and especially the rapid development of modernization in recent decades has meant not only a degree of deodorization of the upscale areas of modern cities, but also that the traditional handcraft of distilling *attars* (oil-based perfumes from rose, jasmine, and other flowers) has declined. This is due not only to their far higher cost than alcohol-based synthetics, but to the fact that many younger middle-class Indians are drawn to imported Western brands.[8]

India was not the only Asian region where the aristocracy developed a sophisticated olfactory culture that permeated almost every aspect of life. As E. H. Schafer remarks in *The Golden Peaches of Samarkand*, "In the medieval Far East there was no clear-cut distinction between medicinal plants, spices, perfumes and incense, i.e. between substances that nourish the body and those that nourish the spirit, those that attract the beloved and those that attract the deity."[9] In China the aristocracy, as Georges Métailié puts it, "appear to have lived in an atmosphere full of fragrances." Not only was incense central to religious ritual, especially Buddhist practice, but among the aristocracy, bathing waters were perfumed with lemongrass and peach blossom, perfume sachets were sewed into clothing, cloves were chewed to perfume the breath, and there were recipes for saves to make the skin smell sweet. The Chinese were also known for their incense clocks, and the highest social classes built scented pavilions, usually of aromatic sandalwood with walls plastered in mortar mixed with camphor and musk.[10]

In 2018, the Paris exhibition *Perfumes of China: The Culture of Incense in the Imperial Period* traced the history of perfume/incense use in China from the Han dynasty beginning in 206 B.C. through the end of the imperial period in 1911.[11] In addition to exhibiting large, intricately wrought perfume burners along with paintings showing incense/perfume use in a variety of social situations, the exhibition had four smelling stations where visitors could sample reconstructions of the actual odors based on surviving incense formulas. For over two thousand years, then, incense/perfume has never stopped playing a role in many aspects of Chinese life, some forms of it reaching all social levels. From the Tang dynasty on and especially during the Song, the literati accompanied their poetic, painting, philosophical, and social activities with incense rising from beautifully wrought ceramic or bronze burners. As Frédéric Obringer remarks, these refined incense sessions "took on the aspect of a superior aesthetic experience shared by a few friends who were knowledgeable in poetry, calligraphy . . . tea and, of course, perfume."[12] Some writers even began to speak of a new "Four Arts of Life": incense, tea, flower arranging, and painting.[13] Although the special association of incense with the literati, aristocracy, and imperial court disappeared after Revolution of 1911, the daily use of incense to honor ancestors in the home and deities in temples, or to scent clothing and accompany meditation, and writing, playing music, and so on, has never completely died out in either of the Chinas. The survival of many of these practices in today's mainland China—despite the Cultural Revolution's attack on tradition—may be one reason the major Western perfume houses have had difficulty penetrating the contemporary Chinese consumer market. The one exception appears to be the interest in perfumes on the part of a wealthy elite, for whom possessing perfumes like Chanel No. 5 is a status symbol.[14]

When we turn to Japan of the Heian period (794–1185), we also find that perfume/incense was being used daily not only in temple worship and monastic practice, but also to scent both clothing and living quarters. And in an aristocratic culture where men and women often did not see each other before marriage, but were separated by screens and shutters, scent was a critical way of revealing one's taste. Moreover, the aristocracy made perfume/incense creation an intellectual challenge and held competitions to see who could create the most elegant and interesting scent combinations. In *The Tale of Genji*, for example, Genji organizes a contest among a small group of intimates to determine which perfume would accompany his young daughter to her initiation and entry into court life.[15] By the fifteenth century these kinds of aristocratic competitions had evolved into a formal ceremony or "art" (*do*) called *kodo*, paralleling the tea ceremony, or *cado*, and flower arranging, or *kado*. By the nineteenth century kodo had spread until it was enjoyed by almost all but the poorest classes. But, as we will see in a later interlude devoted to its contemporary revival and interpretation, kodo's more sophisticated versions became what some aestheticians see as a veritable art form, involving both meditative and literary aspects along with the aesthetic appreciation of odor qualities. As for the more general uses of incense and perfume in contemporary Japan, the situation appears similar to China; incense is omnipresent in temples and homes, but personal use of Western-style perfumes for adornment is making only modest headway.

Of course, from the nineteenth century on, Western influences, often backed by imperialist armies and navies, contributed in all three Asian cultures to the decline of some of the complex olfactory practices we have described. As interesting as it would be to explore further both the influences of Western "deodorization" on Asian cities and the survivals of elements of traditional olfactory culture, we have seen enough to conclude that just as odors and the sense of smell once played a crucial role in Western culture, smell has had and continues to have an even more pervasive and aesthetically sophisticated role in Asian cultures.

In the light of this evidence, the Western intellectual tradition's denigration of the sense of smell, or Darwin's view that it of little use to "civilized" as well as "savage" peoples, seems astonishingly parochial. The previous chapter's survey of the important uses of scents and of the sense of smell in Western history combined with the evidence we have just reviewed from East and South Asia should not only drive the final nail in the coffin of the idea that the human sense of smell is of little use but also challenge claims that all humans are by nature unconscious of smells except in threatening circumstances. Moreover, contrary to the claim that humans intrinsically lack the capacity to identify, name, and linguistically express smells, McHugh's study of India, for example, shows that Sanskrit writers were able to discuss odors and perfumes using highly sophisticated literary devices. There are also a number of serious older treatises on perfume/incense and their enjoyment in Chinese and Japanese.[16] In the next chapter, we will see that a consideration of several contemporary non-Western languages and cultural practices will cast even greater doubt on a supposed universal "poverty of language" for expressing smell.

Notes

1. James McHugh, *Sandalwood and Carrion: Smell in Indian Religion and Culture* (Oxford: Oxford University Press, 2012), 97–100. For a different perspective there is David Shulman's "The Scent of Memory in Hindu South India," in *The Smell Culture Reader*, ed. Jim Drobnick (Oxford: Berg, 2006), 411–26.
2. McHugh, *Sandalwood and Carrion*, 244–45.
3. McHugh, *Sandalwood and Carrion*, 144.

4. McHugh, *Sandalwood and Carrion*, 145.
5. McHugh, *Sandalwood and Carrion*, 121–28.
6. McHugh, *Sandalwood and Carrion*, 148–49.
7. McHugh, *Sandalwood and Carrion*, 153–56. See also David Howes, "Future of Scents Past," in Smith, *Smell and History*, 210.
8. Faisal Fareed, "Scent of a City," *Indian Express*, March 27, 2016, https://indianexpress.com/Lifestyle, 7.
9. Edward H. Schafer, *The Golden Peaches of Samarkand* (Berkeley: University of California Press, 1985), 155.
10. George Métailié, "Fragrances in Medieval China," in Grasse, *Perfume*, 120.
11. Held at the Museé Chernuschi from March 9 to August 26, 2018.
12. Fédéric Obringer, "Introduction," in *Parfums de Chine: La Culture de l'encens au temps des empéreurs*, ed. Éric Lefebvre (Paris: Paris Musées, 2018), 11.
13. The traditional four arts of the literati were playing the *guqin*, mastering Go, calligraphy, and painting. See the announcement for the exhibition at the National History Museum in Taipei, Taiwan, from March 22 to May 5, 2013, *Distilling the Soul's Fragrance: Traditional Chinese Incense Culture*, 11, https://www.nmh.gov.tw/en/exhibition_2_3_22_721.htm?.
14. Chandler Burr, "Fragrance Market Is Establishing a Foothold in China," *New York Times*, May 10, 2008, https://www.nytimes.com/2008/05/10/business/worldbusiness/10perfume.html.
15. Aileen Gatten, "A Wisp of Smoke: Scent and Character in the *Tale of Genji*," in Drobnick, *Scent Culture Reader*, 336–37.
16. Obringer, *Parfums de Chine*, 23; Gatten, "Wisp of Smoke," 337–39.

6
Language, Culture, and Smell

> What's in a name? That which we call a rose
> By any other name would smell as sweet
> —*Romeo and Juliet*, act 2, scene 2

Sorry, Juliet, you got it wrong—as far as the scent of roses is concerned. According to contemporary psychology, a different name can easily change the way things smell, making the sweet sour and turning the smell of Parmesan cheese into the reek of vomit. Of course, where roses themselves are concerned, as we noted earlier, not many of them have much smell in our vision-centric culture. But therein lies a problem for a good deal of our philosophy and psychology of smell. As we have seen, nearly all the philosophical reflections as well as the neuroscience and psychology research that we have followed so far have been developed within modern Western societies that have been hygienically deodorized to a large degree for over a century, societies, moreover, whose intellectuals have long disregarded or depreciated the sense of smell. In this chapter we will continue to pursue our biocultural approach to the question of whether smell has the cognitive capacity for genuine aesthetic experience and judgment by examining the mantra about the essential "poverty of language" to express odors and the belief that humans by nature have an extremely poor ability to name and discuss smells. Here again, as in the cases of the emotionality and unconsciousness of our experience of smell, we will show that a partial truth has been grossly exaggerated. Apart from the intrinsic interest of this foray into anthropology and linguistics, the philosophical payoff will be to demonstrate that when viewed from a cross-cultural perspective, the human sense of smell does have the linguistic resources to make serious aesthetic discussion possible. We will begin by considering olfactory language learning, labeling, and classification, along with the basic vocabulary of Western languages, before focusing on anthropological and psycholinguistic evidence from non-Western languages and cultures.

Art Scents. Larry Shiner, Oxford University Press (2020) © Oxford University Press.
DOI: 10.1093/oso/9780190089818.003.0014

Learning, Labeling, and Classification

One obvious reason most Westerners have trouble identifying and discussing smells is that whereas we teach very small children their colors or even the harmonic scales, we seldom offer any explicit education of the sense of smell (other than an occasional scratch-and-sniff game). As a result, what learning does take place is generally haphazard and leads to idiosyncratic associations. By contrast, Western children who show exceptional gifts for music, art, dance, or sports are likely to be given special training from an early age. Indeed, many of today's celebrated composers and performers got their start in early childhood, like those of the past—Mozart was playing in French salons at age five. The case of professional perfumers today differs in their much later starting date than students of other arts. With the exception of a few children of perfumers, most perfumery students begin studying in their early twenties, although from the time of their first training, they too tend to work with odors almost daily, often into their later years. It is no wonder, then, that perfumers, like pianists or violinists, show increased structural brain plasticity (more gray matter) and decreased functional brain plasticity (more efficient processing) that releases space for artistic creation. But, to my knowledge, there are no celebrated olfactory child prodigies (except the fictional Jean-Baptiste Grenouille) perhaps for the simple reason that in our culture, a child's precocious sense of smell is likely to be either ignored or discouraged in favor of more culturally acceptable pursuits. We still await the Mozart of smell.

One handicap for teaching people their smells is the fact that, as Plato and Aristotle already noted, there exists no widely agreed-upon system of abstract smell terms of the kind that exists for colors or sounds. Various scientific attempts at the classification of smells have been made from Linnaeus to the present, but each of the three major ways of classifying odors—by odor receptor response, by odor molecule structure, or by perceptual experience—has serious drawbacks. Classifications based on receptors in the nose run into the problem that there are not just three types of receptors, as there are in the eyes for color, but over 320 receptors, a number that also varies slightly from person to person. Classifications based on chemical structure, whether molecular shape, weight, or vibration frequencies, have so far failed to predict reliably people's odor perceptions. Classifications based on the olfactory perceptual experience itself suffer from the problem that even experts may disagree on the application of some terms. Thus, there seems to be no general taxonomy of odors that could be taught in elementary schools alongside the color wheel and the tonic scale. And to some people this seems yet another reason to disparage the cognitive potential of the sense of smell, with negative implications for the possibility of serious discussion of aesthetic issues. But a closer look at the problem of odor classifications will show the situation to be not nearly as bleak as often claimed.

Kaeppler and Mueller's 2013 survey of twenty-eight classification experiments from the past fifty years concludes that although the scientific community is still far from an agreed-upon universal system of classification, many descriptive classifications for particular olfactory domains have shown considerable consistency and overlap and have also proved useful tools for professional communication. After all, classifications do not exist in a vacuum; they are made for a variety of purposes and differ accordingly.[1] Thus, as the psychologist Peter Köster has noted, we should not disparage the relatively successful classifications that have been developed within particular fields whose practices involve an important role for smell, such as wine and other beverages (brandy, beer) and certain foods (cheddar cheese, fish), each of which has its own specialized olfactory terminology on which there is relatively broad agreement despite differences in detail.[2]

Of particular interest for our purposes are two of the best-known perfume classifications, the "Odor Effects Diagram" of Peter Jellinek, and the "Fragrance Wheel" developed by Michael Edwards. The Jellinek map (1951, 1992) is based on his years of work as a perfumer and educator and describes the *effects* of perfumes and their materials according to two axes, an "erogenous" to "refreshing axis," and a "narcotic" to "stimulating" axis. Odor quality descriptors can be distributed along the two axes, such that floral scents are associated with the narcotic pole and spicy or woody with the stimulating pole, and so on.[3] Thirty some years after Jellinek's first map, Michael Edwards arranged fragrances in a wheel form to emphasize that categories blend into one another, as do color categories. Although under constant revision since it first appeared in 1983, Edward's wheel has four standard families of fragrance types: floral, oriental, woody, and fresh.[4] Importantly, Jellinek's Odor Effects Diagram and Edward's Fragrance Wheel overlap at many points; for example, floral notes as a group stand opposite the woody notes on both. And the two maps also resemble several other perfume classification systems developed independently by particular perfume houses. Moreover, in 2009, two other researchers, Zarzo and Stanton, decided to statistically compare the semantic classifications of Jellinek and Edwards with the large Bohlens-Haring database of 309 odor materials. The B-H database was numerically derived, that is, was based on experts' perception of odor similarities and differences from a group of thirty reference odors, rather than on verbal responses, which tend to be more subjective and variable. Using a method called principal component analysis, Zarzo and Stanton were surprised to discover that, despite the small size and personal origin of the Jellinek and Edwards classifications, there was statistically a good deal of overlap of the two semantic-based maps with the large, numerically derived Bohlens-Haring database. This result, they conclude, supports the "hypothesis that . . . the effect of a given odorous material is basically the same ... *under a similar context and convention.*"[5]

Of course, many professional perfume classifications tend to mix qualitative perceptual terms (floral) with chemical ones (aldehydes) and are obviously tailored to the working perfumer needing a way to organize knowledge and communicate with other perfumers. Yet the Fragrance Wheel is relatively user friendly, and J. Stephan Jellinek has also come up with a modified version of his father's diagram aimed at being of more practical use to consumers. Moreover, a number of studies in which people have been questioned about their perceptions of similarities and differences among perfumes or other scented products have shown that many people group scents into such contrasting categories as heavy-light, sensual-cool, or floral-nonfloral, polarities that roughly parallel the classifications of Jellinek and Edwards.[6] There is a similar phenomenon in the even more complex case of wine tasting, a cultural practice that has engaged increasing numbers of people. The well-known Wine Aroma Wheel of Ann C. Noble, for example, actually has striking similarities to various perfume schemas, with categories like floral, woody, vegetative, spicy, and various types of "fruity."

Thus, even within the context of Westernized, developed countries, there is evidence that in addition to the very small body of olfactory professionals, there exists a larger group of people who are aware of the odors around them, especially of scents in the perfumes and everyday hygienic products they use. Like those who take the trouble to learn something about wine and develop their sensitivity for it, people who are interested in perfumes and scents are often able to find ways of giving a more nuanced linguistic expression to odors than the rough hedonic responses that typically show up in randomized experimental studies of the general population. Moreover, for our purpose of demonstrating the possibility of an olfactory aesthetics, the question is not whether human language can provide an "objective" classification of odors or even the degree to which the average person can correctly name odors, but whether people who make the effort can learn to rise above merely hedonic or source-based expressions to articulate the qualitative distinctions requisite for the creation of artworks and for aesthetic discussion and judgment.

Yet despite this more positive evidence from the worlds of perfume and wine, a major support for the poverty-of-language claim remains the fact that in laboratory studies the average person seldom gets much beyond expressing a basic hedonic reaction to odors: good/bad, pleasant/unpleasant, aroma/stench, with a guess or two about specific qualities. And if they do manage to come up with a qualifying term, they tend to name odors by their source, using similes: it smells *like* or smells *of*. But we should be cautious in drawing negative conclusions about human linguistic ability to express smells based on some of these naming studies. After critically examining the linguistic assumptions behind many researchers' claims about people "failing" to get the names of odors right, the linguist Danièle Dubois and the neuroscientist Catherine Rouby jointly concluded that many research designs seem to assume that odors actually *have* a single "true" label and that it is in fact the label of the specific conventional source. But, of course, the "odor object" as we established earlier is distinct from the "odor source" (odorant); it is an experienced entity, and that experience can be generated by an actual lemon, or by an artificial lemon scent, or by a closely related chemical compound. And although a few experimenters take more expansive category answers such a "citrus" or "fruit" as near misses, other responses such as "fresh" or "pungent" might also be argued to be reasonably correct and could change the outcome of a study. Moreover, various experiments have shown that certain brand-name products such as Johnson's Baby Powder, Crayola crayons, Play-Doh, and Bazooka Bubble Gum are correctly identified and named with considerable success. Dubois and Rouby conclude that in many cases the supposed "correct" answer or "veridical label" is just "the name the experimenter expects."[7]

One of the things lacking in many naming studies is that they overlook the important role that context and community play in labeling. Earlier we cited the comment by the neuroscientist André Holley that perfumers, who are trained in a particular tradition of "notes" and "accords," when confronted with an identification test that is completely outside those professional habits, are likely to perform only a little better than the inexperienced. Kevin Sweeney reminds us that experienced wine tasters also follow an "accepted critical discourse" that has developed among groups of tasters, such as that reflected in the terms used in the Wine Aroma Wheel.[8] Yet even those designations often have some basis in the chemical structure of the wine, just as perfumer's verbal labels may reflect the actual molecules used in a perfume. One of Sweeney's wine examples is that of finding a green apple taste in a young Chardonnay; the taste comes from the fact that the young wine has not undergone the full process of fermentation in which malic acid (the same acid that gives green apples their crisp taste) is transformed into softer, lactic acid, which has a buttery taste. Sweeney calls this kind of taste analysis and labeling "analytic realism," since the taster's experience is primarily based on a stimulus agent in the wine. He points out that Hume's example of Sancho's two kinsmen, one who tastes iron, the other leather and who later find an iron key attached to a leather thong in the barrel, is a case of "analytic realism."[9]

But most writing about wines and also about perfumes is not of the "analytic realism" type, which names a particular substance in a wine or perfume that is the "cause" of the taster's description. Although one experienced taster of a mature Chardonnay, Sweeney points out, may describe it as buttery, others might use terms like butterscotch, caramel, or even honey. Yet few would use "minty" or "metallic." Thus, there are appropriate ranges of terms that are used by serious students of wine—or of perfumes. Sweeney calls this way of describing qualities "analytic interpretivism," since "the taster must come up with an imaginative interpretation that is apt, that fits within the correct sensory category, but within that category there is room for interpretation."[10] Those verbal descriptions often do have a basis in the wine or perfume, even if that basis cannot always be so graphically identified as leather thong with an iron key or a particular molecule. My conclusion from these wine-tasting examples is that in the parallel case of identifying odorant characteristics, although the labels that experienced people use may vary, responses that fall within an appropriate range should not be considered purely subjective and erroneous.

Odor Terms in Western Languages

Yet even if experienced laypeople's performances in domains such as wine or perfume labeling and discussion are not as poor as some researchers suggest, the fact remains that Western lexicons for odors are indeed restricted and contain hardly any abstract odor terms. English, French, German, Italian, and Spanish have only about a dozen-plus active terms each for odors, and the average person's everyday operational vocabulary is typically even smaller. Consider the main terms for odors in English. We have the roughly interchangeable "odor" and "smell," the latter available both as a noun (meaning either "an odor" or "the sense of smell") and as a verb ("to smell"). Both terms can be more or less neutral, although each of them is often used negatively, for example, "What's that smell (odor)?" Then come more positive terms like "aroma" (generally associated with food odors) and "scent" or "fragrance" (both associated with flowers or perfumes). "Perfume" has both a verbal form, "to perfume something," and a more often used nominative form, which in earlier epochs, as we have seen, covered all "aromatics." Perfume's dominant contemporary nominal reference today, of course, is to alcohol-based compounds intended to be applied to the skin. Because of its commercial associations, some people avoid the term "perfume" in favor of "fragrance" or "scent." The closely related "bouquet" is largely confined to expressing the aroma or fragrance of a wine. "Incense," of course, is closely identified with the religious use of scents, usually of burned wood, although as we have seen, it has historically overlapped with perfume. Whereas aromas, scents, incense, perfumes, and fragrances make up the core of the primarily positive side of the olfactory semantic complex, the negative consists primarily of "stink," for more limited unpleasant smells, and the more potent "stench." The verbal form for a stench or stink, cognate with German term for smell (*riechen*), is "reek," and when something reeks badly enough, we may be impelled to "fumigate" it. Of course, sometimes we only get a "whiff" of an odor, something that is largely involuntary, although it may encourage us to actively "sniff" in order to identify the source.

These are some of the major nominal and verbal terms of ordinary conversation, although they can be supplemented by several rather more literary usages, such as "exhalation" on the positive side and "malodor" for the negative, and one might want to include the names of aromatic substances such as resins (frankincense) or fibers (sandalwood), or animal secretions (musk) that exist primarily as odor terms. As for adjectives, in addition to those borrowed from general discourse such as "rank" for a particularly strong smell, there is also a set of typical adjectives that are more closely associated with smells, such as "redolent" for something penetrated by odors, "aromatic" for a source emitting an odor, and "pungent" for an odor especially sharp. On the negative side there is the stinging sensation, as of smoke, we call "acrid," and for stale odors "fetid," with "putrid" for decay and "musty" and "fusty" suggesting something moldering.

Taken together these most often used terms may not seem like a very large arsenal for expressing the multitude of possible odors and smell experiences, but of course, many adjectives for characterizing smell, as Aristotle noted long ago, are regularly borrowed from other sensory domains such as taste or touch to describe odor effects, such as sweet or sour, soft or harsh, fresh or stale. Researchers at the University of Düsseldorf have offered evidence that in ordinary German the scope of synesthetic borrowings is much more complex than once believed, and they suggest that even in a Western language like German, ordinary speakers can comprehend, and potentially use, a much larger repertoire of terms for smells than the "poverty of language" claims imply. The Düsseldorf studies focused on "cognitive accessibility," that is, to what extent a particular adjective-noun combination is intuitively comprehensible. Subjects were given a list of fifty-some adjective-noun pairs that had been randomly formed and asked to rate each according to cognitive accessibility. The pairs included such things as "yellow silence" (*gelbe Ruhe*), "sweet darkness" (*süsse Dunkelheit*), and "gloomy smell" (*düster Geruch*). The researchers discovered striking differences in the accessibility of certain combinations. For example, none of the 107 respondents rated "yellow silence" meaningful, but 93% found "pale sound" (*blasser Klang*) comprehensible. Similarly, although most mappings of adjectives from taste onto smell, such as "sweet smell" and "bitter smell," or from touch terms, such as "soft smell" and "sharp smell," were found intuitively comprehensible, and over 60% of respondents found "quiet smell" (*stiller Geruch*) comprehensible, most other mappings of sound and color adjectives onto smells were rated not accessible.[11] Although more research needs to be done on the issue of cognitive accessibility, the Düsseldorf study suggests that even in the case of Western languages, ordinary speakers may not be as tongue-tied when it comes to smells as some researchers have intimated.[12]

Smell in Non-Western Cultures and Languages

But if we look at everyday linguistic practices outside Western societies, we find several cultures where not only is the sense of smell more highly valued than in the West, but the linguistic expressions for odors are often subtler and more complex.[13] A particularly striking recent study in this respect has looked at the use of synesthetic metaphors in everyday Chinese. Qingqing Zhao and Chu-Ren Huang, basing their study primarily on the Sinica Corpus, compared fifteen commonly used Chinese sensory adjectives from three realms: taste (bitter, sweet, sour, spicy, salty), touch/temperature (cold, hot, icy, warm, moderately warm), and smell (smelly, fragrant, fishy, urinous, mutton-like). They found that four of the five Chinese taste terms could be used to modify "fragrance," resulting in "light, bitter fragrance," "sweet fragrance," "sour fragrance," and "tangy, spicy fragrance." Only "salty" taste did not transfer to form a metaphor with "fragrance." In the case of touch/temperature, all five source terms transferred to "fragrance," as in "a piece of cold fragrance," "flocks of hot fragrance," "warm fragrance of coffee," "tangy, warm fragrance."

Of the five main Chinese odor terms—smelly, fragrant, fishy, urinous, and mutton—the last three do not transfer at all to other senses, and none of the five smell terms transfers to taste. The other transfers, however, are quite distinctive. Thus, the one transfer from smell to the domain of touch is from "fragrant," with the result that a fragrant touch means a "kiss." The one transfer commonly made from smell to the target domain of emotion is "smelly," which joins with three other Chinese characters to signify "unhappy." Finally, if one thinks of "smelly" and "fragrant" as forming opposites, their effect when joined to the term for "word" in the target domain of hearing is "bad words" and "good words," respectively. And when the characters for "smelly" and "fragrant" are joined to the character for "appearance" in the realm of vision, we get "disgusting appearance" and "beautiful appearance" respectively.[14]

As impressive as this bit of Chinese evidence of the linguistic potential for expressing smell in ordinary non-Western languages is, the French linguist Charles Boisson has discovered even more possibilities in his survey comparing nine language families and sixty specific languages. A little less than a third of the languages were Western, the bulk coming from Africa, Asia, and the Pacific as well as some of the indigenous peoples of both North and South America. Although Boisson regards his survey as only a rough beginning of what is needed, he was impressed by the variety of terms for smells in many non-Western languages. "The result is less negative than one might fear from reading various . . . psychologists, physiologists, chemists, and perfumers" who base their conclusions on "a quick examination of a small number of familiar languages such as French, English and German."[15]

Although Boisson discovered that a hedonic polarization does exist in many languages, his evidence does not support the implication of Yeshurun and Sobel's theory that no one ever perceives and expresses genuine qualitative characteristics. On the contrary, Boisson discovered numerous qualitative terms in a wide span of languages, some of the terms semantically highly complex. Hawaiian, for example, has separate terms for heavy and light fragrances as well as for stronger and weaker negative smells. Moreover, it has a specific term, *puîa*, to express the idea of a pleasant smell that spreads, translatable as "sweet smell, diffused" or in other contexts meaning "permeated with perfume." Then there are the special terms for how smells are carried on the air, for example *mâpu* for "wind-blown," and *moani* for "wafted fragrance." Finally, there is a single term, *anuhea*, that can embrace the complex relation of a smell to temperature, touch sensation, and environmental context all at once, signifying "cool, soft fragrance, as of an upland forest."[16]

Boisson notes that most terms for body smells in the languages he studied tend to be more extensive in cultures that place a high value on cleanliness. Cleanliness, of course, is "next to godliness" according to an old saying in English, but in Arab-Muslim cultures cleanliness is even more important both socially and religiously than in the West. Thus, several other French scholars have suggested that although the total number of terms for odors in Arabic is similar to that of Europeans languages, in some Arabic-speaking societies smell plays an explicitly coded role not only in religion, but also in rites of passage and social relations, leading to a different lexical emphasis than in Western languages. Thus, in more traditional Arab societies, a newborn is subject to fumigations and aromatic massages, and the skin is rubbed with scented potions. At marriage both the bride and groom must be properly scented.[17] And among all Muslims, one purpose of washing the hands, feet, ears, nose, and mouth before entering a mosque is to purify the body of foul odors, especially the breath since, as a traditional saying has it, the angels cannot stand garlic or onion breath anymore than humans can. This leads to a lexical emphasis on terms for the smells of the body, with four terms alone for the breath.[18]

Taken together, the results of Zhao and Huang's study of synesthetic metaphors in Chinese, Boisson's comparison of sixty languages, and the French scholars' study of Arabic usages represent an astonishing rebuke to those who glibly talk of a general poverty of human languages for expressing smell. But even more striking differences from Euro-American social practices and linguistic usages regarding odors and the sense of smell are to be found in several small-scale traditional societies (the supposed "savages" whom Darwin claimed make little use of the sense of smell). For example, the Ongee, a hunting and gathering people of the Andaman Islands off the coast of India, use smell as an organizing principle of thought, constructing their calendar after the succession of odors from plants that come into bloom at different times of the year. Among the Ongee, people are said to be born without odor but to gradually gain olfactory strength as they grow, only to lose it at death, after which their odorless spirit seeks out the odors of the living in order to be born again. Accordingly, the Ongee word for "growth" means literally "a process of smell," and the word for hunter means literally "one who has his smell tied tightly."[19] In addition to these linguistic extensions of the smell vocabulary among the Ongee, David Howes points out that the Sereer Ndut of Senegal recognize five abstract odor categories (i.e., none of them based on a single source reference), the Bororo of Brazil recognize eight, and the Kapsiki of Cameroon fourteen.[20]

Ernest Gell's study of the pig-hunting magic of the Umeda, a small group in the West Sepik region of Papua, led him to some thought-provoking speculations on the interrelation of smells, magic, and dreams, and he drew out some general implications of his speculations for the problem of language and smell. He was struck by the fact that Umeda hunters carry in their shoulder bags an aromatic sachet whose odor, they believe, will powerfully enhance their hunt for the much sought after wild pig. Hunters even sleep with the bag before going on a hunt to allow the odor to influence their dreams. In his reflections on why such perfumes are believed to have the power to inspire helpful dreams, Gell suggests that we should think of the "restricted language of smells" as standing "somewhere between the stimulus and the sign." The fact that smells "only partly detach themselves from the world of objects to which they refer" is analogous to the way magical signs operate in traditional cultures, since any expression of magic both "refers to and *alters* the world" at the same time.[21] From a linguistic point of view, Gell believes, it is no accident that the Umeda word for "dream" (*yinugwi*) and the word for "smell" (*nugwi*) are etymologically close.[22]

Several psycholinguists interested in anthropology have also begun studying the smell terms used by hunter-gatherer societies in Southeast Asia. The Maniq speakers of the mountains of southern Thailand, for example, have a smell lexicon made up of fifteen abstract terms.[23] But as telling as the cases of the Maniq, Ongee, and Umeda are for exposing the Eurocentrism of claims about the poor possibilities of human language for expressing smell, a more direct challenge to the received view comes in a comparative study done by Asifa Majid and Niclas Burenhult, entitled "Odors Are Expressible in Language, as Long as You Speak the Right Language."[24] The right language Majid and Burenhult have in mind is Jahai, the language of a group of nomadic hunter-gatherers in the rainforest on the border between Malaysia and Thailand. Like Maniq, the Jahai language has a lexicon of over a dozen terms for smell that are not individual source-descriptors, but general classification terms. One term, *ltpit*, for example, describes a smell the Jahai find common to certain flowers, ripe fruits, soap, Aquilaria wood, and bearcat. But Majid and Burenhult were interested not just in the size and abstract nature of the lexicon, but in whether those who spoke it could actually use it with ease to identify smells, unlike the typical Westerner's difficulty in naming. They devised an experiment in which ten native Jahai speakers residing in a resettlement village (although they still pursued foraging) were tested in their native language and their responses were compared to ten English-speaking Americans of comparable age residing in Austin, Texas. Both groups were given the same set of color chips and odorants in a free-naming task. The odorants were cinnamon, turpentine, lemon, smoke, chocolate, rose, paint thinner, banana, pineapple, gasoline, soap, and onion. Respondents were asked simply: *What color is this? What odor is this?*

Majid and Burenhult coded responses in three ways, according to (1) how much agreement there was within each language group studied, (2) how long the description was (shorter meaning greater certainty), (3) what type of response was given (abstract, source-based, or hedonic). The English speakers' responses on odors showed low agreement with each other, and the subjects often provided long, hesitant descriptions (five times longer than their descriptions for color). Moreover, this greater length was despite the fact that all the odors used in the experiment were common smells that should have been relatively familiar to the English speakers. But the most important result concerns the response type. The English speakers gave abstract color names, but mostly source-based or hedonic descriptions for odors. The Jahai speakers, on the other hand, not only showed far more agreement in naming odors and quicker response time, but also gave abstract terms from their own odor lexicon 99% of the time for both odors and colors. Masjid and Burenhult conclude: "Contrary to the widely-held belief that people universally struggle to describe odors, Jahai speakers name odors with ease." In fact, the researchers go on, "We suspect the current results underestimate the expressibility of smells in Jahai. Smell features prominently in everyday communication, as well as in the indigenous ideology and rituals of the Jahai."[25]

This study of the Jahai is only one small experiment, but taken together with the other evidence we have presented concerning complex lexicons for smell in other non-Western societies, it raises serious doubts as to whether studies and theories based solely on subjects from modernized Western countries give results that hold universally. Of course, more studies of non-Westerners need to be done, but if we put the existing evidence that several cultures are capable of generating languages with complex smell vocabularies containing a variety of abstract terms, together with the evidence we discussed earlier that showed Western olfactory experts can easily imagine and name odors in the domain of perfume creation, the outlook for cultivating the cognitive dimension of the human sense of smell seems even more promising than before. Moreover, if odors can be the basis of cosmologies and social organization and play a role in ideology and ritual, the promise of odors and the sense of smell for creating art works and developing a reflective aesthetic criticism also seem much brighter. Thus, anthropologists, and psycholinguists with a cross-cultural focus, have given us reason to believe that the generally poor ability of Westerners to accurately identify and name odors in the artificial setting of the laboratory may be partly a cultural artifact that we might begin to overcome by cultivating our sense of smell. Moreover, just as we should not be too quick to assume on the basis of a cursory examination of the vocabulary of a few Western languages that human languages in general lack the capacity to offer sophisticated expressions of smell experience, neither should we overlook the evidence of those Westerners who *are* adept at using language to offer nuanced and penetrating descriptions of olfactory experience. What Jenifer Robinson says of the novel's power to represent and express emotion applies to olfaction as well: "Novels . . . introduce us to characters and emotional states for which there are no one-word descriptions in folk psychology."[26] Thus, if we turn to Western poets and novelists, we can find abundant evidence that there are ample resources in Western languages to give compelling expression to the experience of smell.

Notes

1. Kathrin Kaeppler and Friedrich Mueller, "Odor Classification: A Review of Factors Influencing Perception-Based Odor Arrangements," *Chemical Senses* 18, no. 3 (2013): 189–209.
2. Egon Peter Köster, "The Specific Characteristics of the Sense of Smell," in *Olfaction, Taste, and Cognition*, ed. Catherine Rouby et al. (Cambridge: Cambridge University Press, 2002), 35–36.
3. Paul Jellinek, *The Psychological Basis of Perfumery*, 4th ed., trans. J. Stephan Jellinek (London: Blackie Academic and Professional, 1997). This edition has a long appendix by Paul Jellinek's son, J. Stephan Jellinek, "The Psychological Basis of Perfumery: A Reevaluation," that proposes several useful modifications of his father's diagram.
4. The version of Edward's wheel can be found at https:// www.fragrancesoftheworld.com/fragrancewheel.
5. Manuel Zarzo and David T. Stanton, "Understanding the Underlying Dimensions in Perfumers' Odor Perception Space as a Basis for Developing Meaningful Odor Maps," *Attention Perception and Psychophysics* 71, no. 2 (2009): 245, https://doi.org/10.3758/APP.71.2.225.
6. Zarzo and Stanton, "Understanding the Underlying Dimensions," 239.

7. Danièle Dubois and Catherine Rouby, "Names and Categories for Odors: The Veridical Label," in Rouby et al., *Olfaction, Taste, and Cognition*, 50.
8. The Wine Aroma Wheel developed by Ann C. Noble of the University of California Davis is available at https://www.winearomawheel.com/.
9. Sweeney, *Aesthetics of Food*, 172–74.
10. Sweeney, *Aesthetics of Food*, 175.
11. The researchers hypothesized that what made "quiet smell" accessible was that the adjective was *scalar* (quiet, loud) rather than *qualitative* (yellow, black). Thus, "quiet" in the realm of sound stands at the low end of the sound intensity scale, analogous to the way "mild" stands at the low end of the smell intensity scale, whereas the respondent must make a considerable effort to figure out what a "red smell" or a "yellow silence" might signify. Markus Wernung, Jens Fleischhauer, and Hakan Beseoglu, "The Cognitive Accessibility of Synesthetic Metaphors," in *Proceedings of the 28th Annual Conference of the Cognitive Science Society*, ed. R. Sun (London: Lawrence Erlbaum Associates, 2006), 2365–78.
12. It is worth remembering that color terms themselves are often used metaphorically, so that "green" is a standard category in most perfume classifications, signifying a particular type of "fresh" scent alongside citrus and fruity, and ordinary perfume users can learn the application of the term from salespeople, magazine articles, perfume blogs, advertisements, etc.
13. Although it is not specifically focused on linguistic structures, Susan Rasmussen has offered an illuminating account of the use of smells as a medium of communication in "Making Better 'Scents' in Anthropology: Aroma in Tuareg Sociocultural Systems and the Shaping of Ethnography," *Anthropological Quarterly* 72, no. 2 (April 1999): 55–73. I am grateful to Rebecca Shereikis of Northwestern University for calling my attention to this article.
14. Qingqing Zhao and Chu-Ren Huang, "A Corpus-Based Study on Synesthetic Adjectives in Modern Chinese," in *Chinese Lexical Semantics: 16th Workshop*, ed. Qin Lu and Helena Hong Gao (Cham, Switzerland: Springer International, 2016), 535–42, https://www.springer.com/cn/book/9783319271934.
15. Claude Boisson, "La dénomination des odeurs: Variations et régularités linguistiques," *Intellectica* 1, no. 24 (1997): 31.
16. Boisson, "La dénomination des odeurs," 37.
17. Françoise Aubaile-Sallenave, "Bodies, Odors, and Perfumes in Arab-Muslim Societies," in Drobnick, *Smell Culture Reader*, 391–99.
18. Sophie David, Melissa Barkat-Defradas, and Catherine Rouby, "A Contrastive Study of French and Arabic Olfactory Lexicons," in *Words for Odours: Language Skills and Cultural Insights*, ed. Melissa Barkat-Defrades and Elisabeth Motte-Florac,. (Newcastle upon Tyne, UK: Cambridge Scholars Publishers, 2016), 167–188.
19. David Howes, "Nose-Wise: Olfactory Metaphors in Mind," in Rouby, *Olfaction*, 72–73.
20. David Howes, "Nose-Wise," 75. Another anthropologist has found a complex odor vocabulary containing abstract terms among the Boholano of the Philippines. Betina

Beer, "Boholano Olfaction: Odor Terms, Categories, and Discourses," *Senses and Society* 9, no. 2 (2014): 151–73.
21. Alfred Gell, "Magic Perfume, Dream . . ." in Drobnick, *Smell Culture Reader*, 401–2.
22. Gell, "Magic, Perfume, Dream," 406–7.
23. Ewelina Wnuk and Asifa Majid, "Revisiting the Limits of Language: The Odor Lexicon of Maniq," *Cognition* 131 (2014): 125–38, https://doi.org/10.1016/j.cognition.2013.12.008.
24. Asifa Majid and Niclas Burenhult, "Odors Are Expressible in Language, as Long as You Speak the Right Language," *Cognition* 130 (2014): 266–70, https://doi: 10.1016/j.cognition.2013.11.004. Their title is an intentional retort to Yeshurun and Sobel's essay we considered earlier, "An Odor Is Not Worth a Thousand Words."
25. Majid and Burenhult, "Odors Are Expressible in Language," 269–70.
26. Robinson, *Deeper Than Reason*, 159.

7
Writing Smell

Late in a severe Midwest winter a few years ago, my wife and I rented a small house for a week in the hills above Santa Barbara. The first night, as I stepped out onto the patio to admire the lights of the city below, I was met by a powerful fragrance more compelling than the splendid view. At once sweet and pungent, the scent seemed to conceal a whole world of subtle elements. I had to find the source. It didn't take long to come on a bush at the corner of the house covered with delicate white flowers shining in the faint light. Jasmine.

Mandy Aftel, known for creating unique fragrances from natural essences, has written of jasmine's "deeply floral, warm, rich and highly diffusive odor, with a peculiar honeylike sweetness and tealike, fecal undertone." It is "the intensely narcotic aura that strikes you most," she goes on, "a feeling of intoxication" that comes from indole, "a major element in jasmine, tuberose and orange flower, that is also found in human feces." It is the indole that "lends jasmine the putrid-sweet, sultry-intoxicating nuance."[1] When I came across Aftel's evocative analysis, I understood better why I have never forgotten my nightly rendezvous with jasmine that winter. And I also thought: those theorists who keep harping on "the poverty of language" for expressing smells have not looked very hard. Aftel writes about odors as an olfactory expert, of course, but we find just as rich a linguistic field if we look to those whose expertise is not olfaction itself, but literature. In this chapter I will draw on a few examples from poets and novelists, moving from the way poets and some novelists use fine-grained metaphorical devices to express smell experiences to writers notable for bringing olfactory experiences into language in more discursive ways.

Poetry

Among the crucial devices used by poets and novelists for putting odors into language are the synesthetic metaphors that we briefly discussed in the previous chapter, such as a "sweet smell" or "soft fragrance," usages now so common as not to be noticed as metaphorical.[2] Indeed, as Stephen Ullmann points out, in many languages—ancient Chinese, Sanskrit, Persian, Egyptian—similar devices go back for millennia, and in the West they begin turning up with some frequency from the Renaissance on. John Donne's "The Perfume" (Elegy

IV: 39–42) uses a synesthetic metaphor to describe how the perfume he was wearing as he sneaked into his lover's house betrayed him to her father: "But O! too common ill, I brought with me / That, which betray'd me to mine enemy / A *loud* perfume." But it was the nineteenth-century Romantics, Ullmann suggests, who finally elevated the synesthetic metaphor to a standard poetic device, and passed it on to those later nineteenth-century movements known as Symbolism, Decadence, and Aestheticism several of whose writers gave particular attention to smell.[3]

Among those who have written with great feeling and insight about scents, it is surely Charles Baudelaire who has produced the most dazzling body of work that reveals the creative possibilities for giving odors linguistic expression. Indeed, he devotes the entire second half of his most famous and influential poem, *Correspondences*, to the sense of smell.[4] In the opening stanza of the poem, Nature is likened to a temple from whose pillars emerges a confusion of word-like sounds, a veritable forest of symbols. The second stanza is one long sentence ending in the poem's most often quoted phrase: "like echoes that merge from afar, in a dark and profound unity, vast as the night and brilliant as day, scents, colors, and sounds correspond" (*Les parfums, les couleurs, et les sons se répondent*). Then Baudelaire turns to perfumes themselves in the last two stanzas.

> There are perfumes fresh as the flesh of infants
> *Il est des parfums frais comme des chairs d'enfants*
> Mellow as oboes, green as meadows
> *Doux comme les hautbois, verts comme les prairies,*
> —And others, corrupted, rich and triumphant
> —*Et d'autres, corrompus, riche et troumphants,*

> Having the expansiveness of infinite things
> *Ayant l'expansion des choses infinies*
> Like amber, musk, benzoin and incense,
> *Comme l'ambre, le musc, le benjoin et l'encens,*
> Which sing the raptures of the spirit and the senses.
> *Qui chantent les transports de l'esprit et des sens.*[5]

Although the poem as a whole begins with the *sounds* and *sights* of the symbols through which Nature addresses us, already by the end of the second stanza *smells* emerge and by the third stanza perfumes take over. In the famous line "scents (*parfums*), colors and sounds correspond," I translate *parfums* by "scents" since in French the term *parfum* is often used not only for a perfume to wear, but also for either a scent or a flavor. Moreover, the context here clearly calls for a term of parallel generality to colors and sounds, although as the next verse makes clear, perfumes in the narrower sense are also anticipated. Indeed, this last line of the second stanza—"scents, colors, and sounds correspond"—forms a kind of hinge uniting the two halves of the poem. It joins the visual images (pillars, forest, darkness, light) and auditory images (words, echoes) of the first half of the poem via the notion of scents (*parfums*) in general to the idea of perfumes (*parfums*) in the more specific meaning of the term that dominates the second half of the poem.

It is in this second half that Baudelaire shows us most concretely how to transcend the ordinary reliance on simple source-based similes in favor of synesthetic and other types of metaphors. The fact that there are scents as "fresh as infants' flesh" does not mean that the perfumes in question simply smell *like* babies' skin. Rather, the language suggests that certain perfumes have the freshness or coolness typical of an infant's smell or the feel of an infant's flesh, thus invoking touch as well as smell. That reading is confirmed by the next metaphor Baudelaire uses, which invokes the mellowness or softness (*doux comme*) of an oboe's sound. Similarly, although the phrase "green as meadows" (*verts come les prairies*) might be taken as a bland simile, that is, there are perfumes that "smell like grass," the "green" here is not a narrow source simile, but a more general signifier of freshness and new growth. Moreover, a "*prairie*" is not merely an expanse of green grass, but a field made up of many kinds of grasses, herbs, wildflowers, and so on. (In his own way, Baudelaire anticipates here the twentieth-century perfumers' "Green" category.)

But Baudelaire's finest achievement in giving odors a name comes with the next lines that invoke a series of strikingly different kinds of scents or perfumes, those that are "corrupted, rich and triumphant." This is a metaphoric complex unlike anything one might normally expect in the description of a perfume; "corrupt," "rich," and "triumphant" are moral and social qualities we normally ascribe to persons. In contrast to the previous line, which likened certain perfumes to the feel of babies' flesh, or the gentle sonority of oboes, or to the refreshing air of meadows, the scents that are called corrupt, rich, and triumphant are not directly connected to something we can visualize, touch, or hear. But, as the list suggests, these scents are known not only for their intensity, complexity, and costliness (they are "rich" in two senses), but also for their darker, more sensual aspect, perhaps partaking of decay ("corrupt"), and they are even a little menacing ("triumphant"). Baudelaire piles their names against each other ("amber, musk, benzoin and incense") in a kind of recital that builds momentum toward the climactic closing line in which they "sing the rapture of the spirit and the senses."

What a powerful lesson in how to bring odors into language! Baudelaire shows us not only how to make them speak but even how to make them sing in both a lighter and darker key. (Incidentally, you may have also noticed that Baudelaire's general contrast of the fresher and lighter scents with richer and darker ones remarkably parallels aspects of both Jellinek's and Edwards' scent maps.)

A hundred years after Baudelaire, Seamus Heaney used a series of synesthetic metaphors in "Digging" with equally telling effect. Each metaphor plays a key role in the poem's development, with the last and crucial metaphor from smell. The conceit of the poem is that the poet is sitting with pen in hand, while outside under his window there is a *"clean rasping* sound" each time his father's spade sinks into ground. The sounds of his father's digging remind Heaney of his youth when they scattered seedlings for the next crop, loving the feel of the little potatoes' *"cool hardness."* Then, after several lines in which Heaney remembers his grandfather, who dug potatoes as deftly as his father does now, comes the poem's climactic stanza, growing out of a synesthetic smell metaphor as Heaney remembers the *"cold smell* of potato mold." Together the three synesthetic metaphors set up the poem's closing metaphor as Heaney muses that *he* will dig with his "squat pen."[6] Heaney's "cold smell" is surely as cognitively accessible to ordinary speakers and readers as the synesthetic metaphors in German that we considered in the previous chapter. But here the "cold smell" of potato mold near the end of the poem powerfully evokes and unites not only a world of labor and deep family ties across generations, but also the cold smell of death embedded in the memory of the potato famine.

Novelistic Devices

At this point, rather than pursue the way other poets have shown us how to express odors and the sense of smell through suggestive synesthetic metaphors, I want to turn to the novel. The novel, of course, is a much more discursive genre, and my concern will be less with the fine grain of linguistic usages than with the many ways novelists have vividly conveyed the total experience of smell and its role in human life. Even so, I will start with some examples of novelists who have in fact made impressive use of both synesthetic and other unusual metaphors in descriptive prose. In *Death Comes for the Archbishop*, Willa Cather describes Father LaTour's experience of the smell of New Mexico as not only dryly aromatic, but as "soft and wild and free," thus uniting a synesthetic modifier from touch with metaphors belonging to the moral realm as compelling in their own way as Baudelaire's "corrupt, rich and triumphant."[7]

At the other end of the prose spectrum from Cather's compressed line, we have Proust's long, resonant sentences, piling metaphor on metaphor, sometimes taking up almost a page, as when he evokes the smells of his aunt's rooms at Combray that emanate from the "virtues, wisdom and habits of the village."

Among other things, they are "weather-tinted like those of the neighboring countryside, but already humanized, domesticated, snug . . . lazy and punctual as a village clock, roving and settled, heedless and provident, linen smells, morning smells, pious smells."[8] In this passage, of which I have quoted only a few phrases, Proust uses a number of the kind of source adjectives that so many theorists have claimed are the only device language has for smell, but he gives them life by making them emanate from the wisdom and habits of the village. And what a wonderful series of contrasts he employs: smells that are "weather-tinted" yet already "humanized, domesticated, snug," that are both "lazy" and "punctual," "heedless and provident." Like Cather's naming the smells of frontier New Mexico "wild and free," Proust's calling the village odors at once lazy and punctual, heedless and provident pricks our olfactory imagination by inviting us to make unaccustomed associations.

Where Proust evokes the character of a single household and village, in *Notebooks of the Malte Laurids Brigge* Rainier Maria Rilke uses a description of imagined odors to give the impression of a whole class of people in a city. Malte, a young German on a stroll through Paris, is looking up at a half-demolished tenement where he can see the insides of what were once apartments, but are now only dirty streaked walls with peeling paper and shadows left where pictures once hung, yet walls that he imagines still exhale the smell of the "tough life" that had been lived in them.

> The breath of these lives stood out—the clammy sluggish, musty breath which no wind had yet scattered. . . . There stood the tang of urine and the burn of soot and the grey reek of potatoes and the heavy, smooth stench of ageing grease. The sweet, lingering smell of neglected infants was there, and the fear smell of children who go to school.[9]

Since Rilke was a great poet, it is no surprise that his Malte uses so many synesthetic metaphors, the "grey reek of potatoes" linking smell and color, just as Heaney's "cold smell" of molding potatoes linked smell and temperature. Urine has a "tang" (taste) and soot has a "burn" (touch), the stench of aging grease is "heavy" and "smooth" (weight and touch).

Virginia Woolf rarely invoked odors in her novels, except for one charming novella called *Flush: A Biography*. Here we find a different but equally effective way of describing smells, and they are the smells experienced by a dog! *Flush* is a tour de force that tells the story of Elizabeth Barrett's elopement and marriage to Robert Browning from the perspective of Elizabeth's beloved cocker spaniel, Flush. When Flush, who grew up in the country, first enters the Barrett's fine house on London's Wimpole Street and is taken on his first walk, Woolf finds striking adjectives for the odors that assault him: "the swooning smells that lie in the gutters; the bitter smells that corrode iron railings; the fuming, heady smells that rise from basements—smells more complex, corrupt, violently contrasted and compounded" than any he had smelt in his earlier life in the countryside.[10] And when Elizabeth finally elopes to Italy, Flush discovers a marvelous new olfactory world as he wanders about Florence enjoying "the rapture of smell . . . goat and macaroni were raucous smells, crimson smells." He even follows the "sweetness of incense" into dark cathedrals.[11] Although Woolf at one point echoes the poverty of language claim, lamenting the dearth of words for expressing smell as compared to those for vision, she shows an extraordinary ability to give odors a name: fuming, heady, bitter, swooning, corrupt, queer, crimson, raucous. When it comes to expressing smells, perhaps it is not words that fail most of us, but imagination.

Character

So far I have focused on novelists' uses of specific literary devices that complicate or exceed the supposed inability of language to articulate smell experiences. But as recent scholarship has shown, even novelists who do not use synesthetic and other metaphorical devices are able to vividly deploy odors to establish character, further plot, and explore themes.[12] As a transition to some examples of such practices, I want to look at James Joyce's uses of smell in *Ulysses* since Joyce's style superbly combines a jocular/serious poetic inflection of language with a brilliant display of the deep connection between smell and character.[13] Smell is present in the novel from the first lines describing Leopold Bloom as especially liking mutton kidneys, which gave his palate "a fine tang of faintly scented urine" (4:3–6) to the last line of Molly's soliloquy that ends the book, when she remembers drawing him down "so he could feel my breasts all perfume . . . yes I said yes I will Yes" (18:1606–9).

The "Nausicaa" chapter of *Ulysses*, for example, sets up a complex and subtle contrast of sacred and profane smells. As the smell of incense wafts out over Sandymount Strand from a church temperance meeting, Bloom lecherously eyes the young Gerty. who enjoys his ogling her from a distance as she leans back so that he can see up above her knee (13:695–700). The first part of the chapter is told from her naively romantic perspective, the second from the perspective of Bloom, who has been masturbating as he stares at her and later sniffs for traces of the smell of semen (13:1042). As Gerty departs, she waves her wadded up handkerchief that is impregnated with a perfume, and Bloom imagines he can smell it, which sets him off on one of his most scent-filled reveries:

> Wait. Hm. Hm. Yes. That's her perfume. Why she waved her hand . . . What is it? Heliotrope? No. Hyacinth? Hm. Roses, I think. She'd like scent of that kind.

Sweet and cheap: soon sour. Why Molly likes opoponax. Suits her with a little Jessamine mixed. Her high notes and her low notes. At the dance night she met him. . . . Clings to everything she takes off. . . . Know her smell in a thousand (13:1007–24).

The "low notes" of Mollie's perfume and the "night she met him," of course, refer to Mollie's lover, Blazes Boylan. And when Bloom finally reaches home at the end of his long day that began with the pungent smell of a slightly burned kidney, and is now musing about his moral and legal options for dealing with Mollie's adultery, he notices, among other things: "a pair of outsize ladies' drawers . . . redolent of opoponax, jessamine and Muratti's Turkish cigarettes . . . new clean bedlinen, additional odours, the presence of a human form, hers, the imprint of a human form not his" (17:2094–95, 2123–24). After examining his feelings of envy, jealousy, and abnegation, Bloom finally settles on equanimity, and sensing the stirrings of an erection at the sight and smell of Mollie's behind, "he kissed the plump mellow yellow smellow melons of her rump" (17:2241). And she, in her dreaming reverie, pulls him down to "feel my breasts all perfume" (18:1610). As Frances Devlin-Glass puts it, "Just as on the Hill of Howth, Bloom had fallen in love with Molly at first smell, Joyce makes it clear that for Bloom at this moment her 'female hemispheres' have become heavenly 'islands of the blessed . . . redolent of milk and honey.'"[14]

Although William Faulkner did not engage in the kind of wordplay that marks *Ulysses*, he did make complex use of odors for expressing character through smell.[15] Thus, the "idiot," Benjy, repeats many times over in the first chapter of *The Sound and the Fury* that his sister Caddy "smelled like trees," and his brother Quentin's veiled incestuous impulses are repeatedly expressed later in the novel in terms of her smelling like honeysuckle. Indeed, several younger scholars have argued that Faulkner's oeuvre as a whole is pervaded with the theme of smell, something that seems especially true of *The Sound and the Fury*, which conventional criticism might characterize as partly "seen through Benjy's eyes" but could with equal justice be characterized as "smelled through Benjy's nose."[16] In the first chapter alone, in addition to the numerous repetitions of "Caddy smelled like trees," Benjy can "smell the cold" (6), smell "the clothes flapping, and the smoke blowing across the branch" (14), "the pigs" (20), the covers that "smelled like T.P.," the grandmother's death ("could smell it") (34), "the sickness . . . on a cloth folded on Mother's head" (61), and "Father," who "smelled like rain" (64). Versh also "smelled like rain. He smelled like a dog, too" (68). At the end of the chapter when the children are all in bed in one room, Benjy can "hear . . . the darkness, and something I could smell" (75). Faulkner has Benjy's nose tell the reader so much that other characters in the novel fail to see or hear.

My final novelistic examples of reflecting character through smell come from two works notable for reversing the conventional association of positive fragrances with love and life and of stench with sickness and death: Toni Morrison's *Sula* and Jamaica Kincaid's *Autobiography of My Mother*.[17] For example, when Sula's baby, Plum, is gagging in pain from constipation, she takes him into the smelly outhouse, where "deep in its darkness and freezing stench she squatted down and shoved the last bit of food she had in the world up his ass . . . softening the insertion with a dab of lard."[18] In contrast to this association of motherly love with the stench of the outhouse is the scene in which Sula's lover decides to leave her when he finds her "lying on fresh white sheets, wrapped in the deadly odor of freshly applied cologne."[19] A similar revalorization of the conventional idea of negative versus positive odors occurs in Kincaid's *Autobiography of My Mother*. Xuella, whose "human form and odor," we are told, "were an opportunity to heap scorn on her," affirms her dignity by embracing the strong smells of her body: "I loved the smell of the thin dirt behind my ears, the smell of my unwashed mouth, the smell that came from between my legs, the smell in the pit of my arm, the smell of my unwashed feet."[20]

Clearly, neither Morrison nor Kincaid makes any effort to avoid the simple "smell of" usage that some have derided as the only way language can express smell. We have surely by now exploded that myth, but it is worth noting that in the right hands, like Morrison's or Kincaid's, even the phrase the "smell of" can be powerfully expressive of character and action. The same is true when character and action are reflected simply in terms of the effect of a smell. At the end of Norman Mailer's *An American Dream*, for example, Rojack, who has been driving across the country during a heat wave, is watching his old army buddy perform an autopsy on a man with cancer who had suddenly died from a burst appendix and peritoneal gangrene. The smell from the incision is so powerful "it called for the bite of one's jaws not to retch." The smell keeps following Rojack for two days; he experiences it every time he passes a fertilized field, a dead animal on the road, or a rotting tree stump. There is "madness in it . . . come in with breath." In Rojack's rumination, Kant's fear of inhaling the other becomes palpably threatening.[21]

I have drawn these varied ways of expressing smell in the novel from the works of canonical or established authors, but, of course, there are interesting explorations of smell in many contemporary works, from Tom Robbins's *Jitterbug Perfume* to Radhika Jha's *Smell*.[22] In a later interlude preceding the discussion of the ethics of perfume wearing, I will consider two other novels in which smell and perfume play a central role, J.-K. Huysmans's *Against Nature* and Patrick Süskind's *Perfume: the Story of a Murderer*. Yet the passages from the authors we have just surveyed should be enough to show what an impressive span of linguistic devices, narrative techniques, and thematic variety modern literature has developed to bring smells to life in ways that provoke reflection, awaken insight, and stimulate the imagination, the very requisites of thoughtful aesthetic experience.

Notes

1. Mandy Aftel, "Perfumed Obsession," in Drobnick, *Smell Culture Reader*, 211–13.
2. For other kinds of metaphorical devices see Rémi Digonnet, *Métaphore et olfaction: Une approche cognitive* (Paris: Honoré Champion, 2016). Digonnet offers a highly detailed analysis of olfactory metaphors drawing on three novels.
3. Stephen Ullmann, *Language and Style: Collected Papers* (Oxford: Basil Blackwell, 1966), 86.
4. I have made my own more or less literal translation from the first French edition. Charles Baudelaire, *Les Fleurs du Mal* (Paris: Poulet-Malassis et de Broise, 1857).
5. Baudelaire, *Fleurs du mal,* 19. I am grateful to Rosina Nejinsky for comments on my translation.
6. Seamus Heaney, *Opened Ground: Selected Poems, 1966–1996* (New York: Farrar, Straus and Giroux, 1998), 3–4.
7. Willa Cather, *Death Comes for the Archbishop* (New York: Alfred A. Knopf, 1927), 275–76.
8. Marcel Proust, *Remembrance of Things Past*, vol. 1, trans. C. K. Moncrief and Terence Kilmartin (New York: Random House, 1981), 53.
9. Rainer Maria Rilke, *The Notebooks of Malte Laurids Brigge*, trans. M. D. Herter (New York: Norton, 1949), 47–48.
10. Virginia Woolf, *Flush: A Biography* (New York: Harcourt Brace, 1933), 37.
11. Woolf, *Flush*, 139–40.
12. A pioneering work here is Hans Rindisbacher's *The Smell of Books: A Cultural-Historical Study of Olfactory Perception in Literature* (Ann Arbor: University of Michigan Press, 1992), which focuses on nineteenth-century continental literature. For English literature there are several works, including Janice Carlisle's *Common Scents: Comparative Encounters in High-Victorian Fiction* (Oxford: Oxford University Press, 2004), 4–6, and Emily Friedman's *Reading Smell in Eighteenth-Century Fiction* (Lewisburg, PA: Bucknell University Press, 2016).
13. James Joyce, *Ulysses* (New York: Random House, 1986). All references in parentheses are to this edition. See also Laura Frost, *The Problem with Pleasure: Modernism and Its Discontents* (New York: Columbia University Press, 2013), 33–62.
14. Frances Devlin-Glass, "Armpits and Melons: An Olfactory Reading of James Joyce," *The Conversation*, June 15, 2017, http://www.theconversation.com/armpits-and-melons-an-olfactory-reading-of-james-joyce-78832.
15. William Faulkner, *The Sound and the Fury* (New York: Random House, 1984).

16. See Terri Smith Ruckel, "The Scent of a New World Novel: Translating the Olfactory Language of Faulkner and Garcia Márquez," PhD diss. (Louisiana State University, 2006).
17. Danuta Fjellestad, "Towards an Aesthetics of Smell, or, the Foul and the Fragrant in Contemporary Literature," *Cauce* 24 (2001): 637–51, https://cvc.cervantes.es/literatura/cauce/pdf/cauce24/cauce24_37.
18. Toni Morrison, *Sula* (New York: Alfred A. Knopf. 1973), 34.
19. Morrison, *Sula*, 137.
20. Jamaica Kincaid, *Autobiography of My Mother* (New York: Farrar, Straus and Giroux, 1996), 32.
21. Norman Mailer, *An American Dream* (New York: Dial Press, 1965), 265–67. I am grateful to J. Michael Lennon, author of a major biography and many other works on Mailer, for calling my attention to this passage.
22. Tom Robbins, *Jitterbug Perfume* (New York: Bantam, 1984); Radhika Jha, *Smell* (New York: Viking, 1999).

8
Odor, Memory, and Proust

One of my fondest childhood memories is the smell of the screen wire on my bedroom window as I fell asleep on summer nights. I was about eleven, and we lived in an old frame house in a scruffy neighborhood of Topeka, Kansas, where my bedroom was at the back of the second floor. In those days before air conditioning, I would put my pillow on the sill with my nose almost touching the screen so I could catch any faint breeze that might be stirring. Of course, it was a multisensory experience; I could feel the cool night air entering my nostrils and caressing my face, hear the sound of rustling leaves on the big elm outside, and peer into the darkness, barely making out shapes. But it was the slightly dusty, metallic smell of the screen that was at the center of my nightly experience. I don't know when or in what circumstances I first recalled it as an adult. Whenever it was, I know the experience was not like Proust's experience of tasting the madeleine: a sudden rush of bliss that opens out onto an entire past. The Proustian involuntary experience is what psychologists call an "odor-triggered memory"; mine, I think, was merely what they call the "memory of an odor," though I find it no less meaningful for that. The smell of the window screen and its attendant sensations was, and still is, for me the heart of an experience that mingled a child's sense of security with the simple pleasures of falling asleep in an open window on a summer night.

Voluntary versus Involuntary Memories

I don't doubt the existence of the Proustian type of highly charged, involuntary memoires. There are many accounts of how a sound or smell in the present suddenly takes one back to a similar sensation in the past and a whole world of time regained dramatically unfolds. Even Darwin records an involuntary smell-induced memory in his 1838 *Notebooks*. One day he visited the National Gallery in London and, "not feeling much enthusiasm, happened to go close to one [painting] & smelt the peculiar smell of Picture. Association with much pleasure immediately thrilled across me, bringing up old indistinct ideas of FitzWilliam Museum."[1] Before turning to Proust's famous literary version of such memories, I think it will be helpful to examine a few findings from a subdiscipline of social psychology that focuses on long-term autobiographical memory. The first

Art Scents. Larry Shiner, Oxford University Press (2020) © Oxford University Press.
DOI: 10.1093/oso/9780190089818.003.0016

finding concerns age and affectivity. Researchers have long known about what they have called the memory "bump" (a clustering of memories from a particular developmental period), and they have found that for most adults, the "bump" concerns memories from one's twenties and thirties, whether the cues triggering the bump are verbal, visual, or olfactory. More recent studies that focus on the memories of older people, ages sixty-five to eighty, have found two "bumps." Memories elicited by verbal and visual cues still peak in the early years of adulthood, but older people's odor-evoked memories cluster in childhood, ages six to ten. One explanation for this is that associative learning begins early in life and that most memories from childhood seem to be highly affective.[2]

A second finding from long-term autobiographical memory studies concerns the basic distinction between voluntary and involuntary memory. Voluntary memories, as the name suggests, emerge from a deliberate search process that is consciously monitored and draws on a person's general autobiographical knowledge; involuntary memories emerge spontaneously by an associative process that is not consciously monitored, although once a memory emerges, it can be consciously elaborated and extended. As two of the leading experts on long-term autobiographical memory, David Rubin and Dorthe Berntsen, point out, involuntary memories in general are actually quite frequent; many people have as many as five a day. Of course, few of these are likely to be either remarkable or from the distant past and are quickly forgotten. Moreover, the bulk of *both* voluntary and involuntary memories tend to be recent, to have a strong emotional component, and to posses some novel aspect that makes them stand out and give them relevance to the current context. This last characteristic is particularly important since without it we would be constantly flooded, even overwhelmed, with involuntary memories.[3] Although we experience involuntary memories of various kinds every day, several experiments have shown that odor-evoked memories are likely to be less rehearsed (talked about and thought of) afterward, and that fewer odor memories are typically elicited than memories evoked by verbal or visual cues.[4]

Finally, although many popular discussions of involuntary memories evoked through the senses tend to treat them as benign, smell-evoked involuntary memories can also be unwanted and deeply disruptive, especially those related to post-traumatic stress disorder. The psychiatrists Vermetten and Bremner have written about a Vietnam veteran who avoided driving behind big diesel trucks because the smell brought on bouts of guilt and nausea. One day the source of the recurrent trauma emerged when a neighborhood fire, involving the smell of diesel, evoked vivid thirty-year-old images of a furiously burning army vehicle with open doors, inside which the man could see fellow soldiers he was helpless to save.[5]

There are also plenty of literary examples of both voluntary and involuntary memories that are not so benign, such as Faulkner's Benjy, whose smell memories are in equal measure painful and comforting. My impression is that in most literary works the line between the voluntary and involuntary is not always clear, and surely voluntary memories such as those described in Joyce's *Ulysses* can be as richly suggestive and meaningful as the involuntary ones of Proust's novel that are so often vaunted. As Laura Frost remarks, whereas with Proust "a sensuous experience in the present triggers the journey to the past, for Joyce, even the mere contemplation of odors from the distant past can prompt a profound reverie."[6] Thus, both Bloom and Molly, when remembering the moment of their embrace on Howth Hill, recall the smell of her perfume: "A warm human plumpness settled down on his brain. His brain yielded. Perfume of embraces all him assailed" (8:637–39); "I put my arms around him yes . . . so he could feel my breasts all perfume" (18:1608–9). Clearly, the Proustian type of involuntary smell-triggered memory is hardly the dominant kind of meaningful remembering of smells in modern literature, although many psychologists seem fixated on Proust and his madeleine.

Proust, Psychology, and Transcendence

Indeed, since the year 2000 alone, there has not only been a book-length neuroscience study centered on Proust's ideas, *The Proust Effect* (2015), but dozens of scientific articles alluding to the madeleine or even claiming to prove Proust's insight.[7] So it's time to look more closely at that famous episode near the beginning of *Swann's Way*, a passage more often alluded to than carefully read, keeping in mind that *Swann's Way* is only the first of the eight novels that make up the three thousand pages of *Remembrance of Things Past*.[8]

The novel's narrator is having tea with his mother, who offers him a madeleine, which he dips in his tea. "No sooner had the warm liquid with the crumbs touched my palate than a shudder ran through me . . . an exquisite pleasure had invaded my senses . . . the vicissitudes of life had become indifferent to me" (I, 43). The narrator takes another sip, then a third, but the effect diminishes each time; he can't seem to call forth that first shudder of joy. But just as he is about to give up, the memory suddenly reveals itself. "The taste was that of the little piece of madeleine which on Sunday mornings at Combray . . . my aunt Leonie used to give me, dipping it first in her own cup of tea" (I, 50). And with that memory, he tells us, there came a flood of visual images from the past: the village, the church, the houses, the gardens, the water-lilies on the river Vivonne (I, 51).

The first thing to notice here is that, contrary to the impression one might get from the pervasive use of this episode in discussions of smell by psychologists and others, it is the *taste* of the madeleine, not the *odor* (the orthonasal smell) that does the work. Of course, since we now know, as Proust did not, that *retronasal* smell forms a large part of what we ordinarily call taste (flavor), olfactory psychologists can perhaps be forgiven for treating this passage as if it described a smell experience. Yet it is only *after* the sudden re-emergence of the moment when he tasted the bit of madeleine that the narrator even mentions the term "smell," and then he uses it not for the smell of the tea-soaked madeleine, but as part of a comment, albeit a lovely and often quoted line, about the general relationship of taste and smell to memory. "But when from a long distant past nothing subsists ... taste and smell alone, more fragile but more enduring ... bear unflinchingly, in the tiny and almost impalpable drop of their essence, the vast structure of recollection" (I, 50–51).

Given the merely indirect role odor plays in Proust's account of the madeleine, and olfactory psychologists' constant reference to the episode as exemplifying smell-evoked memory, it is not surprising that a few olfactory psychologists have struck back. Avery Gilbert became so incensed at the constant allusion to the madeleine episode that he set out to debunk the idea that Proust should be considered *the* great literary exponent of odor-evoked memories. Gilbert not only points to evidence suggesting that odor memories decay at the same rate as visual and auditory ones, he also points to literary critics who have shown that auditory, visual, and tactile memories each independently far outnumber those of smell in Proust.[9] "Perhaps it's time," he snickers, for "Proust boosters ... to set aside the soggy Twinkie."[10]

Although I am sympathetic with Gilbert's irritation at the ritual invocation of the madeleine episode, Proust, in fact, makes telling use of both voluntary and involuntary memories of odors throughout the several novels that make up *Remembrance of Things Past*, even if smell is less frequently thematized than the other senses.[11] Consider this involuntary odor-induced memory from a late novel called *The Captive*, in which the protagonist is now an adult lying in bed in his Paris apartment. He catches the smell of the exhaust from a car passing beneath his open window, and it instantly brings him back to summer afternoons in the beach town of Balbec, where he used to take drives into the countryside. The smell of the car exhaust "called into blossom now on either side of me ... cornflowers, poppies and red clover," yet the smell was also "a symbol of elastic motion and power" since it reminded him of his drives to visit a lover (III, 418–19). I much prefer this example of an involuntary, odor-evoked memory to the overworked madeleine story, not only because it clearly is an *odor*, not a taste, memory, but also because its trigger is the smell of car exhaust, upending our usual hedonic expectations of the kind of odors that might remind us of flowering meadows. Moreover, we can see the narrator's typical move from an initial, spontaneously triggered sensation to the more ruminative process of recollection, so that the involuntary blends into the voluntary.

In order to more fully understand the role of involuntary sense memories in Proust's set of novels, and their implications for an olfactory aesthetics, we need to take a careful look at the long, climactic passage near the end of the final novel, *Time Regained*. There the narrator describes three sensory epiphanies that occur in rapid succession that force him to probe why these and other involuntary memories, including the earlier madeleine episode, have filled him each time with a deep feeling of happiness and certitude, an experience that has liberated him from the anxiety of death and now gives him once and for all the confidence to become the artist who is writing the very novel we are reading.

Arriving late for an afternoon concert at the Guermantes' Paris townhouse, the narrator trips on the uneven paving stones of the courtyard. Recovering his balance, he feels a sudden happiness come over him as he remembers experiencing the same sensation years before in Venice as he stumbled on the paving stones of St. Marks. Why, he asks himself, do these kinds of experiences always fill him with joy, making death "a matter of indifference?" But, at that moment, he is ushered into a small sitting room to wait until the first musical selection is over (III, 898–900). As he continues to reflect on the paving stone experience, a second epiphany occurs when a servant accidentally knocks a spoon against a plate, provoking the similar feeling of happiness as had come from the experience of uneven paving stones, although this time accompanied by "a whiff of smoke and relieved by the cool smell of a forest" (III, 900–901). He immediately recognizes the past setting as a forest where a train he was on had stopped and a noise identical in quality to the noise of the spoon on the plate had come from the distant sound of a workman hammering a wheel. But no sooner has this auditory epiphany occurred than a butler arrives with food and hands him a linen napkin. As the narrator wipes his mouth, he experiences another epiphany: an identical tactile sensation had occurred during a visit to Balbec when he dried his face on a towel with the same stiffness as this napkin, and he again experiences a feeling of happiness and certainty against death, affirming his vocation as a writer (III, 901).

Proust's narrator now concludes that when a sensation in the present suddenly evokes an identical sensation from the past, it brings with it a host of associated feelings and experiences, and because they are involuntary, entering us through the senses, the narrator is convinced that they reveal the nature of things in a way that our conscious memory and intellect always distort. But "let a noise or a scent, once heard or once smelt, be heard or smelt again . . . and immediately . . . the essence of things is liberated and our true self . . . is awakened" (III, 904–6). A few pages later, the narrator even characterizes these fugitive moments as belonging to "eternity" (III, 908). The narrator's interpretation of the three linked epiphanies offers a compelling description of the power of a sensation, in the present, whether vestibular, auditory, tactile, or olfactory, to transcend time. Thus, even if smell is not given a privileged place among the senses, as some olfactory psychologists' frequent invocations of the madeleine passage mistakenly suggest, smells, as much as the objects of the other senses, can be a vehicle for epiphanies of eternity.[12]

But there is an even more serious misreading of Proust's madeleine episode than mistaking it as primarily about smell, a misreading that would actually undermine the case we have been building for the possibility of an olfactory aesthetics. As the anthropologist David Howes remarks, although "the madeleine incident might seem like a celebration of smell," the way it has been interpreted by most psychologists and literary critics means that

> it was actually a demotion, which compounded the Kantian devaluation of olfaction on cognitive and aesthetic grounds. No good for thinking, at least smell is good for emoting and remembering, the doctrine insinuates. And so smell has come to be known as the "affective" sense, with the result that the gap between it and the intellectual and aesthetic senses of sight and hearing grows even wider.[13]

Howes adds that he is not aiming at the novel itself, but at a certain way of reading Proust. I agree. Those who see *Remembrance of Things Past* as reinforcing the view of smell as highly emotional and as most valuable as a spur to memory have missed the novel's incorporation of an intellectual dimension *within* its multisensory understanding of involuntary memory.

For Proust, the emotional and evocative qualities of odor-triggered memories are not ends in themselves, but what the narrator calls "signs," whose meaning must be deciphered.[14] Although these resurgent, sensory experiences of the past may be superior to voluntary intellectual discovery for Proust's narrator, the experiences themselves are still "what I had merely felt," whereas the narrator believes his real task as an artist is to convert these feelings into something that is also in part cognitive, namely, a work of art. (III, 912). And he can only convert them into an intelligible work of art because the emotional experiences *already* contain within themselves cognitive elements. When the narrator says he cannot rest content with describing merely "what I felt," he is asserting that the revelatory experiences contain an intelligible dimension that it is the artist's task to bring to expression, a position reminiscent of Collingwood's view of the artist's role. In Proust's novel, we have not only a magnificent display of how to put the experience of the lowest-ranked senses of touch, taste, and smell into language, but, at the same time, a demonstration of how to bring emotion and cognition together in a unique aesthetic creation.

What a wonderful paradox! Smell, the supposedly most animal and emotional of senses, the sense that Hegel thought could have nothing to do with Spirit, the sense that Kant, Darwin, Freud, Adorno, and others have thought irrelevant to developed humanity, becomes in Proust's novel—along with the other "lower" sense of taste—a vehicle for the experience of eternity in time. Yet this should not come as a complete surprise if we remember that the smell of burning sacrifices has pleased the gods and connected us to them since ancient times and that the smell of incense is still associated with spiritual aspiration and devotion around the world.

Yet there are some other implications of the narrator's claims for sensory epiphanies in *Remembrance of Things Past* that, despite their beauty and resonance, I find troubling. The narrator describes such aesthetic epiphanies as the "only genuine and fruitful pleasures" in life—compared to which love, friendship, and society are "unreal" (III, 908). "Art," we are told, "is the most real of all things . . . the true last judgment" (III, 914). Proust, of course, was hardly alone in his time in embracing the religion of Art. Yet as important and spiritually rewarding as great works of art can be, it is surely in our ordinary human relations, in love and friendship and our social and political responsibility for one another, that we are more likely to find whatever eternity there is. Given Proust's illness and his race to give permanent form to his discovery of the paradise of "lost time," perhaps his narrator's valuing of art above human relationships made sense for him at that moment, and it resulted in a great gift to us. And whatever his *narrator* says, Proust himself was deeply attached to his family, to his servant Françoise, to his chauffer/lover Albert, and to many others in his life.

Proust's *Remembrance*, Baudelaire's *Flowers of Evil*, Joyce's *Ulysses*, Woolf's *Flush*, and the other literary works we have visited contain linguistic expressions of olfactory experience that make the complaint about the "poverty of language" for expressing smell seem almost ludicrous. But a skeptic might still object that even if a few olfactory experts like perfumers, or the members of some exotic tribes in Southeast Asia, or a handful of exceptional literary artists can easily articulate smell, this still leaves most of us tongue-tied and would render an olfactory aesthetics possible for only a tiny elite. But that is false. There is abundant evidence that one need not spend two years in perfumery school or be a celebrated literary artist in order to convincingly articulate experiences of smell. In fact, as beautiful as the sensory epiphanies described by Proust may be, I think we may draw equally profound lessons from odor memories that have been accessed more directly, above all those odor memories that *connect* us to others rather than provide private epiphanies waiting to be transformed into art. Moreover, directly accessed memories are not always the arid, abstract, intellectualist phenomena that Proust's narrator caricatures, as if a brittle rationality were the only alternative to emotional spontaneity. Nor, as I suggested earlier, is there an absolute difference between voluntary and involuntary memories.[15] Here I have in mind the kind of sensory memories that emerge naturally when people recount an important moment of their past, whether distant or recent, and whether the experience was satisfying, spiritual, poignant, painful, or even horrific.

We can see this by briefly visiting two deeply moving memoirs of the horrific. As Hans Rindisbacher points out in his fine discussion of these two Auschwitz memoirs, smell plays a small but exemplary role in each of them. Consider Olga Lengyel's *Five Chimneys: The Story of Auschwitz*.[16] In 1944, she and her husband, their children, and her parents had been crammed into a cattle car with ninety-six other people for what turned out to be a terrifying eight-day ordeal taking them from Transylvania to Auschwitz. They were denied food and a place to evacuate, given water only in exchange for whatever valuables they still had, and, worst of all, forced to live on top of the many who died on the way, constantly breathing in the nauseating stench of decaying corpses (20). Then, after the survivors entered Auschwitz and the children and the elderly were sent directly to the gas chambers, a new smell assaulted their senses, a "strange, sickening, sweetish odor" that a guard assured them was simply the camp bakery (30–31). Of course, they would soon learn otherwise, and this odor blanketed the camp day after day. Then another smell began to torment the newly arrived, the smell of their own bodies, powerful odors that blended into a mass of human stench. Lengyel writes, "The herd of dirty, evil-smelling women inspired profound disgust in their companions and even in themselves" (45). Torn from the routines of civilized cleanliness in the deodorized cities of Europe, the prisoners intensely felt the degradation of their smell, which at once separated them from each other and united them in a common misery. Their stink, as much as their tattered rags, defined them as degraded outcasts in the midst of guards and functionaries who wore clean uniforms and were able to bathe and scent themselves. The women were especially tormented by the presence of the feared "blond angel," a beautiful SS woman whose daily rounds and roll calls included selection for the gas chamber (147). The prisoners, writes Lengyel, were entranced by her smell, since she wore a rare perfume and sprayed her hair with "tantalizing" scents. The SS woman's heavy use of perfume

> was perhaps the supreme refinement of her cruelty. The internees . . . inhaled these fragrances joyfully. By contrast, when she left us, and the stale, sickening odor of burnt human flesh, which covered the camp like a blanket, crept over us again, the atmosphere became even more unbearable. (147–48)

Primo Levi's account of his time in Auschwitz is less visceral, yet equally telling.[17] Because he was a chemist, after his first year in Auschwitz he was sent to work at the Buna factory laboratories on the campgrounds during the day and was spared some of the worst of what Lengyel describes. He offers this odor-triggered flashback that occurred the moment he first stepped into the Buna labs. The smell made him "start back as if from the blow of a whip: the weak aromatic smell of organic chemistry laboratories." Momentarily, he was taken to a spring day at his university in Italy, but the images and sensations quickly vanished (139). Although Levi had such advantages as a heated workspace and water to drink when he was in the labs, he and the other two prisoners assigned there were constantly ridiculed for their smell, and when he happened to ask one of the regular women employees a question, she turned her back on him and told a nearby male staffer not to let the *Stinkjuden* bother her again. Among the tasks Levi had been assigned during the year before he entered the labs was to help a couple of other prisoners clean out an underground fuel tank. One day as he emerged from the darkness, he found that "it was warmish outside" and "the sun drew a faint smell of paint and tar from the greasy earth, which made me think of a holiday beach of my infancy" (111). But there was no time for him to savor this moment or explore any associations; he and his fellow prisoners had to keep moving.

It is hard to think of a greater olfactory and literary contrast than between these smell memories from the concentration camps and Proust's' descriptions of involuntary memories that flooded his narrator with happiness and a glimpse of eternity. In the Auschwitz memoirs, there is no joyous feeling of the coincidence of two sensory experiences, but nearly always a wretched reminder of the gulf separating the world outside from the inhuman conditions inside where the sickly smell of burning flesh hangs continually in the air. And because these accounts of the camps do not aim at transforming sensory memories into works of high art, the directness of the writing powerfully evokes not only the experience of smell, but the moral pain that accompanied it. In the camps, the odor of burning bodies and of stinking prisoners is a stifling daily reality, but it too is a sign: not a sign of eternity in time needing to be given lasting form in art, but a sign of cruelty, degradation and death, a sign that becomes in these memoirs witness and warning. Even the experiences of positive scents, whether a female guard's tantalizing perfume or the remembered odor of a university chemistry lab, are additional reminders of the awful actuality of captivity. Only for the briefest of moments might something like the smell of "paint and tar on the greasy earth" bring a faint intimation of a long-ago holiday and the bonds of family and friendship, before the oppressive reality of imprisonment and death quickly closes back in. Above all, in the camps, what glimpses of eternity there were lay more in rare acts of human solidarity than in sensory epiphanies. Yet although Lengyel's and Levi's concentration camp memoires are meant to bear witness, not to be "literature," their simplicity and directness show that nonprofessional writers are often just as capable as great literary artists of expressing smell in powerful and convincing language.

Notes

1. Darwin, *Charles Darwin's Notebooks*, 539. The Fitzwilliam is the art and antiquities museum at Cambridge University.
2. Rachel S. Herz, "Odor Memory and the Special Role of Associative Learning," in *Olfactory Cognition: From Perception and Memory to Environmental Odors and Neuroscience*, ed. Gesualdo M. Zucco, Rachel S. Herz, and Benoist Schaal (Amsterdam: John Benjamins Publishing, 2012), 95. As Herz points out, although some odor-induced memories may be emotionally compelling and vivid, this does not mean they are more accurate than visual or verbal memories.
3. Dorthe Berntsen, "Spontaneous Recollections: Involuntary Autobiographical Memories Are a Basic Mode of Remembering," in *Understanding Autobiographical Memory*, ed. Dorthe Berntsen and David C. Rubin (Cambridge: Cambridge University Press, 2012), 290–310.
4. For a useful overview of research up to 2014 see Maria Larsson, Johan Willander, et al., "Olfactory Lover: Behavioral and Neural Correlates of Autobiographical Odor Memory," *Frontiers in Psychology* 5, no. 214 (2014): 1–4, https:// doi:10.3389 /fpsyg .2014.00312.
5. Eric Vermetten and James Douglas Bremner, "Olfaction as a Traumatic Reminder in Posttraumatic Stress Disorder: Case Reports and Review," *Journal of Clinical Psychiatry* 64, no. 2 (2003): 202–7. My friend Gene Brodland, who has worked with Vietnam veterans in his therapy practice, tells me he too has encountered cases of smell-associated PTSD.
6. Frost, *The Problem with Pleasure*, 46.
7. Chretien van Campen, *The Proust Effect: The Senses as Doorways to Lost Memories* (Oxford: Oxford University Press, 2014). Typical of the many psychology articles celebrating Proust for his psychological insights into smell is Simon Chu and John Joseph Downes, "Proust Nose Best: A Scientific Investigation of a Literary Legend. Odors are Better Cues of Autobiographical Memory," *Memory and Cognition* 30, no. 4 (2002): 511–18. I did find one title claiming to *refute* Proust: John H. Mace, "Involuntary Memories Are Highly Dependent on Abstract Cuing: The Proustian View is Incorrect," *Applied Cognitive Psychology* 18, no. 7 (2004): 893–99, https://doi.org/10.1002/acp.1020.
8. Marcel Proust, *Remembrance of Things Past*, vols. I, II, III, trans. C. K. Moncrief and Terence Kilmartin (New York: Random House, 1981). I will henceforth cite this translation in the text with volume and page number.
9. Gilbert, *What the Nose Knows*, 197–99. See the autobiographical memory study Larsson et al., "Olfactory Lover."
10. Gilbert, *What the Nose Knows*, 200.

11. In the previous chapter I quoted Proust's description of a voluntary memory of the smell of his aunt's rooms at Combray. There is also a poignant example of a voluntary odor memory that occurs *before* the madeleine episode in *Swann's Way*, when the narrator recalls how as a child he hated the smell of the varnish on the stairway he went up each night. Proust, *Remembrance*, I, 29–30.
12. None of the three epiphanies we have just examined is evoked solely by an odor, although the memory of the sound of a hammer on the rail car wheel, which is triggered by the spoon on the plate, is at least accompanied by the memory of the smell of smoke and a "cool forest background."
13. David Howes, "Introduction," in Henshaw et al., *Designing with Smell*, 7.
14. Gilles Deleuze, *Proust et les signes* (Paris: Presses Universitaires de France, 2014).
15. For a similar interpretation of Proust's madeleine episode see Emily T. Troscianko, "Cognitive Realism and Memory in Proust's Madeleine Episode," *Memory Studies* 6, no. 4 (2013): 437–56, http://mss.sagepub.com/.
16. Olga Lengyel, *Five Chimneys: The Story of Auschwitz* trans. Clifford Coch and Paul P. Weiss (Chicago: Zoff-Davis, 1947). Subsequent quotations in the text are from this edition.
17. Primo Levi, *Survival in Auschwitz and the Reawakening: Two Memoires*, trans. Stuart Woolf (New York: Summit Books, 1985). Subsequent quotations in the text are from this edition.

Postlude

Is an Olfactory Aesthetics Possible?

We began this book with examples of works of olfactory art ranging from installation and participatory works to hybrids of odors with theater and music, and to perfumes exhibited as fine art. Clearly, olfactory art and its aesthetic appreciation are *actual*, so why ask if an olfactory aesthetics is *possible*? The reason, as we have seen, is that there is a long philosophical tradition going back at least to Kant and Hegel and still embraced by a number of contemporary philosophers for whom olfactory aesthetics is *impossible* in the sense that smell is not considered an adequate vehicle for genuine artistic creation or genuine aesthetic experience and judgment. And that philosophical tradition has been underpinned by a broader negative intellectual tradition, exemplified by Darwin, Freud, and others who reduce smell to a nearly useless evolutionary vestige. Because elements of those traditions are still in circulation, it has seemed important to provide counterarguments and evidence that shows the sense of smell does have the intellectual potential to form the basis of a reflective olfactory aesthetics.

Chapter 1 offered a series of arguments against the claims that the sense of smell is disreputable, defective, deceptive, and dispensable, and I reinforced those arguments in Chapter 2 with evidence from contemporary neuroscience and psychology, showing humans are very good at detecting, discriminating, and learning smells. But in Chapter 3, we examined empirical evidence claiming to show that smell is primarily emotional and hedonic, that it tends to operate unconsciously, and that the average person has an extremely poor ability to name and describe smells due in part to a supposed poverty of human languages for expressing smell. At that point, things looked bad for a cognitively informed olfactory aesthetics. But we found evidence from other neuroscience studies that these deficits may not be biological universals, since olfactory experts (specifically perfumers and wine connoisseurs) can make fine cognitive discriminations and qualitative judgments regarding odors. We also drew on the psychology and philosophy of emotions to refute those who would set up a stark opposition between smell as emotional and vision and hearing as intellectual, thereby *removing a first major barrier to the possibility of an olfactory aesthetics.*

But given the small number of olfactory experts in the world, I turned from arguments based primarily on neuroscience and psychology to arguments based

Art Scents. Larry Shiner, Oxford University Press (2020) © Oxford University Press.
DOI: 10.1093/oso/9780190089818.003.0017

on evidence from the social sciences and humanities to show the possibility of an olfactory aesthetics viable for society at large. First, I answered the claim that smell is an evolutionary vestige of little use to humans by citing recent evidence for the importance of smell in human evolution and tracing the social and cultural importance of smell in both Western and non-Western history. By refuting the charge that smell is a vestige of little use, I *removed a second major barrier to the possibility of an olfactory aesthetics*.

But the most serious threat to an olfactory aesthetics comes from the contention that the sense of smell is essentially mute. This position is based on research studies that seem to show the average person is unable to name odors accurately or to express olfactory experiences linguistically, due to a presumed universal physiological deficit and a general poverty of human languages for expressing smell. I devoted three chapters to refuting these two claims. Chapter 6 presented evidence from anthropology and linguistics showing the complexity of smell vocabularies and linguistic usages in many non-Western cultures, including the existence of languages that have abstract odor terms and whose speakers can quickly identify and name smells. Then we examined the way works of Western literature (Baudelaire, Rilke, Woolf, Joyce, and others) cast doubt on the claims that there is a universal poverty of language for expressing smells and that humans intrinsically lack the ability to articulate smell experiences, and we ended our literary survey with Proust's articulation of epiphanies of involuntary memory and two powerful representations of smell in Holocaust memoirs. Taken together, this evidence shows that humans and human languages are able to articulate the experience of smell well enough to sustain aesthetic discussion. Thus, I *removed a third major barrier to the possibility of an olfactory aesthetics*.

All three of the barriers toppled have in common an underlying dualistic pattern of thinking that grossly exaggerates the cognitive and linguistic gap between the so-called higher or intellectual senses of vision and hearing and the lower or bodily senses of touch, taste, and smell. Hence we have shown that although smell is not as intellectually powerful as vision or hearing, it nevertheless has far greater cognitive and linguistic capacities than has been assumed by those who have disparaged it and denied that it can sustain genuine aesthetic experience and judgment. Moreover, we only occasionally mentioned that vision and hearing are themselves not so purely rational, self-conscious, and linguistically transparent as the traditional polarity of "higher" and "lower" senses implies. Murray Smith's book on film, for example, emphasizes the host of visual and auditory mechanisms that operate below the threshold of consciousness and cognitive control in viewing films, and concludes that, in the case of both vision and hearing, "our discriminative capacities outreach our identificatory capacities," precisely the point made in many psychological studies of olfaction.[1] To that point we could add such things as "inattentional blindness," as in the famous experiment in which people who have been primed to focus on a certain aspect of a crowded scene don't notice the figure in a gorilla suit who waltzes through the room. I conclude that neither are smell's cognitive and linguistic limitations different in kind from the limitations of vision and hearing, nor is the degree of difference nearly as great as many of those who have denied the capacity of smell to support genuine artistic creation and aesthetic appreciation have assumed.

Of course, a really determined skeptic might still bring up the objection that although an olfactory aesthetics is humanly possible, it is not a practical option worth pursuing since it would require everyone to become either an olfactory or a literary expert. We may answer that many Westerners besides professional perfumers, and members of exotic tribes, have developed considerable olfactory sensitivity in relation to perfumes and other scented products, as well as to wines and food. Moreover, some nonprofessionals have deliberately sought to cultivate their sense of smell and written convincingly about it. Consider just one recent case, Barney Shaw's *The Smell of Fresh Rain: The Unexpected Pleasures of Our Most Elusive Sense* (2017).[2] Shaw is a retired British civil servant who set out to explore the ordinary smells of his environment and find a way to describe them. Although he read some neuroscience and psychology along the way, the heart of his book is an account of his own adventures, notebook in hand, seeking out smells, whether on the streets or in the stores and markets of London and environs or on field trips to Portsmouth Harbour, the Dorset woods, or a French garden of fragrant plants. Most importantly, in each place he visited he engaged people in a discussion of odors. Some suspicious boatbuilders at Portsmouth, seeing him taking notes, accosted him and at first made fun of his experiment, but soon joined in a discussion of how to describe the smell of tar, rope, their boat crane, diesel fuel, synthetic rubber. At a small charcoal company in the Dorset countryside, he talked with workers about the smell of the different woods that are packed into kilns to be reduced to charcoal, how they smell when they burn and the smell of the cold ash afterward. What struck me about these brief conversations in workplaces, stores, and markets was not that his interlocutors were especially articulate, but that despite their difficulty in finding words, there *was* discussion. Toward the end of his book Shaw also makes several suggestions about how one ought to go about identifying smells and describing them in everyday language without resort to special literary devices.[3] Thus, in addition to the two Holocaust memoires we discussed previously, Shaw's ethnographic and linguistic efforts are proof that one need not be trained as either an olfactory professional or a literary artist in order to give odors a convincing linguistic expression and engage in intelligent discussions of them.

Notes

1. Smith, *Film, Art and the Third Culture*, 30.
2. Barney Shaw, *The Smell of Fresh Rain: The Unexpected Pleasures of Our Most Elusive Sense* (London: Icon Books, 2017).
3. Shaw, *Smell of Fresh Rain*, 267–86. Shaw even offers short descriptors of some two hundred ordinary smells; it is a heroic effort, although I found many of the descriptions unsatisfying. The value of his book lies in its personal testimony and informal ethnography.

PART III
DISCOVERING THE OLFACTORY ARTS

Overview
What Is Olfactory Art?

Around the turn of this century, the artist Helgard Haug won a prize to create a public art piece for the Berlin Alexanderplatz U-bahn station, once at the center of the former East Berlin. The piece consisted in a distillation of the scents of the Communist-era station that was put in little souvenir glass vials and dispensed from a vending machine during the year 2000. The scent, called *U-deur*, was created for Haug by a professional perfumer based on his perception of how the station once smelled, including cleaning agents, oil, and electricity, along with the smell of bread from a bakery stand. Haug invited people to write their responses, which turned out to be extraordinary. People wrote that the little sniff bottle brought to mind memories and associations with the smells of divided Berlin, for instance, of the "dead" stations that West Berlin subway trains went through after the Wall was built, as well as thoughts about the Stasi (secret police) archive with its jars of socks, handkerchiefs, and other items saturated with the body odor of East German dissidents and criminals.[1]

Haug's work is fairly typical of contemporary art that uses installation and participatory strategies. Although some conservative aesthetic theorists might complain that works like Haug's seem more like sociological experiments than serious art, most aestheticians today accept what David Davies calls the "pragmatic constraint" on theorizing about art: our aesthetic concepts need to be consistent with the best practices of the art world itself. From that perspective, works like Haug's *U-deur* or Tolaas's *Fear of Smell and the Smell of Fear* are indeed instances of art given the profound conceptual and "postmedium" turn that has developed in the (fine) arts since the 1960s, opening the way to the use of all sorts of new forms, techniques, and materials, including odors. And despite vigorous debate within philosophical aesthetics on just how the concept of fine art should be defined in the light of that conceptual turn, nearly all the major definitional alternatives currently in play attempt to accommodate "postmedium" and "conceptual" art experiments.

As for the role of aesthetic experience and judgment in the appreciation of such art, one might think it problematic since some contemporary artists have intentionally made what critics like Hal Foster have called "antiaesthetic" art, works that do not aim at evoking sensory pleasures, let alone "beauty." Yet an

Art Scents. Larry Shiner, Oxford University Press (2020) © Oxford University Press.
DOI: 10.1093/oso/9780190089818.003.0018

antiaesthetic intention is hardly characteristic of all contemporary art, including olfactory works. Moreover, the idea of the antiaesthetic sometimes unfairly assumes a rather old-fashioned formalist idea of what an interest in "aesthetic" properties entails, narrowly associating it with traditional ideas of exalted beauty or pleasure in pure form.[2] A complicating factor in the case of olfactory art is that most olfactory works are hybrids that either use odors to enhance traditional fine art forms like drama, film, and music or else integrate odors into multisensory installation, performance, or participatory pieces like Haug's *U-deur*. In the cases of drama, film, and music, the question of art status and the relevance of aesthetic values is largely moot since the genres of which olfactory works are hybrids have been long accepted as (fine) art and been the subject of much aesthetic theorizing. And since there are established critical traditions to guide aesthetic responses to the parent genres, such works' olfactory aspect may require some adjustment from the theorist or critic, but not starting from zero. But is an "aesthetic" approach to conceptual installations or participatory works like Haug's *U-deur* an appropriate way to engage such art? To the extent that one thinks aesthetic appreciation is based on an immediate sensory response to perceptual properties, the answer might seem to be no. But, just as contemporary philosophers of art have put forth many competing definitions of art (see Chapter 11), so there are many competing concepts of aesthetic properties, aesthetic experience, and aesthetic judgment, among which some make a place for conceptual as well as sensory and formal aspects. As Elisabeth Schellekens has argued with respect to the aim of conceptual works in general, "We ought to 'undergo' the idea rather than merely think of it." Doing so "will involve all of the idea's experiential qualities, amongst which aesthetic ones are included."[3]

But even if we take a somewhat traditional view of aesthetic experience that identifies it with an immediate sensory response to perceptual properties, there is a way in which appropriate responses to conceptual works like Haug's could be considered to have an aesthetic component. After all, Haug's *U-deur* afforded an experience inseparably uniting the ideas and memories related to the old Alexanderplatz station with the sensory/perceptual experience of inhaling the odor that prompted the memories. Although visitors who participated in the Alexanderplatz work might have been more consciously focused on the conceptual than on the perceptual aspects of it, the two aspects were inseparable not only in the work itself, but also in the experience of it. For even if most people normally pay little attention to the smells around them or consciously register their qualities, the thrust of conceptually oriented olfactory artworks like *U-deur* is precisely to call attention to certain smell qualities that, in this case at least, evoke a particular set of associations from the past.

Given this general argument in favor of taking olfactory art hybrids seriously as objects of aesthetic appreciation, there still remain a number of difficult preliminary issues that need to be clarified. One such issue concerns the differences between the majority of olfactory artworks that are hybrids and the relatively "pure" olfactory works such as perfumes, whether created by artists or professional perfumers. Of course, it would be difficult to create any work of olfactory art that consisted solely of odors since odor molecules have to have some sort of vehicle or container until they are released—just as there could be no "pure" work of paint (even a monochrome) without a surface for the paint, although someone, somewhere has probably exhibited some cans or tubes of paint as a "painting." Even Clara Ursitti's *Self-Portraits in Scent* from the early 1990s, which involved simulations of her body scents, required a vehicle (alcohol) for the scent molecules and either a bottle sitting on a table or a dispenser inside a box hanging on the wall. Haug's *U-deur*, although it also involved odor molecules suspended in alcohol, is more clearly a hybrid since it was both installation art (the vials dispensed in a subway station) and participatory art, (the audience needed to take an action, namely, retrieving one of the vials, smelling it, and perhaps commenting on it).

In addition to "pure" and "hybrid" olfactory artworks, a third possible kind of olfactory art might be works of painting, music, or literature that simply *represent* odors. For example, should literary works like those of Baudelaire, Woolf, Joyce, or Proust that give an important place to smell experiences be considered in some sense works of olfactory art? It would seem not, since just because a poem or novel (or a painting or musical composition) represents, evokes, or expresses the *idea* of odors it is not necessarily a work of olfactory art, any more than a poem or novel about music is a work of music. There would seem to be a need for actual odors. As Roland Barthes once put it: when written, "shit" no longer smells.[4] Yet since some works of literature, film, or music by their *effect* on our imagination can evoke genuine emotions in us, as several philosophers have argued, might not a particularly vivid literary passage of the kind we cited from Baudelaire also make us imagine or react to its language as if we were inhaling real odors? G. Gabrielle Starr has surveyed some of the neuroscience evidence that shows that when images are evoked by literature, some of the same areas of the brain are activated and "function in similar patterns for imagined sensation as during actual perception."[5]

Moreover, there are also a few cases of literary works that have actually been part of olfactory art hybrids. For example, the artist Brian Goeltzenleucter developed a collaborative project called *Olfactory Memoirs* (2015) that involved diffusing scents as writers read aloud works expressing childhood smell memories. He followed this by creating a scent to accompany the reading of a poem by Anna van Suchetelen as part of *Volatile! A Poetry and Scent Exhibition*, held at the Poetry Foundation in Chicago from December 11, 2015, to February 19, 2016.[6] Not long before that exhibition opened, an issue of the Foundation's *Poetry*

Magazine came out that included a microencapsulated scent commissioned from perfumers D.S. & Durga intended to provide a scent to complement to a poem by Jeffery Skinner.[7] Two obvious issues raised by such hybrids that combine a poem with an actual smell are, first, the identity of the work—for example, should we speak of three works, a poem, a scent, and a poem-scent combination? Second, does the olfactory aspect actually enhance the poem involved, or does it distract from the literary experience? No such problems were raised by the Brazilian artist Eduardo Kac's work *Aromapoetry* (2011), which was included in the *Volatile!* event, since Kac's work is an artist's book with twelve custom aromas embedded in a nanolayer of mesoporous glass forming the "pages" of the book. Thus, his "poems" (consisting of one to a dozen molecules) are emitted from each page as the reader/smeller turns the pages and sniffs, making this a more or less "pure" olfactory artwork. Of course, someone might ask whether Kac's use of the term "poetry" is anything more than a questionable metaphor.[8] In any case, the existence of both "pure" olfactory poetry such as Kac's and the various hybrids that were presented at the Poetry Foundation's *Volatile!* seem to raise even more urgently the question: "What *is* "olfactory art?"

There is no better indication of the current ambiguity of the term "olfactory art" than the contrast between two major exhibitions that have been mounted in the last decade. I have already mentioned *The Art of Scent*, held at the Museum of Arts and Design in 2012–2013, which presented a dozen classic commercial perfumes as if perfumes were the only kind of olfactory art. Neither the curator, Chandler Burr, who held the title director of olfactory art at the time, nor the exhibition literature even mentioned the existence of works such as those of Sissel Tolaas, Clara Ursitti, or Peter de Cupere, let alone works like Helgard Haug's *U-deur*. By contrast, the 2015 exhibition of "olfactory art" called *Belle Haleine: The Scent of Art*, at the Museum Tinguely in Basel, showed only conceptual, installation, performance, and participatory works involving odors but included no commercial perfumes.[9] One can find similar contrasts in the writings of theorists and art critics. The philosopher Chantal Jaquet, in her introduction to the symposium *L'Art olfactif contemporain* (2015), gives perfume a central place in olfactory art alongside both theatrical works that feature odors and multimedia olfactory works shown in galleries, whereas Jim Drobnick's extensive writings on olfactory art typically do not discuss either traditional perfumes or dramas accompanied by odors, but focus on multimedia works for visual art galleries and museums.[10]

One way to bring greater clarity to the discussion would be to use "olfactory arts" in the plural as an umbrella term for any artworks (sculptures, installations, plays, poems, perfumes) that intentionally give a distinctive place to actual odors and to adopt some other term for the narrower set of olfactory arts created by professional artists and intended for galleries, museums, and art expositions.

Accordingly, as a working definition for the rest of this book, I will characterize the olfactory arts or "art scents" in general as "the *intentional* use of *actual* odors as a *distinctive-making* feature of an artwork," whether that work consists of "pure" odors like perfume and incense or is a hybrid of odors and some more established medium. The presence of the odors should be intentional not accidental since almost any material or activity used to make an artwork is likely to give off some odor, but if a work is to be included in a discussion of "olfactory arts," its creator should have intended that the audience notice its smell, whatever other senses may also be addressed. This leads to the second requirement: that the odors be *actual*. This characteristic excludes works of art such as poems, novels, or paintings that only refer to or represent odors and the sense of smell rather than directly or indirectly stimulating our olfactory system. The third criterion, that the odors play a role in giving a work its *distinctive* character compared to otherwise similar works that lack a distinctive-making odor, means that one cannot adequately understand and appreciate such a work without considering the role played by intentionally present scents. This last criterion, of course, requires interpretive application since there will be a range of such works, some in which the olfactory aspect is dominant, others where it plays a less prominent or even minor role. In the case of theater, for example, in the 2016 musical *Waitress*, the smell of apple pie drifting through the auditorium from a convection oven hidden in the orchestra could be dropped from the production with only a minor effect on the audience's total experience, whereas in Violaine de Carné's *The Scents of the Soul* (*Les parfums de l'âme*), which we will discuss in the next chapter, the audience's experience would be totally different without the actual odors that give the play its distinctive character. The many kinds of actual odors that give the various olfactory arts their distinctive character are the "art scents" mentioned in the title of this book. Obviously, the three characteristics of the olfactory arts that I have suggested do not add up to what philosophers consider a classic "definition," namely, necessary conditions that taken together are sufficient to clearly separate out every instance of "olfactory arts" from other groupings of arts. But for purposes of circumscribing the general phenomena we are about to explore in the chapters that follow, this characterization will have to do.

But if we don't restrict the term "olfactory arts" to hybrids shown in galleries and museums, what name might we use to distinguish the latter as a subgroup within the "olfactory arts" in general? Of course, we could still use "olfactory art" in the singular for such works since that usage is now well established. After all, we do something similar for "art" in the singular, sometimes using it to mean the visual arts (painting, sculpture, architecture) in contrast to literature, music, and drama, at other times using it in both the singular and the plural to refer to all the (fine) arts together. But there are several other possible names for museum and gallery works worth considering that would create less ambiguity, among them "scent art," a term that has occasionally been used by the olfactory art critic Jim Drobnick and also serves as the title of an important blog about olfactory art by Ashraf Osman.[11] The art historian Francesca Bacci has even suggested the term "scent-ific art."[12] One problem with either "scent art" or "scent-ific art" is that "scent" tends to suggest light and pleasurable odors, whereas the odors in many of the multimedia artworks we will be considering in Chapter 10 are anything but light and pleasurable, for example, de Cupere's *Tree Virus* (2008), which caused many visitors to flee its enclosure, or Wim Delvoye's *Cloaca Professional* (2010), with its smell of mechanically produced feces. Other terms that come to mind are "smell art" or "odor art." Yet "smell," like "odor," often carries unwanted negative overtones, for example, "What's that smell?" Since "scent art" already has some traction in contemporary writing about olfactory art, I will use it interchangeably with "olfactory art" in the singular for olfactory artworks that (1) integrate odors with other art materials or with established media such as sculpture and installation, (2) are typically made by professional artists, and (3) are created to be shown in established art venues such as galleries, museums, or expositions.[13]

The alternative rubric "scent art" would also have the advantage of paralleling the category "sound art" that is widely used in the contemporary art world. Both sound art and scent (olfactory) art are experimental directions within the broad arena of the contemporary fine arts that have come to prominence since the 1990s. Thus, just as we don't refer to sound art as "aural art," but regard both sound art and music as part of a broader category of "aural arts," so it would make sense on some occasions to refer to works produced in the context of the world of galleries and museums as "scent art" and regard both scent (olfactory) art and perfume, along with hybrids of odors with theater, music, or film, as part of the larger umbrella category, "olfactory arts." Naturally, these terminological suggestions are mostly a matter of convenience, a tentative effort to order the multifarious practices of a nascent field for purposes of discussion in this book. My guess is that "olfactory art" in the singular will continue to be the most widely used term for hybrid works intended for the gallery or museum, which is why I will use it most of the time.

The first two chapters of Part III will focus on the aesthetic issues that arise when odors are combined with various established media and art practices; the last two chapters will discuss the art status of perfumes. Since previous chapters have already explored the literary representation of odors and we have already considered a few literary/odor hybrids, the following prelude will focus on the representation of odors in the pictorial arts along with some rare attempts at hybrids of actual odors with paintings. We will save olfactory hybrids with sculpture for Chapter 10, "Sublime Stenches: Contemporary Olfactory Art," since sculpture is a major partner in olfactory artworks intended for the gallery, museum, or art festival circuit.

Notes

1. Margaret Morse, "Burnt Offerings (Incense): Body Odors and the Olfactory Arts in Digital Culture," conference paper, 2000. I am grateful to Margaret Morse for sharing a copy of her paper. See also Jim Drobnick, "The City, Distilled," in *Senses and the City*, ed. Mădălina Diaconu et al. (Vienna: LIT Verlag, 2011), 259–61.
2. Arthur Danto, *The Abuse of Beauty* (Chicago: Open Court, 2003).
3. Elisabeth Schellekens, "The Aesthetic Value of Ideas," in *Philosophy and Conceptual Art*, ed. Peter Goldie and Elisabeth Schellekens (Oxford: Oxford University Press, 2007), 86.
4. Roland Barthes, *Sade, Loyola, Fourier*, Richard Miller (Paris: Éditions du Seuil, 1971), 140.
5. G. Gabrielle Starr, *Feeling Beauty: The Neuroscience of Aesthetic Experience* (Cambridge, MA: MIT Press, 2013), 75.
6. The exhibition was curated by Debora Riley Parr.
7. "The Bookshelf of the God of Infinite Space," *Poetry Magazine*, 207, no. 3 (December 2015).
8. Kac claims his book is the world's first book "written exclusively with smells." See https//www.ekac.org/armapoetry.htlm.
9. See the collection of essays from the symposium that accompanied the exhibition and contains a list of the works shown. Lisa Anette Ahlers and Annja Müller-Alsbach, eds. *Belle Haleine-The Scent of Art, An Interdisciplinary Symposium* (Basel: Museum Tinguely, 2015, 148–151.
10. Chantal Jaquet, "Introduction," in *L'Art olfactif contemporain*, ed. Chantal Jaquet (Paris: Classiques Garnier, 2015), 7–16. Jaquet's earlier discussion of perfume as an art form can be found in Jaquet, *Philosophie de l'odorat*, 223–303.
11. See "Scent Art.net: Stop and Smell the (Olfactory) Art," https://www.scentart.net/tag/ashraf-osman/.
12. Francesca Bacci, "Scent-ific Art in Context: Developing a Methodology for a Multisensory Museum," in Ahlers and Müller-Alsbach, eds. *Belle Haleine: The Scent of Art*, 126–36.
13. This characterization is similar to Jim Drobnick's definition of "olfactory art" in his essay "Smell: The Hybrid Art," in Jaquet, *L'Art olfactif*, 173–89.

Prelude
Picturing Smell

The 2015 olfactory art exhibition *Belle Haleine: The Scent of Art* at Basel's Tinguely museum included one contemporary picture: Louise Bourgeois's 1999 drypoint, *The Smell of the Feet,* a work that juxtaposes a profile self-portrait with the bottom of a pair of feet under her nose. It seems that, as a child, she was made to remove her father's shoes each evening, leaving the smell of his feet indelibly imprinted in her memory. Pictorial representations of the sense of smell are relatively infrequent in the history of the visual arts with the exception of cartoons and the older tradition of creating painting series portraying the five senses. In Western Europe, these series typically showed a female figure in a symbolic pose to suggest each scent, for example, an elegant young woman holding a flower to signify smell. In the famous *Lady and the Unicorn* tapestries from around 1500, however, the lady is shown weaving a wreath made out of flowers, while a monkey nearby smells a flower he has stolen from her bouquet. Among the best-known series of works in the tradition of using a female figure to symbolize the senses is Jan Brueghel and Peter Paul Rubens's series, *Allegory of the Five Senses* of 1617–1618. The panel on smell shows a finely dressed young woman in a flower garden, holding a cutting to her nose; instead of a monkey, there is a perfume-distilling apparatus nearby.

Within a decade, seventeenth-century Netherlandish paintings of the five senses began to depict the senses not only through allegorical female figures, but also through genre scenes. Among the most interesting of the genre scenes are the set of small panels created by the young Rembrandt and considered among his earliest surviving works. The age-darkened panel on smell was rediscovered in a New Jersey basement in 2015 and when cleaned not only revealed its similarity to the already known Rembrandt panels, but his signature confirmed the attribution. The painting shows an elderly woman holding a cloth, likely soaked in smelling salts, under the nose of an unconscious young man.

In addition to the tradition of serial depictions of the five senses, the theme of smell also turns up as an aspect of narrative subjects, for example, in depictions of the New Testament story of the raising of Lazarus (John 11:1–44). The dramatic moment of Lazarus's emergence from the tomb has been portrayed countless times, as in Giotto's panel of the Scrovegni Chapel frescoes in Padua, which shows the figures nearest Lazarus covering their noses at the stench given off by his body. In Duccio's version, one of the bystanders is actually holding his nose with his fingers. The use of so obvious a gesture invites us to imagine what the smell must have been like. The same is true of some of the Dutch genre depictions of everyday activities that emit strong odors, such as smoking. A work attributed to Adriaen Brouwer, *Interior of a Tavern* (c. 1630), graphically portrays two men blowing smoke as other denizens of the tavern carouse around them. The English writer William Hazlitt remarked on seeing the painting that it "almost gives one a sick headache."[1]

Of course, for those of us brought up on formalist principles for looking at paintings, our imagination may not summon vicarious experiences of smell or taste, even when we are confronted with older works that are clearly intended to evoke a response to their olfactory content. Yet just how vividly one might imagine the odors represented in a painting can be seen in the passage of Huysmans's *Against Nature,* where the protagonist Des Esseintes imagines not only what Herod sees of Salome in one of Gustave Moreau's famous paintings of her, but also what Herod must have smelled of her, "maddened by the nakedness of this woman soaked in musky scents, steeped in balms, and smoked in the fumes of incense and myrrh."[2]

It is hard to tell whether Moreau intended his Salome to evoke such intense olfactory responses, but there was one painter at the end of the nineteenth century who did try to capture the poetry of scent through painting: Paul Gauguin. Although Gauguin's Tahiti is largely a mythical creation, it is meant to be a multisensory myth in which smell plays a key role. His famous account of his first stay in Tahiti is entitled simply *Noa Noa,* the Maori term for scent or perfume, and he wrote that it would "embody the scent that Tahiti gives off."[3] Jim Drobnick has pointed out that not only *Noa Noa* but Gauguin's journals and other writings at this time are replete with references to odors and perfumes. As part of his mythicizing of Tahitian women, he invokes their closeness to unspoiled nature, especially in their smell, which he likens to that of "healthy young animals." It is "a mingled perfume, half animal, half vegetable . . . the perfume of their blood and of the gardenias—*tiaré*—which all wore in their hair."[4] Some of Gauguin's musings suggest that he embraced something like Baudelaire's notion of correspondences among colors, sounds, and smells. As Chantal Jaquet points out, for example, when Gauguin describes his painting *Vahine no tiare* (*The Woman with the Flower*) in *Noa Noa,* he focuses not on the visual aspects themselves, but on what she evokes in him, "her perfumed aura symbolized by the flower."[5] And a powerfully odorant flower it was; the Tahitian gardenia was often used in perfumery. With Gauguin we come about as close as a visual art like painting can do to expressing smell without resorting to some tactic like having figures holding their noses, although as we will argue in the next chapter, film may be able to generate a kind of mental olfactory equivalent through the combined effects of visual motion and sound.

Not many years after Gauguin's death, the Futurist Carlo Carra issued his 1911 manifesto titled, the "Painting of Sounds, Noises and Smells." He begins by declaring: "As artists we have already created a love of modern life in its essential dynamism—full of sounds, noises and smells." He is no doubt referring to the Futurist painters' celebration of the speed, noise, and machinery of the modern city by filling their pictures with dynamic lines of force, but in the case of smell, it seems Cara is not thinking of "representing" smells, but of finding in strong smells an impetus to paint. "We are not exaggerating when we claim that smell alone is enough to determine in our minds the arabesques of form and color which could be said to constitute the motive and justify the necessity of a painting." A Futurist painting, Carra proclaims, will be "*total painting*, which requires the active cooperation of all the senses . . . you must paint as drunkards sing and vomit: sounds noises and smells!"[6] Like so many Futurist pronouncements, Carra's manifesto is full of hyperbole verging on incoherence. But it also contains a genuine aesthetic insight, namely, that painting as a visual art can mostly give us indirect evocations of smell by reflecting the impact of odors on the painter or on the human subjects in the painting.

One possibility that we have not yet examined is for the painter either to juxtapose a painting with an odor source or to incorporate actual odors in a painting. Of course, when a painting is "fresh," it will give off a certain odor, and even older paintings may still emit a distinctive scent (likely varnish), as Darwin indicated on his visit to the National Gallery. But more apropos of the intentional use of actual odors in paintings are variations on the kind of hybrid works that make up the category I am calling "scent art" or "olfactory art" in the singular. Although there are several kinds of smell-sculpture hybrids, as we will see in Chapter 10, smell-painting hybrids are of two main types. One type simply juxtaposes a painting or photograph with an odor source. For example, the artist Andrew Marvick once exhibited a series of abstract paintings each of which had what he called an "olfactory predella" below it: a jar containing a perfume. Although one could enjoy the painting without leaning over to experience the fragrance from the jar, the total artwork was intended to consist of the painting plus the olfactory predella. Marvick's hybrid works not only invited reflection on the interrelation of the particular scents and the colors and shapes in each painting-perfume juxtaposition, but also made one more aware of our general conventions for experiencing paintings and perfumes.[7]

Peter de Cupere's various *Soap Paintings* (1996–2002) have also played on the conventions for exhibiting paintings, but in a more metaphorical way than Marvick's works. Most of de Cupere's soap "paintings" consist of bars of scented soaps arranged in rectangular formats that are hung on the wall (one is actually in a frame). Although they allude to the tradition of geometric abstraction in painting, they involve no paint and could just as well be considered a kind of sculptural assemblage and thus belong more properly with the other kinds of olfactory art that will be discussed in Chapter 10.[8]

The second major way of using of paint with smells, of course, is in works like Sissel Tolaas's *Fear of Smell and the Smell of Fear* (2006) that involve the micro-encapsulation of odors into a paint that is spread on the gallery walls. Peter de Cupere used a similar technique in a work called *Invisible Scent Paintings* presented in 2014 at the Marta Museum in Herford, Germany. De Cupere has made a video of visitors going along the gallery walls guide in hand, feeling and sniffing the work. That same year Sean Raspet created *Micro-encapsulated Surface Coating* as part of his show *Residuals* at the Jessica Silverman Gallery in San Francisco. The odors were a distillation and concentration of his capture of all the odors—of gallery surfaces, artworks, cleaning materials, bodies—emitted into the air of the gallery over the course of a full week. Tolaas's, de Cupere's, and Raspet's works lack a conventional frame or other visible separation of each of the areas representing a different odor, and Raspet's work presented only a single, continuous odor. In this way they differ from the recent monochrome tradition, although one might consider them monochrome murals, a reductio ad absurdum of Clement Greenberg's idea of painting as paint on a flat surface. But the major difference, of course, is that none of these monochrome murals seem to be intended primarily as extensions of the medium of painting, nor are the audience experiences those we normally associate with paintings directed at our visual sense. Tolaas, de Cupere, and Raspet are concerned neither with the appearance of the painted surface nor with engaging the tradition of viewing paintings, but with providing a smell experience that comes through touching and sniffing. This makes these kinds of work primarily olfactory installations that solicit physical engagement; at most, they distantly allude to painting rather than constituting a deliberate extension of the medium of painting.

Traditional painting and drawing seem even more limited in their ability to represent or evoke smell than poetry and the novel. One reason, of course, is that odors are constituted of volatile molecules typically invisible to the human eye, so that, at best, they can only be inferred to be present when borne in mists or vapors. In the case of traditional figurative painting or drawing, as we have seen, there are a few situations that can be visualized that will suggest odors to the imagination, such as cases in which figures hold things up to the nose (flowers, smelling salts), or where the nose is covered or pinched (the raising of Lazarus), or where some well-known odor source is emphasized (smokers in a tavern, feet). In the case of cartoon drawings there are also a few conventions for indicating odors. In many cartoons, vertical wavy lines, sometimes with scattered dots, are placed above a cup of coffee to signify aroma or above feet or excrement to indicate stench. But we should not underestimate the suggestive power of "comics" or graphic novels to communicate a simulacrum of smell experiences, especially given the point made earlier about the way in which even purely literary works can evoke in us images that activate the same areas of the brain that are involved in actual smell perception. The graphic novelist Chris Ware has remarked that by combining words and actual images a cartoonist is able to bring alive not only sounds but "sometimes even smells."[9]

Another kind of visual representation of smell is the use of scientific photographs of streams of exhalations that bring into view volatile molecules that are invisible to the naked eye. Volatiles can also be given two-dimensional digital representation, as in the charts showing the results of a gas chromatograph, but these charts are even less likely than paintings, drawings, or photographs to arouse an odor image in the mind of anyone but a specialist. Then there are what are called "smellmaps." Contemporary cartographers have not been left behind in the sensory turn of the sciences of recent decades. One contemporary artist/designer, Kate McLean, deserves special mention for her creative smellmaps, which are based in part on data gathered from leading "smellwalks," something we will look at more closely in Chapter 14 on the smell of cities.

Even under the broad definition of "olfactory arts" that I have given, few of the historically prominent two-dimensional depictions of smells qualifies as part of "olfactory arts" since they do not use actual odors. Among the few uses of paintings with scents that might be included in the category of olfactory arts are hybrids like Marvick's juxtaposition of abstract paintings with fragrances. To adapt Barthes's comment, shit merely pictured doesn't smell any more than it does when written, sculpted, or, in the case of Piero Manzoni's famous *Merda d'Artista*, even canned. And this would apply to any nonodorous sculptural representations of smelly or fragrant objects. On the other hand, as we will see in Chapter 10, hybrids of actual odors with sculpture have become a major genre of contemporary scent or olfactory art. Before we turn to them, however, we first need to consider hybrids of actual odors with three other long-established fine art forms, one of which is also pictorial, namely, instances where odors are used to enhance theater, music, and film.

Notes

1. Quoted from the etiquette next to the painting in the Dulwich Gallery, UK.
2. J.-K. Huysmans, *Against Nature* (Sawtry, UK: Daedalus, 2008), 86.
3. Paul Gauguin, *Noa, Noa: A Journal of the South Seas*, trans. O. F. Theis (New York: Farrar, Strauss and Giroux, 1957).

4. Quoted in Jim Drobnick, "Towards an Olfactory Art History: The Mingled, Fatal, and Rejuvenating Perfumes of Paul Gauguin," *Senses and Society* 7, no. 2 (2012): 196–208. Drobnick's suggestive essay traces three themes in Gauguin's reflections on the smells of Tahiti: rejuvenating power, a mingling of animality and flowers, and their "fatal" erotic power.
5. Jaquet, *Philosophie de l'odorat*, 212.
6. Carlo Carra, "The Painting of Sounds, Noises, and Smells," in *Futurism: An Anthology*, ed. Lawrence Rainey, Christine Poggi, and Laura Wittman (New Haven: Yale University Press, 2009), 156–59.
7. The works were shown in 2013 at the art gallery of Southern Utah University, where Marvick is on the faculty of the art department. I am grateful to Andy Marvick for discussing his ideas about this work with me during a visit to the SUU art gallery.
8. For images of most of these works see Peter de Cupere, *Scent in Context: Olfactory Art* (Antwerp: Stockmans Art Publishers, 2016), 430.
9. From a personal communication with the artist, October 2018.

9
Toward a Total Work of Art
Smell in Theater, Film, and Music

The classic ideal of the "total work of art" envisions the highest work of art as a reunion of all the major art forms. It would be a reunion, since Richard Wagner and others who first espoused the *Gesamtkunstwerk* ideal believed that in the past the arts *were* united. The tragedies of Aeschylus, for example, integrated poetry and dramatic action with a chorus that chanted and danced, sometimes in front of painted scenery. But *Gesamtkunstwerk* theorists complained that in the modern period the individual arts had increasingly gone their separate ways. Since Wagner's day, the dream of the total work of art has experienced many permutations, including the one I want to consider in this chapter: its extension to the "lower" senses to include smell.[1]

But first we should note that in some non-Western cultures smell has often played a role in the performing arts. In contemporary Balinese culture, for example, "welcoming dances, for either gods or tourists, use smells combined with action, music and dance to provide a strong sense of joy to both participants and audience members."[2] In a work like *Panyembrama*, which was developed from traditional forms and intended for outsiders (partly to preserve sacred performances from profane eyes, ears, and noses), the olfactory elements come from the incense used to consecrate the performance space as well as from the scents of the flowers worn by the dancers and the petals thrown on the audience during the performance. In the West the practice of scenting theater performances goes back at least to the time of Shakespeare. Although some critics have dismissed the practice of adding scents to films or plays as a "gimmick," when thoughtfully done it may help focus audiences' attention on important aspects of a work as well as add another channel of interest and pleasure.

Theater

Although the decision to use odors with Renaissance dramas was often made by the theater companies, in other cases they were following explicit stage directions. Webster's *White Devil*, for example, specifies that two characters enter and "draw a curtain where Bracciano's picture is . . . and then burn perfumes afore the picture," and Ben Jonson gives an even more elaborate description of the ritual censing of an altar to be done in *Sejanus*.[3] In *Lingua: or, the Combat of the Tongue and the Five Senses for Superiority* (1607) each of the five senses appears before a jury led by Common Sense. Olfactus, or the Sense of Smell, is accompanied by seven boys carrying perfume bottles, censers, flowers, herbs, and ointments and by a page who recites a recipe for creating a pomander. In his plea to the jury, Olfactus claims that smells, among other things,

> clear your heads, and make your fantasy
> To refine wit, and sharpen invention
> And strengthen memory, from whence it came
> That old devotion, incense did ordain
> To make mans spirits more apt for things divine.[4]

The most striking use of odors in English Renaissance theaters came from the stench of "squibs" employed to simulate thunder and lightning. Composed of sulfurous brimstone, coal, and saltpeter, the explosion that resulted from igniting them not only produced a startling noise and flash of light, but also a smell reminiscent of gunpowder. Ben Jonson alludes to the "stink" of squibs in one of his plays and explicitly called for them in the stage directions for *Dr. Faustus*. In Shakespeare's *The Tempest*, when Ariel speaks of "the fire, and cracks / Of sulfurous roaring," he could be alluding to how the players made the "tempestuous noise of Thunder and Lightning" called for in Shakespeare's stage directions. Squibs were probably used at the opening of *Macbeth*, whose stage direction also calls for "Thunder and Lightening."[5]

Several scholars have argued that the experience of odors not only lent verisimilitude, but also worked on audiences' perception of political and religious issues. In the case of *Macbeth*, for example, the smell generated by the squibs at the beginning would have likely conjured thoughts related to the Gunpowder Plot that had occurred not long before. The use of incense in plays like *Sejanus* and *Lingua* could also have touched a religious nerve, since the conflict over the use of incense in the English church was still very much alive in the early 1600s.[6]

If odors seemed to have been used in English Renaissance productions to enhance the illusion of reality, Sally Banes points out that they were still showing up this way in the late nineteenth and early twentieth centuries. David Belasco was famous for using such things as strewing pine needles on stage for a forest setting, burning incense for a Chinatown, or having pancakes cooked onstage.[7] But late nineteenth century Symbolist productions on the Continent tended to use odors in more "suggestive, mysterious, ways," as Mary Fleischer puts it, hoping to evoke "a hidden reality."[8] Often the action would take place behind a gauze curtain, and scents would be diffused in the auditorium to create a dreamlike mood. An

1891 drama based on the biblical Song of Songs went much farther. The director staged the play as a total artwork by appealing not only to the visual and aural senses but also to smell. In fact, as Kristin Shepherd-Barr notes, the production attempted an almost literal enactment of Baudelaire's line "sounds, scents and colors correspond."[9] As the audience entered, people were given a set of program notes that not only summarized the action and the "mystical meaning" of each scene, but also listed the corresponding sounds, colors, music, and scents that would be used. The notes for the first scene tell us that the actors' intonation will emphasize the vowel " 'i' illuminated by 'o,' " the music will be in the "key of C" and will feature the "viola," the color scheme will be "pale purple," and the scent will be "frankincense."[10] Unfortunately, the scents accompanying each of the five scenes were diffused from atomizers held by assistants stationed at the proscenium and in the balcony, but, of course, they did not reach everyone at the same time and in the absence of a modern ventilation system, gradually built up. As a result, the "bewildered audience was doused with perfume and left choking in fumes of incense."[11]

When it comes to twentieth- and twenty-first-century productions, theater and dance historians such as Sally Banes have usually grouped the uses of odors in two broad, often overlapping categories: to illustrate and authenticate or to establish atmosphere and mood.[12] Examples of atmospheric odor include the 2014 Minneapolis Theater in the Round production of *Treasure Island*, which used the ventilation system to deliver four ambient scents corresponding to scenes involving the pirate ship, the old tavern, sea breezes, and Skull Island jungle.[13] The use of odors for illustrative or authenticating purposes comes in many varieties and manners, often simply drifting out from the stage. In *Balti Kings* (2000), a British drama about an Indian restaurant, there was a functioning on-stage kitchen in which actual cooking was done; and since the production took place in a theater-in-the-round, most of the audience could easily inhale the spicy odors.

But sometimes, as Stephen Di Benedetto says, odors are more than "a mere artifice to make a bit of naturalistic mimesis authentic."[14] The philosopher Susan Feagin, for example, describes how odors can be used not only to make the audience feel a part of the action on stage, but also to configure the acting space and guide audience attention. In a 2015 adaptation of Tennessee William's *The Glass Menagerie*, when Amanda sprays herself with perfume, a moment later the scent reaches the audience. Here, as Feagin says, the perfume scent was not just illustrative and mood setting, it also had the effect of "shifting the audience's experience of their own bodies" to feeling as if they were "in the same space and place as the characters."[15] Feagin also gives an example of smell being used as an adjunct to guide spatial attention in a 2016 adaptation of *Macbeth*, entitled *Til Birnam Wood*, a production for which the audience wore eye masks in order to intensify nonvisual sensory experience. Late in the play, as Malcolm's army is approaching Macbeth's, carrying pine trees for disguise in fulfillment of the witches' prophecy that Macbeth will not be vanquished until "Great Birnam wood to high Dunsinane hill / Shall come against him," the audience not only hears a murmur of voices coming from the left, but smells a strong pine odor, which it assumes is coming from the same direction as the voices.[16] But as Feagin points out, without the cue of the voices, the pine smell by itself may not have directed attention to the left (she cites Clare Batty's emphasis on the sense of smell's general lack of ability to locate direction on its own). Feagin concludes that given our difficulty in locating the direction of odor sources, especially in a darkened theater, a smell may only represent space in special cases.[17]

Perhaps the most creative use of odors in a contemporary theater production has been in Violaine de Carné's *The Scents of the Soul* (*Les parfums de l'âme*) (2012), which actually made the characters' "odor signatures" the subject of the play. Carné's play offers one of the best examples to date of a genuine integration of odors with the dramatic action and shows both the potential and limitations of odors in the theater. Carné also combined the odors with a musical score and occasional video images in a metaphoric rather than merely illustrative way. She had long been interested in the possibility of a "total sensory experience" in the theater. But it was only after spending several months visiting a hospital to observe brain-injured patients whose treatment included exposure to odors as a way of reclaiming memory that she began writing and directing plays focusing on the sense of smell. For *Les parfums de l'âme*, she worked with a perfumer for a year, perfecting thirteen odors that were diffused to the audience, most from under the seats, a few from the back of the theater.[18]

The play is set in a future where there is an institute that can recreate the odor signature of a departed loved one based on scents impregnated in the loved one's clothes. Seven people gather in the institute's waiting room and tell their stories as each waits to be presented with a vial containing the scent of the departed. From time to time the voice of one of the dead resonates from offstage, or their faces appear on a giant video screen. Among the seven characters are Melissa, part African, trying to make contact with her African grandfather, and Takuni, a Japanese student who wants to reclaim the smell of his beloved Ophélie, who disappeared without a trace. As the sensuous mouth of Ophélie appears on the screen in a giant close-up, she speaks of her red shoes (Takuni had brought them along with one of her T-shirts) and at that moment, the pungent odor of her smelly feet is delivered to the audience.[19]

Despite such humorous touches, Carné meant her play not only to deepen the audiences' awareness of the importance of smell, but to raise the question of whether the desire to obtain the scent of a departed one is not a way of short-circuiting the grief process and refusing to face the finality of death. If one is to believe the results of questionnaires administered to 319 audience members (including personal interviews with 35 of them), Carné and her team largely achieved her aim of raising olfactory awareness. Most respondents were satisfied by the diffusion technique, some commenting on how quickly the theater was cleared between tableaux. As for detection and impact, nearly all said they could smell the odors well enough, and at least half said they were prompted by some of the smells to recall personal memories. As one might expect, the individual odor most remarked on was the foot smell, which several claimed was too strong and lasted too long, but many others thought the joke worth it.[20]

Despite Violaine de Carné's success with *Les parfums de l'âme*, and the earlier success of Bellasco and others, smell remains only an occasional experience in theater. There are obviously a number of practical obstacles. First, there is the fact that until recently, the technology for the delivery of odors at the right time, the right place, and right amount has been inadequate to support a nuanced coordination with lighting, sound, and other effects. Then there is the problem of formulating and calibrating the odors themselves and the readiness of actors to perform olfactory works, and of critics and audiences to understand and appreciate their aesthetic value. Finally, given most people's generally low level of olfactory knowledge, it is not surprising that some critics and audience members may be bewildered or put off by the prominent use of odors, especially by anything that exceeds the hedonically pleasant and conventionally illustrative. After studying a number of critical reviews of plays involving odors, Matthew Reason has concluded that although most critics respond well to implied but inexistent smells, when real odors are involved, many critics seem unable get beyond either irony or ridicule.[21]

A key factor inhibiting positive critical and audience responses is probably the "fourth wall" convention, the assumption that there is an invisible wall separating the audience from the stage, a wall that permits sights and sounds to reach us but is impervious to most other sensory information. But even audience members accustomed to more traditional theater or opera are likely to accept mild violations of the fourth wall, for example, when cooking smells or characters' perfumes drift out to the audience. On the other hand, when a fuller range of odors connected to specific stage actions are diffused into the theater, some people find it disturbing, since it breaks too sharply with the habit of following dramas primarily with eyes and ears.

Resistance to the fuller range of odors probably also reflects the prevalence among Westerners of hedonic reactions over qualitative perceptions that we examined in Chapters 2 and 3. Thus, it is not surprising that the use of negative smells in the theater is particularly risky. Romeo Castellucci's play *On the Concept of the Face of the Son of God* (2011) follows a young man's dialogue with his dying father, who lies beneath a picture of Christ. The father's uncontrollable diarrhea and flatulence are not only seen and heard, but the smell reaches audience through a finely calibrated diffusion system. As Dominique Paquet remarks, the use of fecal odor in the play profoundly reinforces the reality of degradation and death. Yet despite Paquet's general enthusiasm for the use of odors in the theater, she wonders if, in this case, "the color of the liquids and the sound of the flatulence would not have been enough to evoke dereliction and produce the simulacrum of excremental odor by the power of suggestion."[22] This is an important observation to which we will return when we discuss the use of odors with films.

So far we have mostly considered practical obstacles to an "olfactory theater." Now we need to consider some of the more general aesthetic principles for dealing with the issues raised by smells in the theater. Let's look at the issue first from an Aristotelian perspective. For Aristotle, the heart of dramatic theater is plot, with character, thought, and diction playing a supporting role, and what he called "embellishment," music/dance and spectacle (masks, costume, sets, etc.) bordering on the dispensable. Smells on this theory would surely be an irrelevancy, at most a part of spectacle, and probably considered a distraction from experiencing the plot in a way that leads to catharsis. Even Violaine de Carné has said that by including so many odors in her play, she risked losing some traditional spectators since the diffusion of odors is likely "to prevent catharsis."[23]

Moreover, adding odors to most traditional plays would seem to run afoul of the spirit of the Aristotelian derived principle of unity, which concerns the integrity of the dramatic action and its presentation as a unified conception. This suggests that the fourth-wall convention we have considered could be taken to mean that once you let one or two odors through, you not only break with the integrity of drama as a visual/auditory medium, but it will be hard to justify not attaching odors to everything. If we are to smell the lightening at the beginning of *Macbeth*, it might be argued, why not smell the witches' potion? And if we smell the witches' potion, why not smell Macbeth's horses and the sweaty bodies of his retainers? And what about Lady Macbeth's famous little hand? "Here is the smell of the blood still; all the perfumes of Arabia cannot sweeten this little hand" (act 5, scene 1). I suppose one might try to argue for a difference among the various cases, for example, that were we near the witches' cauldron, we could smell it, but the retainers might be too far away, and Lady Macbeth's hand is so small, and so on. Such trains of reasoning can quickly become absurd, yet they seem inevitable if one starts adding odors to plays. The sensible and principled choice, the conservative theorist might argue, is to leave smell entirely out of staging traditional dramas since odors will inevitably raise such puzzles.

Yet since odors have, in fact, been used and accepted off and on for centuries, is there any principled way to answer these worries? One answer in relation to more traditional realist plays might be that because odor molecules can travel, although more slowly than sound waves, it is reasonable to expect to smell something happening on stage if one sits near enough, for example, the odor of curry that floated out to the audience of *Balti Kings*. But what of the plethora of specific scents in plays like Carné's *Les parfums de l'âme*, scents that were diffused from beneath the audience's seats, yet were taken as coming virtually from the stage? Clearly, Carné's play as a whole breaks with traditional realist theatrical practices so that its use of odors should be evaluated in the way we approach other works of experimental theater, whether Brechtian, postmodern, or "postdramatic." But for avant-garde theater, we lack a dominant philosophically grounded aesthetic approach similar Aristotle's. Even so, I think we can delineate some general aesthetic guidelines for the use of odors in experimental theater.

One such guideline we could call "proportional effect," the rather obvious idea that any use of odors must be proportionate to the effect sought. This would apply to the use of the excremental odors in Castellucci's play *On the Concept of the Face of the Son of God*, powerful smells that were intended to intensify the audience's experience of the dying man's degradation. But confronting us with a fecal smell in such a case seems less likely to make us feel more keenly the abjection of the dying man than it is to trigger visceral repulsion and make us wonder why we need to be subjected to it. The point isn't that fecal or other negative odors should never be used in the theater, but that an artist has to carefully consider the proportionate effects. If, instead of deepening our understanding and empathy, the odors shock, sicken, or occupy our whole attention, little has been gained and much may be lost.

A second aesthetic guideline concerns what we might call "cognitive affect," namely whether our emotional response to odors is of a kind that can sustain a publicly articulable aesthetic experience and judgment. Given the fact of each individual's somewhat idiosyncratic and partly unconscious emotional associations with odors, it is apparent that writers, composers, or choreographers concerned to use odors will be working with considerable uncertainty. Of course, some artists embrace this variability in olfactory responses as having the advantage that strong emotional associations afford the possibility of connecting with the audience on a deeper, more individual, level. Violaine de Carné, for example, has written that olfactory theater cannot be addressed to an abstract "public" but only to individuals who will experience the odors by moving back and forth between what is happening on the stage and the personal memories aroused by the odors. Yet Carné also speaks of communicating meanings that can be shared across her audience, even though each person's experience of those meanings will be individually inflected. According to the interviews I mentioned earlier, a majority of those questioned about *Les parfums de l'âme* understood the appropriateness of the odors as evocations of the departed loved ones, and even got the joke about the beautiful woman with smelly feet. Thus, although some respondents seemed to be mostly expressing purely subjective reactions, others seemed capable of giving considered responses and able to discuss the play's intended aesthetic effects.

My tentative conclusion is that the occasional use of odors in more traditional drama is aesthetically justified to the degree that its violations of the fourth-wall convention seem "natural," for example, such things as odors drifting out from the stage. In works of experimental theater, on the other hand, a much wider span of odor uses is artistically and aesthetically justifiable if one allows for proportionate effect and cognitive affect. But to develop olfactory performance arts farther than Carné has done will require writers, directors, critics, and audiences with the courage and patience not only to explore the opportunities but to ponder the limits set by the basic characteristics of olfaction.

Film

Obviously, many of the issues we have discussed with respect to the live performing arts such as theater apply to the use of odors with film, so that our treatment of smell in film can be much briefer. Accordingly, I will first mention some highlights of past attempts to "odorize" films, and then discuss why I believe the artistic and aesthetic limitations of the use of odor in standard narrative films may currently outweigh the artistic opportunities to authenticate or modulate actions in narrative films. Even so, given that films were once silent but now give sound an indispensable role, we should keep an open mind about the possibility of scent scores, especially in experimental films of the future.

Since the late 1950s when the rival experiments of Smell-O-Vision and AromaRama ended in commercial disaster, there have been hardly any major Hollywood efforts to add odors to films, unless one counts *Polyester* (1980) with its ten scratch-and-sniff cards and various kid movies like *Rugrats Go Wild* (2003). The problems with diffusing actual odors have been partly technological and partly economic. In the case of the AromaRama system used in the 1959 travelogue film *The Great Wall*, with a hundred different odors, many of the smells not only seemed off target but were also diffused over the entire theater so that as the film progressed, the smells got out of sync with the images and gradually built up and afterward clung to the seats and people's clothes. Although a few critics found things to praise, most found the smells distracting rather than enhancing, and a part of the press had a field day ridiculing the AromaRama film as a "stinker."[24]

The Smell-O-Vision drama *The Scent of Mystery* of 1960 had more subtly calibrated odors and a far superior delivery system that released the odors from beneath the seats. But it tried to coordinate thirty some different scents with specific scenes and actions of the narrative, still too much for an unaccustomed audience to easily follow. *The Scent of Mystery* also ran into technical difficulties. The scents reached the balconies slightly behind the corresponding events on screen, and some scents were barely noticeable, whereas on the main floor the strength of the odors varied in different areas, and in some places the pipes carrying them made a hissing sound. Although several of the problems were corrected after the first few showings, critical and public perception had already been soured by the previous failures of AromaRama. The poor reception of *The Scent of Mystery* so discouraged Smell-O-Vision's producer, Mike Todd ,that he gave up the project rather than keep pouring money into improving it. The olfactory psychologist Avery Gilbert, who has closely studied the competition of the two systems, thinks Smell-O-Vision might have caught on if Mike Todd had persevered.[25]

These days, each year seems to bring more sophisticated, digitally controlled machinery for releasing odors, so that the idea of odorizing films has cropped up again. Of course, odors are sometimes added to the growing number of productions in 4DX aimed at the seventeen- to twenty-four-year-old market. These films feature such things as vibrating seats, water dripping on your head, forced air blown in your face, and sometimes the release of odors, but the effects require costly mechanisms, and theaters using them will probably be few and confined to large metropolitan areas. So far we hardly need to worry over how to evaluate the artistic and aesthetic achievements of 4DX, although someone might be able to make a more artistically sophisticated use of some of the devices, just as there are now aesthetically interesting video games (which, like virtual reality devices, could also be odorized).

A recent attempt at adding scents to a traditional narrative film was the 2006 partial odorizing of Terrence Malick's *The New World* (2005) by a Japanese distributor and NTT Communications, the latter seeking publicity for the introduction of its new scent delivery technology for home entertainment uses. The odorized version of *The New World* was tried out in two theaters, each of which had several rows outfitted with the large aroma diffusing balls NTT hoped to market. A decision was made to include only a dozen pleasant scents in a suggestive way, for example, woody smells to accompany the virgin American landscape, citrus for the English court, and floral for romantic scenes. The problem with this, as one commentator remarked, is that it left him puzzled at the absence of diffused odors when something appeared onscreen that would have smelled strongly in the film's fictional world, such as boiling leather or gunpowder.[26]

The issue to which the reviewer of *The New World* calls attention shows that, as in the case of live theater, once technological problems of delivery begin to be solved, the artistic problems of making odors in films aesthetically meaningful show up more clearly. It is no longer a question of can we effectively odorize, but should we and how should odors be used? A terminological shorthand developed by film theorists for distinguishing two major kinds of sounds in film may be of some help in discussing the issue. Sounds in a film scene that are part of the narrative world of the film (a television blaring in a living room) are called "diegetic," whereas sounds that are not part of the narrative world (the musical score) are called "nondiegetic." Thus, if the odor of gunpowder had accompanied the sight and sound of a gun firing in *The New World*, it would have been diegetic, whereas the floral odors diffused to the audience during the romantic scenes when there were no flowers present would be nondiegetic, functioning like a musical score. But there is a major difference between nondiegetic odors and a film's nondiegetic music that has relevance for the question of whether to accompany films with a smell track. Most of the time we do not notice a film's musical score and, with some notable exceptions such as Michael Levi's score for *Jackie* (2016), we are not supposed to notice it, but to focus our attention on the story. By not only announcing to the audience of *The New World* that the film would be accompanied by odors (and charging a premium to sit in the seats near the scent-dispensing globes), it was almost inevitable that people in those seats would wonder why some things were accompanied by smells and others not.

A more recent attempt to enhance the experience of an existing film with real odors was a special 2013 screening in Los Angeles of Tom Tykwer's film version of Patrick Süsskind's novel *Perfume: The Story of a Murderer*. (We will consider the novel at the beginning of Part IV.) There is an interesting story about how the special, odorized screening of *Perfume* came about. In the early 2000s, the perfumer Christophe Laudamiel was so taken with Süskind's 1988 novel that he set out in his spare time to create a series of scents that would express various scenes and events in the novel, ranging from the garbage heap on which the protagonist, Grenouille, was abandoned at birth through the virginal scent of the murdered young women, to the supreme perfume Grenouille achieves at the end.[27] When Laudamiel heard that a film version was in the works, he arranged with the producers to have a luxury coffret of fifteen of the scents he had originally designed based on the novel sold to theatergoers by the Thierry Mugler perfume house as part of the film's promotion. Although those who attended the premiers in 2007 could buy the coffrets in the lobby and sniff the odors, no attempt was made to dispense scents along with the film or to market the fragrances individually. Then, in 2013, the Los Angeles Institute for Art and Olfaction (a small nonprofit dedicated to promoting the understanding of perfumes and their use as art), decided to host two screenings of the film for which each of the fifty attendees would be handed, at the appropriate moment, a scent strip that had been dipped in one of Laudamiel's scents.[28]

But for all the inventiveness of Laudamiel's scents compared to the handful of humdrum smells that accompanied *The New World*, the actual cinematic experience raises similar questions. Why just these fifteen scents? And are we to assume we should only smell what Grenouille himself smells (point-of-view issue)? And, of course, the big question, how can one integrate scents into a film in a way that is truly convincing and enhancing rather than a distraction? One way of addressing these kinds of problems would be to take the radical position that the difficulty with odors in past film attempts was not that there were too many, but that there were not enough. Perhaps a filmmaker should go all out and make a true "scent track" to parallel a film's soundtrack (although many films include stretches when there is no music). A full scent track might include not only scents to accompany diegetic on-screen smells, but could have a second scent track that would function like nondiegetic film music and simply accompany what is happening on screen. To create an olfactory parallel of just the diegetic soundtrack itself would, of course, be an enormous undertaking, but it could be done. As with diegetic sounds in most films, particular diegetic smells would be transmitted to the film audience only when the odors were noticeable to the characters on screen, unless a director wanted to foreshadow or make an ironic or humorous comment using odors. An example of the latter was actually tried in *The Scent of Mystery* when a cab driver is shown with a steaming cup of coffee, but the audience smells liquor instead and realizes the driver is tippling. As for whether there could also be the scent equivalent of a musical score, we need to keep in mind that music offers endless possibilities of melodies, timbres, and rhythms for suggesting a general mood, characterizing a person, or accompanying an action, and as Noel Carroll points out, whatever one's philosophical position of whether music can express emotions, there are clearly many established conventions for sad, happy, tense, and soothing film music.[29] Unfortunately, the scent composer would face not only the absence of an established expressive tradition, but also the fact of people's highly idiosyncratic emotional associations with different smells. Moreover, in a realistic film narrative, the atmospheric scents would have to be kept at a sufficiently low volume that they would not be consciously perceived or interfere with the diegetic smells, yet could still supplement the musical score by subliminally modifying our emotional reactions. It sounds not merely daunting but impractical for a standard narrative film. But in the niche genre of the short experimental art house film, given enough financial resources and determination, it would in principle be possible to create a genuine scent track for both diegetic and nondiegetic odors. As we will see in the next section on music, *Green Aria: A Scent Opera* offers a precedent for solutions to many of the problems that have faced scenting films in the past.

Yet, one could argue that even if the creation of comprehensive scent track(s) were not so daunting, it may, in fact, not be necessary—or even advisable. Remember Dominique Paquet's suggestion that the vivid images and sounds of the dying man's diarrhea and flatulence could have been sufficient to suggest the smells? The film theorist Vivian Sobchack has written: "We are in some carnal modality able to touch and be touched by the substance of images . . . to take flight in kinetic exhilaration . . . to be knocked backward by a sound; sometimes even smell and taste the world we see up on the screen."[30] And the philosopher Cynthia Freeland has pointed out that neuroscience studies have demonstrated that the combined images and sounds of food in films activate some of the same brain areas as become active when we taste real foods.[31] This is quite consistent with the notion we have encountered since the beginning of this book, namely, that all perception is multimodal in varying degrees. Although Luis Rocha Antunes's book *The Multi-sensory Film Experience* does not address taste or smell, but focuses on the vestibular and proprioceptive senses, its argument, like Freeland's, draws on contemporary neuroscience. His detailed examination of how film images and sounds depicting bodily movements activate the same brain areas as the vestibular and proprioceptive senses themselves suggests that our visual and auditory experience of a film could trigger sensory experiences of smell at the level of perception and not simply by imagination or inference.[32] Whether one accepts Sobchack's emphasis on imagination, or Antunes's claims for a multisensory perceptual response, or Freeland's blending of both approaches, there seems good reason to believe that film images and sounds are able to trigger a multisensory experience that can include a virtual dimension of taste and smell. Because our responses to films, although provoked by vision and audition, nevertheless actuate other sensory areas of the brain, it might be that vision and hearing are adequate to bring to mind a simulation of the experience of smell. At least some reviewers of the film version of *Perfume: The Story of a Murder*, whose story is primarily about smell, found that the film satisfactorily conveyed the olfactory aspects of the story—which in its original book form, after all, used only language printed on a page to evoke the imaginative experience of smells.[33] Our literary experience of reading *Perfume* would probably not have been improved with scratch-and-sniff cards or even having Laudamiel's coffret of scents at hand.

Does this mean that the dream of creating films with integrated scent tracks should be forgotten? Not at all. To refuse such experimentation would be to give in to what Noel Carroll has called "media essentialism," the idea that the medium of film is singular and itself sets limits on what should even be tried.[34] Kevin Sweeney makes a similar point, suggesting that the term "film" may not name a single medium but a group of related media; narrative, documentary, and abstract films now coexist, and we have gone from silent to sound, black and white to color, 2D to 3D, and now 4DX, and so on.[35] No doubt, just as Violaine de Carné has been able to use actual odors in a play with positive aesthetic consequences, so an avant-garde filmmaker working with a perfumer like Laudamiel may come along and do something similar with an experimental film. It may be that the most promising avenue for an artist wishing to create a film that included actual odors would not be to add odors to the existing art form and its conventions, but to venture into the uncharted waters of creating what would be an essentially a new multisensory art form, an art form in which odors and the sense of smell would be one part of the conception from the beginning, not an embellishment or addition, something that happened with *Green Aria: A Scent Opera*.

Music

At the beginning of this chapter I mentioned the traditional Balinese welcoming dance that is presented in a space redolent of incense and punctuated by the smell of the flower petals showered on visitors. Even contemporary Indonesian musicians and artists such as I Wayan Sadra may incorporate smell into their works. As he describes a performance of his piece *Lad-lud-an*, the gamelan musicians began playing outside the theater so that "the audience only heard the faint indiscernible sound of the gamelan as it gradually approached." After the players entered and continued playing,

> A performer stood up . . . in his hand he held an egg . . . high above a black oval shaped stone . . . the egg was dropped. . . . Then the air circulating in the theatre spread a foul smell. I had deliberately chosen an egg that was rotten—and the audience reacted by holding their noses.[36]

Here we have a musical work that bridges traditional and modern cultures, using smell in a way that would not be out of place in a performance piece for a museum of contemporary art.

When we turn to Western uses of scents with music, we find a few composers and performers have tried to convey odors through musical sound. Among them, Debussy stands out for such works as "Les parfums de la nuit," or the second movement of *Iberia*, or the section of *Preludes I* called "Les sons et les parfums tournent dans l'air," named after a line from Baudelaire's "Harmonie du Soir." In each of these pieces Debussy's titles invite one to listen imaginatively for Baudelaire's correspondences of sounds, smells, and colors. In "Parfums de la nuit" the sinuous rhythmic forms beneath the delicate oboe melody seem to suggest the shimmering and blending of scents on a summer night. Of course, the gentle oboe sound here might also momentarily remind one of Baudelaire's words from "Correspondences," "scents as soft as oboes." That Debussy sought such correspondences is clear from a 1901 comment:

I see the possibility of music constructed especially for the open air ... a mysterious collaboration of the movement of leaves and the fragrance of flowers with the music, which would unite all these elements in a natural accord.[37]

One can find even more pointed correspondences of sound and scent in Debussy's only opera, *Pelléas et Mélisande* (1902). As Chantal Jaquet points out, although the use of music to evoke odors occurs throughout the opera, the effect is particularly marked in act 3 when the jealous Golaud makes his half-brother lean over a stagnant pool at bottom of the castle vault, asking him menacingly, "Do you smell the odor of death?" The music includes a descending bassoon passage over dark orchestral colors and slow tympani effects. Pelléas pleads to leave the stifling atmosphere, and as the two begin to climb up toward the light, the music also ascends and brightens with fluttering strings and harp. Once in the free air, Pelléas exclaims over the "odor of wet roses" rising up along the terraces. That a sensitive listener could in fact respond almost viscerally to such musical expressions of smells we know from Proust's narrator in *Remembrance of Things Past*, who says that the smell of roses at that moment in *Pelléas et Mélisande* seems so real that it triggers his allergies and he starts sneezing every time he listens to the scene.[38] Even if one does not have such extreme physical reactions to the olfactory suggestiveness of Debussy's music, Proust's comment implies that music and words in combination could have an effect similar to the combined effect of images, words, and music in film that we discussed earlier. Yet in music as in film or theater, adding actual odors, unless handled with great care, may distract from the desired experience rather than enhance it.

Although major compositions that call for the use of actual odors are rare, using scents to accompany actual performances is less so in both classical and popular music. Popular music examples of actual odors have ranged from singer Katy Perry's 2016 tour that including cotton candy scents to the Broadway musical *Waitress* with its apple pie aroma. Noteworthy among contemporary classical music performance groups who use smells is the French choral group Les Métaboles, who, in 2015, did a program of American music that was accompanied by scents designed for the occasion by the perfumer Quentin Bisch.[39] That same year, the pianist and composer Laurent Assoulen released an album called *Sentire* that included a series of reusable fragrance patches designed by the artist to accompany each piece.[40] Finally, in 2016, the Australian Art Quartet commissioned the perfumer Carlos Huber to design a series of scents to accompany pieces by composers such as Tchaikovsky and Gurdjieff in a concert called "Scent of Memory." In this case, the scents were provided to the audience on paper scent sticks they held up to their noses. Before each piece Huber described how each scent was constructed to evoke a particular historical moment or mood. Interestingly, in the light of our previous discussion of using scents with dramas or films, the decision to use scent sticks was not for lack of a means to diffuse scents in the hall, but in recognition of the fact that individuals have highly variable responses to both the quality and intensity of smells.[41]

One of the few modern composers who at least planned to use actual odors as part of a musical work was Scriabin, whose visionary last project called for the inclusion of incense. Scriabin died in 1915 with only the long prelude to his projected magnum opus, *Mysterium*, partially written. The work as a whole was not only to be a true *Gesamtkunstwerk* involving all the senses, including smell, but was also to make the audience full participants, with the first performance to take place in a monastery somewhere in the Himalayas. In 2015, a centennial celebration of Scriabin's multisensory vision called "Scriabin in the Himalayas" was actually held at the Thiksey Monastery near Ladakh that involved not only live piano and solo performances of Scriabin works, some accompanied with dances by Buddhist monks and a color light show, but also the diffusion of six ambient scents created for the occasion by the perfumer Michel Roudnitska.[42]

The most complete integration of scent and music in a contemporary *Gesamtkunstwerk* is *Green Aria: A Scent Opera*. Presented in 2009 in an auditorium at the Guggenheim Museum in New York and later at the Guggenheim in Bilbao, *Green Aria* took the integration of scent and music to an entirely new level. Moreover, *Green Aria* was notable for putting the emphasis squarely on scent and integrating it with the music in a way both cognitively and emotionally rich. Yet *Green Aria* was a strange sort of "opera" since it had no human voices, no theatrical actions, and no sets. It consisted of an original score for scents coordinated with a commissioned score of electronic music that followed a loose narrative. Moreover, it was presented in the dark since its creators felt that if the audience were able to look around they would be distracted from focusing on the scents and sounds. Stewart Matthew, who had the idea for the work, originally wanted to produce an opera that addressed all five senses, but soon realized that the project wasn't feasible and settled for this creative hybrid of odors and music. In order of composition, the narrative came first, then the scent score, then the music score to accompany it. The scent score, created by the perfumer Christoph Laudamiel, was based on a vague environmental scenario written by Matthew. This scenario, along with samples of Laudamiel's scents and a crucial timing matrix, were sent to two composers, who created the electronic music track.

The scenario itself is rather nebulous, positing a symbolic conflict between technology and nature in four "movements," featuring the elements Earth, Air, Fire, and Water and eighteen "characters," with names like Base Metal, Evangelical Green, Fresh Air, Funky Green Imposter, Industry, Chaos, and Green Aria. In a way these abstract characters were the equivalent of human figures in a regular opera and their scents were their "voices." In the narrative, technology usurps the powers of nature, Chaos ensues and is only overcome when

Evangelical Green instigates a reconciliation of technology and nature out of which Green Aria emerges. Although the story sounds corny, there was nothing corny about either the scents or the music, both of which many of those who attended found interesting and well coordinated. The classical music critic for the *New York Times*, Anthony Thommasini, characterized the music as "episodic yet subtly flowing, with skittish flights; contrapuntal passages where dueling voices were pushed to wide extremes of register; steely electronic agitation; and calming harmonic writing for dusky sustained strings."[43]

Green Aria's olfactory achievements were both technical and artistic. As we have seen, technical requisites for the successful use of odors with any dramatic production, whether theater, film, or music, are that the odors reach everyone in the audience at roughly the same time, that the scents have the same intensity everywhere, and that they remain distinguishable and not build up and merge into a stew. Laudamiel worked with designers at the international firm Fläck-Wood, which specializes in ventilation, to come up with a "scent organ" to diffuse the odors. It used compressed air to push the odors through flexible Teflon tubes that ended in front of each seat with what looked like a scent "microphone" at the tip. Thus, members of the audience could bring the scent tube as close to their nose as they needed. As a result of this setup, Laudamiel could send a much lighter burst of each odor composition and be confident that its smell would remain relatively localized, thus avoiding the kind of spread and buildup that have doomed many previous odorizing experiments.

Crucially, Laudamiel carefully calibrated the dosage and timing of the bursts, based on the roughly six seconds it takes a person to inhale and exhale. After each six-second exposure to an odor, a two-second burst of plain air was sent through the tubes to clear the nostrils and prepare the way for the next odor. This kind of precision allowed Laudamiel to send thirty different odors in the space of fifteen minutes, with most of the audience able to follow the sequence and perceive the coordination with the musical motifs. The scents themselves were abstract constructions since, as the scent psychologist Avery Gilbert, pointed out in his review of *Green Aria*, if Laudamiel had used familiar scents, people would have been distracted from the work itself and started "guessing the smells or puzzling over their mental associations."[44] In addition, Laudamiel set aside considerations of pleasantness in favor of scent combinations that would evoke the "characters" and themes of the narrative.

Many people entering the Guggenheim's theater were skeptical but came away pleasantly surprised. As Amanda Gefter wrote in *New Scientist*, she expected the scents from her "microphone" to eventually merge into a fog, but was amazed that "each smell was as precise as a staccato note, lasting only a few seconds before giving way."[45] And Avery Gilbert found the experience "totally compelling," noting in particular that "the sensory connections between odors and musical themes were easily perceived."[46] As one might expect, there were considerable differences in people's individual experiences, especially in their reaction to the role of the narrative, a summary of which was presented during the prologue. In addition to a précis of the narrative, the prologue offered a kind of trial run by flashing the name of each protagonist, Absolute Zero, Meretricious Green, Industry, and so forth, on a screen as their scents were transmitted to the audience through the scent tubes and the character's associated musical themes were played. Then the lights were turned out and the opera proper ran for fifteen minutes, ending with a kind of curtain call where each "character," its scent, and its music took a bow.

Interviews with some members of the audience several months later found an interesting division in the way people remembered experiencing *Green Aria*. Although some appreciated being given a sketch of the scenario and a trial run of the scents and music and recalled making a concerted effort to follow the interweaving of the two, others said the directions given in the prologue seemed to demand too much concentration, and these participants simply let the scents and sounds sweep over them, evoking feelings and memories at random. Yet many of those who did try to follow the scenario expressed surprise that they had actually been able to perceive the differences among the scents and could grasp the relation of each scent to the musical themes as well as they did.[47] Although these interviews involved only a handful of people and were well after the fact, they at least suggest that although artists can orchestrate odors and other elements of a production in a way that many audience members will accept and attempt to track, other people are likely to follow their own lead and enjoy just as satisfying an aesthetic experience. After all, audience experiences of symphonies also vary along similar lines with some highly knowledgeable people able to follow the score and perceive interpretive dynamics and at the other end people who simply like to sit back and let the sounds sweep over them.

The two reviews of *Green Aria* that I have already cited found it to be overall a highly satisfying aesthetic experience. Amanda Gefter wrote, "I was awed by the innovative concept and the meticulous execution." And Avery Gilbert concluded that *Green Aria* demonstrates that "an elegant technology in the hands of truly creative people can deliver an original, beautiful, sensory performance that enlivens the mind as well." Surely a sensory experience that is original, beautiful, and cognitively engaging should meet almost any definition of a successful artistic and aesthetic achievement. More specifically, *Green Aria* offers ample evidence to justify setting aside the claims of Beardsley and Scruton that odors and the sense of smell cannot be used to create works that involve tension, climax, and resolution and thus be considered genuine works of art and the experience of them genuinely aesthetic.

Notes

1. David Roberts, *The Total Work of Art in European Modernism* (Ithaca: Cornell University Press, 2011).
2. Zachar Laskiewicz, "From the Hideous to the Sublime: Olfactory Processes, Performance Texts and the Sensory Episteme," *Performance Research* 8, no. 3 (2003): 57.
3. Tiffany Stern, "Taking Part: Actors and Audiences on the Stage at Blackfriers," in *Inside Shakespeare: Essays on the Blackfriars Stage*, ed. Paul Menzer (Selinsgrove: Susquehanna University Press, 2006), 35–53.
4. The Tonkin passage is cited in Jonathan Gil Harris, *Untimely Matter in the Time of Shakespeare* (Philadelphia: University of Pennsylvania Press, 2010), 133. The last line of Olfactus's plea is strikingly parallel to Montaigne's comment on the use of incense.
5. Harris, *Untimely Matter*, 119–39.
6. Holly Crawford Pickett, "The Idolatrous Nose: Incense on the Early Modern Stage," in *The Performance of Religion on the Renaissance Stage*, ed. Jane Hwang Degenhardt and Elizabeth Williamson (Farnham, UK: Ashgate, 2011), 19–37.
7. Sally Banes, "Olfactory Performances," in *The Senses in Performance*, ed. Sally Banes and André Lepecki (New York: Routledge, 2007), 350.
8. Mary Fleischer, "Incense & Decadents: Symbolist Theatre's Use of Scent," in Banes and Lepecki, *Senses in Performance*, 105.
9. Kristen Shepherd-Barr, "'Mis en Scent': The Théâtre des Art's Cantique des Cantiques and the Use of Smell as a Theatrical Device," *Theater Research International* 24, no. 2 (1999): 152–59, https://doi.org/10.1017/S0307883300020770.
10. František Deák, "Symbolist Staging at the Théâtre d'Art," *Drama Review* 20, no. 3 (1976): 120–22.
11. Claude Schumacher, ed., *Naturalism and Symbolism in European Theater, 1850–1915* (Cambridge: Cambridge University Press, 1996), 18.
12. Sally Banes also mentions ironic, contrastive, and distancing uses in "Olfactory Performances," 351–52. Michael McGinley and Charles McGinley, "Olfactory Design Elements in the Theater: The Practical Considerations," in Henshaw, *Designing with Smell*, 219–23.
13. McGinley and McGinley, "Olfactory Design Elements," 224.
14. Stephen Di Benedetto, *The Provocation of the Senses in Contemporary Theater* (New York: Routledge, 2010), 109.
15. Susan Feagin, "Olfaction and Space in the Theater," *British Journal of Aesthetics* 58, no. 2 (2018): 137.
16. Feagin, "Olfaction and Space in the Theater," 141.
17. Feagin, "Olfaction and Space in the Theater," 146.
18. Violaine de Carné, "Théâtre olfactif: Un itinéraire singulier," in Jaquet, *L'Art olfactif*, 255–70.
19. For a discussion of the characters and their signature odors and the challenge of creating them, see the essay by the perfumer for the play, Laurence Fanuel, "'Création

olfactive 'Hors Piste,' L'ordorisation théâtrale des *Parfums de l'âme*," in Jaquet, *L'Art olfactif*, 271–86.
20. Sophie Domisseck and Roland Salesse, "Le spectateur olfactif: La reception de la pièce *Les Parfums de l'âme* par le public," in Jaquet, *L'Art olfactif*, 287–98.
21. Matthew Reason, "Writing the Olfactory in the Live Performance Review," *Performance Research* 8, no. 3 (2003): 73–84.
22. Dominique Paquet, "L'acteur olfactif," in Jaquet, *L'Art olfactif*, 244.
23. Carné, "Théâtre olfactif," 269.
24. For an excellent brief discussion see Gilbert, *What the Nose Knows*, 158–62.
25. Gilbert, *What the Nose Knows*, 152–58, 163–69.
26. Chris Fujiwara, "Wake Up and Smell the New World," *Film Comment*, July–August 2006, https://www.filmcomment.com › article › wake-up-and-smell-the-new-world.
27. See Marian Bendeth's January 2007 interview in her blog "Basenotes." "Interview with Les Christophs: Christophe Laudamiel and Christoph Hornetz: Perfumers of Le Coffret," https://www.basenotes.net.
28. Lauri Pike, "'Perfume: The Story of a Murderer,' to Reshow with Newly Created Scent Track," *Hollywood Reporter*, November 4, 2013, https://www.hollywoodreporter.com/news/perfume-story-a-murderer-reshow-652912.
29. Nöel Carroll, *Theorizing the Moving Image* (Cambridge: Cambridge University Press, 1996), 141–44.
30. Vivian Sobchack, *Carnal Thoughts: Embodiment and Moving Image Culture* (Berkeley: University of California Press, 1995), 65
31. Cynthia Freeland, "Gustatory Film Perception," paper presented at the Annual Meeting of the American Society for Aesthetics, New Orleans, November 2017. I am grateful to Professor Freeland for sharing her paper.
32. Although Antunes does not reject the relevance of association and imagination based approach of Sobchack, he wants to show that multisensoriality is perceptual before it is "intellectual, imagined or remembered." Luis Rocha Antunes, *The Multisensory Film Experience: A Cognitive Model of Experiential Film Aesthetics* (Bristol, UK: Intellect, 2016), 4.
33. See Roger Ebert's review at https://www.rogerebert.com/reviews/perfume-the-story-of-a-murderer-2007.
34. Carroll, *Theorizing the Moving Image*, 49–55.
35. Kevin Sweeney, "Medium," in *The Routledge Companion to Philosophy and Film*, ed. Paisley Livingston and Carl Plantinga (New York: Routledge, 2009), 181–83.
36. Laskiewicz, "From the Hideous to the Sublime," 64.
37. Jaquet, *Philosophie de l'odorat*, 201.
38. Proust, *Remembrance*, II, 842.
39. Michel Grinand, "Une Nuit américaine à écouter et humer," *AvantChœur* 3951 (June 4, 2015), https://www.avantchoeur.com
40. Jennifer Weil, "Album Mixes Music with Scent," *WWD*, May 7, 2015, https://www.wwd.com.
41. Clarissa Sebag-Montefiore, "Scent of Memory: Smell and Classical Music Prove an Intoxicating Combination," *The Guardian*, November 11, 2016.

42. Coady Green, "Scriabin in the Himalayas," *Classical Music.Com*, September 29, 2015, http://www.classical-music.com/article/scriabin-himalayas.
43. Anthony Tommasini, "Opera to Sniff At: A Score Offers Uncommon Scents," *New York Times*, June 1, 2009.
44. Avery Gilbert, "Green Aria—a Scent Opera: Bravo!," on his website *First Nerve: Taking a Scientific Sniff at the Culture of Smell*, June 3, 2009, https://www.firstnerve.com › 2009/06.
45. Amanda Gefter, "Green Aria: An Opera for Your Nose," *New Scientist*, June 5, 2009, http://www.newscientist.com/article/dn17236-green-aria-an-opera-for-your-nose.html.
46. Gilbert, "Green Aria—a Scent Opera: Bravo!" *First Nerve*, June 3, 2009, https://www.firstnerve.com › 2009/06.
47. Sophie Domisseck and Roland Salesse, "Peut-on raconteur une histoire rien qu'avec des odeurs? La gageure de 'Green Aria: A Scent Opera,'" https://prodinra.inra.fr.

Interlude
Smeller 2.0 and the Osmodrama

There is an often-quoted passage in Aldous Huxley's dystopian novel *Brave New World* of 1932 that describes a "scent organ playing a delightfully refreshing Herbal Capriccio—rippling arpeggios of thyme and lavender, of rosemary, basil, myrtle, tarragon."[1] Actual working scent organs finally began to appear in this century. Earlier I mentioned Peter de Cupere's *Olfactiano*, on which he played a scent sonata for a Brussels festival in 2004. The scent organ used in Laudamiel and Matthew's 2009 *Green Aria: A Scent Opera* was clearly another step forward. In 2012, Wolfgang Georgsdorf unveiled an even larger and more elaborate "scent organ" in Linz, Austria, the product of years of research and trials, including an earlier *Smeller 1.0*. But it was not until 2016, when Georgsdorf organized a two-month-long demonstration of *Smeller 2.0* in Berlin under the title *Osmodrama: Storytelling with Scents*, that his creation showed its true versatility by allowing him to accompany poetry readings, films, musical works, and produce pure scent "narratives." In the preface to the program for the 2016 series, he wrote of "the dream that never left me throughout my years of artistic work: to stage smells in sequences."[2]

Smeller 2.0 is an enormous instrument (weighing over a ton) that Georgsdorf regards as both an artwork itself (a "functional sculpture") and as an instrument for "timebased composing and storytelling."[3] And indeed it is a striking assemblage: large, silvery interweaving pipes visible behind a perforated screen with a cluster of 64 bronze-toned outlets in the center, one for each of the smell canisters attached to the pipes. The organ took up the entire wall of a moderately large room at the Martin Gropius Bau in Berlin when I saw it in 2018 and was able to experience a composition called *Autocomplete*. *Smeller 2.0* produces a stream of air at a constant speed, carrying the changing scents and their combinations to every corner of the room, but it is also installed in such a way that the ambient air of the space is "exchanged 1–2 times per minute, so there is no build-up or undesirable mixing in the temporal sequence of the changing chords."[4] Thus, using a different setup than the organ produced for Mathews and Laudamiel by Fläkt-Woods for *Green Aria*—which sent tubes to each seat—*Smeller 2.0* also solves the technical problem of odor buildup and unwanted mixing that doomed earlier attempts at accompanying a film or telling a narrative with smells. A reviewer for the Berlin paper *Die Zeit* found the machine altogether astounding and remarked that the attempt to coordinate scents with a contemporary film during the 2016 event, despite some occasional timing problems, was an impressive effort.[5]

Smeller 2.0 has a MIDI keyboard that allows Georgsdorf to play improvised pieces, something he did at the 2016 *Osmodrama* demonstration series. For example, during the 2016 event he joined in a jam session with the Berlin Improvisers Orchestra following their performance of "Orchestral Whifftracks" (a piece that integrated preprogramed smells from *Smeller 2.0* with music by Stephen Crowe). Georgsdorf also performed live smell accompaniments at the 2016 event for two poets and a novelist reading from their works. In the case of the poems, Georgsdorf played a prologue and epilogue of appropriate scents, but for the chapter from a novel by Julia Kissina, he attempted to keep up with the changing images and actions, a daunting task since smells travel slower than sound, with the result that Georgsdorf had to anticipate the spoken words by several seconds. Among the more affecting experiences during the 2016 series according to some comments was the live performance on the final night of a sound collage by Carl Stone accompanied by smells played by Georgsdorf.[6]

In some ways, the most important work on the 2016 program was not the various hybrids of smell with film, poetry, or music, but Georgsdorf's own osmodrama of pure smells. Called *Autocomplete: Synosmy* (scent symphony), it ran for over fifty minutes and was described by Georgsdorf as

> a pure synosmy, a complex composition of smell sequences. The audience automatically completes olfactory storytelling in a reflex-like reaction to the smells flowing through the room. Spheres of familiarity, trusted smells from childhood can shift into eerie notions within two breaths.[7]

There were no program notes to accompany the 2016 *Autocomplete* so that visitors were left to their own devices to find a story. I had a chance to experience a much shorter (twelve-minute) version of *Autocomplete* as part of an exhibition of contemporary "immersive art" held at the Martin Gropius Bau in Berlin in 2018. Georgsdorf called this synosmy *Quarter Autocomplete (Evolution in 12 Minutes)* and had it run three or four times an hour. Here again, there were no program notes to guide one's experience, and I found myself trying to guess the identity of the odors as they came. I sat through it three times, trying to find or construct a narrative. Unfortunately, I was never able to construct a pattern, although this may have been a result of my own olfactory or imaginative deficiencies. Yet given Georgsdorf's general description for the 2016 version of *Autocomplete*—"the audience automatically completes olfactory storytelling in a reflex-like reaction to the smells"—perhaps his idea of a "story" is closer to simply enjoying a sequence of smells and free-floating personal associations.[8] Indeed, in a short video documentary on the Gropius Bau presentation of *Smeller 2.0* made later in the summer of 2018, snippets of a few audience associations are quoted (burning tires, mushroom in a forest), but no one describes an actual story. One person said *Quarter Autocomplete* was "like an abstract painting, but through the nose," which is perhaps a good way to think of it; when I experienced it last summer, I may have been trying too hard to find a "story."[9]

Yet in one of Georgsdorf's earlier plans for smell sequences, he did offer some indication of what the smell stories in that work were to be about. For example, he said he was working on a narrative called *Childhood* that would include a movement called "My First Circus," consisting of smells of popcorn, elephant droppings, cotton candy, sawdust, caramel apples, and so on.[10] It would have been interesting to try to follow a sequence of smells with the help of such explicit clues. *Autocomplete* seems to be attempting for smell something parallel to pure instrumental music rather than program music that tells a story or seeks to create a connection to a scene or situation. But even many notable composers of program music have provided a descriptive title or even a few program notes (Beethoven's titles for the movements of the Sixth Symphony) to help their listeners, and this would seem even more important with smell compositions, an art form few people have ever experienced.

An interesting nonart use of *Smeller 2.0* occurred in conjunction with the 2018 Gropius Bau presentation. Professor Thomas Hummel of the Interdisciplinary Center for Smell and Taste of Dresden University along with two colleagues from clinics in Berlin set up an experiment to test the effect of *Smeller 2.0* on people suffering from various degree of smell loss by having them undergo repeated exposures to *Quarter Autocomplete (Evolution in 12 Minutes)* for periods of thirty to sixty minutes a day over several weeks.[11]

Smeller 2.0 is a fascinating and impressive machine, beautiful to look at and remarkably versatile given the enormous number of smells and smell combinations it can generate. Taken together with the achievements of Matthews and Laudamiel's *Green Aria*, which used a different technology, we clearly seem to have reached a stage in olfactory musical and narrative experimentation where more complex hybrids of drama and music will be possible, and where "pure" scent narratives can be constructed, although their appreciation will require educating a new audience.

Green Aria and Georgsdorf's *Smeller 2.0* form a natural transition from our discussion of the uses of scents with theater, film, and music to the next chapter on the kinds of scent art hybrids that are intended primarily for art galleries or museums.

Notes

1. Aldous Huxley, *Brave New World* (New York: Bantam Books, 1949), 112.
2. Wolfgang Georgsdorf, "*Osmodrama: Storytelling with Scents*," July 15–September 18, 2016, http://www.osmodrama.com. This website has many sub-sections, one of which is entitled "Osmodrama Festival Berlin 2016."
3. Wolfgang Georgsdorf, "Osmodrama via Smeller 2.0," https://smeller.net › about.
4. Georgsdorf, "Osmodrama via Smeller 2.0." https://smeller.net › about.
5. Rabea Weihser, "Die Wahnsinnlichkeitsmachine," *ZeitOnline*, July 24, 2016, https://www.zeit.de/kultur/kunst/2016-07/smeller-osmodrama-wolfgang-georgsdorf/seite-2. I am grateful to Mădălina Diaconu for calling my attention to this source.
6. One visitor found the smells a bit repetitive and, although pleasant, little more than "a coloristic play of surfaces." Sam Johnstone, "Osmodrama/Review," *The Cusp*, September 21, 2016, https://www.thecuspmagazine.com/reviews/osmodrama-review/.
7. Georgsdorf, "Osmodrama: Molecules Turn to Emotions" https://osmodrama.com.
8. Georgsdorf, "Osmodrama: Molecules Turn to Emotions."
9. Georgsdorf, "'Quarter Autocomplete'/Osmodrama via Smeller 2.0." These comments can also be followed on Video No.16-Osmodrama at Martin Gropius Bau 2018 which can also be watched on You Tube, https://osmodrama.com/videos/osmodrama-im-martin. Georgsdorf dates his comments on the video December 2018/February 2019.
10. Georgsdorf, "Osmodrama via Smeller 2.0." In 2012 at a Linz cultural center Georgsdorf ran a synosmy he called *Childhood* that integrated scent sequences with music, https://osmodrama.com/.../eine-kindheit-a-childhood-2012-wolfgan.
11. Georgsdorf, "Quarter Autocomplete." Georgsdorf has made his machine available for other clinical studies that plan to measure the effect on brain mass in olfactory areas and whether exposure to *Smeller 2.0* can modify the symptoms of dementia or Alzheimer's, https://osmodrama.com/videos/osmodrama-im-martin.

10
Sublime Stenches
Contemporary Olfactory Art

When visitors entered Anicka Yi's 2015 exhibition *You Can Call Me F* at New York's Kitchen Gallery, the first thing they saw in the darkened room was a six-foot-long rectangular glass dish on whose agar bed bacteria and fungus were growing. Yi had created this work, *Grabbing at Newer Vegetables,* using Q-tip swabs to gather mouth samples from one hundred art world women in order to reflect female networks. With the help of a biologist, she put them under Plexiglas so that they would multiply over the course of the exhibition and in that way symbolize the contagion the public fears when women band together. But one could also get a whiff of an unusual scent that drifted on the air coming from a second room, where three large, colorful "quarantine tents," each containing a variety of objects, among them scent diffusers set in motorcycle helmets. The odor combined the somewhat musky smell emitted by the bacteria in the giant dish with a more antiseptic note. The latter came from the famous Gagosian Gallery in uptown New York. The Gagosian's thin smell had been captured by one of Yi's collaborators using a small headspace machine while Yi distracted the guard.[1]

Of including in her work the smell of the Gagosian's largely deodorized white cube space, Yi says, "My interest in smell is very political, critiquing the regime of vision our society imposes on us, re-thinking how art should work on us."[2] Indeed, one of the reasons Yi makes odors a prominent part of her artworks is that they force the viewer to "participate by having to be there physically to smell the work."[3] One of her previous exhibitions, *Divorce* (2014), for example, included two clothes dryers that visitors were encouraged to open and stick their head into. Both dryers held odors that struck a *New York Times* critic as highly unpleasant: "One reeks of fried food and wet cardboard, the other of a peat bog," things that suggested "domesticity gone awry."[4]

Yi's use of odors and the sense of smell in her works is typical of contemporary scent art in two ways. First, her works combine odors with other materials that engage not only the sense of smell but also the sense of touch and sometimes hearing as well as vision. Most contemporary scent or olfactory art, as Jim Drobnick says, is "by necessity a hybrid form. Even when artists seek to isolate smell as a pure aesthetic experience, its deployment depends on other factors,

Art Scents. Larry Shiner, Oxford University Press (2020) © Oxford University Press.
DOI: 10.1093/oso/9780190089818.003.0021

technology, architecture, installation, or performance, to name just a few."[5] The second way in which Anicka Yi's practice typifies contemporary olfactory art is that it is motivated by a desire to engage audiences more directly and intimately than traditional visual arts that more easily lend themselves to distancing and reproduction. Moreover, some of the very characteristics of odors and the sense of smell that philosophers in the past have used to argue that smell cannot be involved in the creation and appreciation of serious art turn out to be advantages in the view of many contemporary artists. As Drobnick pointed out in an influential early essay on smell in contemporary art, artists are drawn to smell as a medium on the one hand because the volatility and evanescence of odors demand physical attention from the audience, and on the other because odors and scents arouse strong, highly personal emotional associations.[6] Moreover, the supposed "animality" of smell and its close association with bodily functions also makes it an ideal expressive medium for many themes important to contemporary performance artists, such as identity and sexuality.[7] Finally, some artists see the very difficulty of exhibiting and collecting scent art as one of its advantages. Brian Goeltzenleuchter has summed up his reasons for using smells this way: "Through its volatility and immateriality olfactory art inherently challenges commodification, collection and archiving. It is this invisible, involuntary nature of smell that prompted me to use it in my art."[8]

Some Types of Olfactory or Scent Art

In order to better address the issue of whether olfactory or scent art constitutes a distinct art form, we need to consider briefly some of the types of artworks that are normally lumped together as olfactory art.[9] These experiments are so multifarious, however, that it is hard to find an organizing principle for sorting them into categories. The art historian Francesca Bacci has made a first attempt at an organizing framework with a three-part matrix for what she calls "scent-ific art": (1) "art with a scent diffused by a scented object in sight," or (2) "art with scent disbursed in the environment [but no] scent emitting object," or (3) "art that represents smell, but does not emit any scent."[10] (She adds one further criterion cutting across these: whether the scent in the artwork is "intentional or accidental.") Although her matrix is a good start, as indicated earlier, I believe it is misleading to include artworks that simply represent smell, unless one adds some further qualification. Moreover, including works of purely visual or auditory representation in a typology of contemporary olfactory art (as opposed to the "olfactory arts" in general) seems to run counter to the many statements by artists such as Yi who claim they use actual odors precisely because they want audiences to be present and interact physically with their works.

My point in raising these questions is not to engage in an academic quarrel over classification schemes, since all such schemes are largely heuristic, but to help us better understand whether or not the works gathered under the category in question, whether it is called scent-ific art, scent art, or olfactory art, finally have enough in common to qualify as a distinct art kind. Thus, rather than propose a competing matrix to Bacci's, I will take a pragmatic approach and organize my brief survey of salient examples of olfactory or scent art after Drobnick's idea that most olfactory art is a hybrid that includes some kind of support, whether it be another material, technology, or an established medium. Of course, cutting across all the differences among olfactory artworks as defined by their type of support are the other criteria I proposed earlier for olfactory arts in general, namely that the actual odors are used intentionally and give the work its distinctive character. Thus, in Anicka Yi's gallery installation, *You Can Call Me F*, the odor is both intentional and essential to the nature of the work, even if it is not the dominant element. As will soon become apparent, many of the following subcategories overlap, and individual works could be placed in more than one (I omit hybrids of actual odors with paintings or poems, which are relatively rare and those "scent organ" artworks such as the *Olfactiano* or *Smeller 2.0*, that have already been discussed).

Scent Sculptures

The best-known sculptures involving smells are those of Ernesto Neto, such as *Mother Body Densities* (2007), in which huge Lycra sacks, filled with aromatic spices, are hung from the gallery ceiling, their aroma often pervading an entire museum. Peter de Cupere's ice sculpture of the Virgin, *The Deflowering*, melted within an hour, but its core contained a synthesized vaginal scent and visitors were encouraged to dip their fingers into the "holy water." Oswaldo Macia makes haunting sculptures that combine scent and sound such as *Transition* (2014), which joins scents from plants about to be extinct with wild animal calls.

Installations

Installation art generally refers to works that often occupy an entire gallery space so that one must physically enter the work. Obviously, many of Anicka Yi's works are conceptual installations, such as her 2011 *Auras, Orgasms, and Nervous Peaches*, a work that invited visitors to enter a small tiled room whose walls oozed an odorous material. Gwenn-Aël Lynn's installations involve scent-sound machines, which he wraps in muslin and, in some cases, as in *Audiofactory Creolization* (2013), to which he adds cast noses and ears that give his installations the feel of a collection of hanging sculptures—whose scents and sounds are triggered by visitors' movements.[11] Nobi Shioya's elaborate 2003 installation, *7S* (for the seven deadly sins), involved generic photographs for each sin, with a cast jug hanging in front of it emitting scents designed by a perfumer, and on the wall opposite each jug, a black canvas impregnated with one of the perfumer's commercial fragrances.

Performance

In Angela Ellsworth's *Actual Odor* (1997), the artist wore a black cocktail dress soaked in her own urine to an art opening, fanning herself with a fan that had the words "Actual" and "Odor" printed on either side. A less public kind of performance was Rachel Morrisson's *Smelling the Books*, in which she began systematically smelling each of the books in the Museum of Modern Art library in 2010 and recording her impressions. In de Cupere's *Black Beauty Smell Happening* (1999), male and female models wore black cat suits with cutouts into which de Cupere sprayed a perfume so that members of the audience had to draw their noses close to the "smell zones" to fully experience the work.

Participation

"Participatory art" means the audience is required to take some kind of action to complete the work. Many of the works I have already mentioned have a participatory aspect, but there are other works whose primary focus is on public interaction, such as Haug's *U-deur*. For *World Sensorium* (2000), Gayil Nalls asked hundreds of public officials around the world to identify their country's most typical smell, then had the smells synthesized and put them on scratch and sniff postcards that were dropped on the crowds in Times Square at midnight of the millennium. Jenny Marketou's *Smell It* (2008) was a mural-size map of Philadelphia on a gallery wall. Visitors were asked to walk around the neighborhood and return to mark notable smell sources on the map with colored markers. In Brian Geltzenleuchter's *Sillage* (2016) at Baltimore's Walters Museum, arriving visitors were asked to identify the area of the city where they resided and a member of the museum staff sprayed a scent on their wrist that the artist had created to represent that residential area.

Perfumes

This category includes works like Clara Ursitti's *Self-Portraits in Scent* (1994–) that we encountered earlier, as well as Janna Sterbak's *Perspiration: An Olfactory Portrait* (1995), based on the body odor of Sterbak's partner (visitors were asked to rub some of it on their own skin). Martynka Wawrzynik's more recent *Smell Me* (2012) is another self-portrait in smell, using odors derived from her own skin, hair, sweat, and tears, which are used to scent a chamber visitors can enter to inhale her smell. In *Lure* (1996) by Maciej Toporowicz, the scents of young Thai prostitutes were combined with incense smells from Buddhist temples to create a perfume that he presented in a gallery with mock advertisements, the work as a whole meant to draw attention to Western fantasies about Asian exoticism that end up supporting sexual slavery.

Atmospheres

These are works where odors are diffused into a largely empty gallery space (corresponding to Bacci's odors with no visible object source).[12] A good example is Kim Jeong A's *Before the Rain* (2011), presented by DIA Foundation at the Hispanic Society in New York and meant to capture the violent atmosphere that precedes a heavy Pacific rainstorm. In Maki Ueda's *Invisible White* (2013) the gallery is semidark and you can barely see so that you switch to smell, touch, and hearing to get oriented. Wolfgang Laib is noted for his claustrophobia-inducing rooms lined with beeswax that give off an intense odor. Matt Morris uses perfumes to create subtle atmospheres, as in a recent group show where all the other works were addressed to vision and Morris had the gallery attendant wear a perfume he had selected and mingle with the visitors.[13]

Olfactory/Scent Art and Sound Art as Art Forms

This brief survey should have given you some idea of the enormous variety of olfactory art. On the other hand, the very variety casts doubt on whether olfactory or scent art constitutes a distinct art medium or art form. The leading olfactory art critic and curator, Jim Drobnick, has written that "olfactory art . . . is not a single-sense phenomenon, it just means that the artworks being discussed have a pronounced olfactory dimension."[14] Accordingly, one could argue that such works are better understood under the rubrics that name the art forms of which they are hybrids. Yet even if one cannot give a classic "real" definition of olfactory or scent art using a set of necessary conditions that are jointly sufficient, the category of olfactory art is characterized by the same three criteria as olfactory arts in general (the *intentional* use of *actual* odors that are *distinctive-making*) plus the proviso that an olfactory artwork is intended for an art gallery or art museum setting or a public art situation. Works of olfactory or scent art, then, involve an *intention* to use *actual* odors in a *distinctive-making* way that typically gives the resulting artwork its effect in a recognized *visual art setting*. I believe such a category of olfactory or scent art can be heuristically useful for reflecting on the way such works affect our aesthetic experience. Moreover, there exists a strong precedent and analogy for such a category of olfactory or scent art, given the widespread acceptance of the category "sound art," whose status as an art form has also been controversial but is now relatively well established among art curators, critics, and historians. Like scent art or "olfactory art" in the singular, which is a subcategory within the olfactory arts in general, sound art is often treated as a subcategory within the general span of all the auditory arts aimed at our sense of hearing.

In his book *Background Noise: Perspectives on Sound Art*, Brandon LaBelle emphasizes that we live in a sea of sounds, coming from every direction, including from our own bodies, sounds that may be a barely audible murmur but at other times may be almost deafening.[15] Yet, as he points out, most of the sounds that fill our every waking and sleeping moment are not "heard" in the sense of listened to but are part of a background that can be brought forward. This is also true of smells; almost everything around us, including our own body, is constantly emitting odors of various intensities from the barely detectable to the overwhelming. And just as most of the ambient sounds of our environment are not consciously heard, so most of the scents that accompany our daily life are not smelled, as the neuroscientist Peter Köster says.[16] A common aim of sound art and olfactory art is to make us notice the unnoticed.

Forerunners of sound art include the Italian Futurist Russolo with his "noise intoners" and his manifesto calling for an "Art of Noises" (1913), although it was John Cage's theories and works such as *4′ 33″* of 1952 with its moments of "silence" inviting the audience to attend to the random sounds of the concert environment that marked a decisive step beyond traditional music. By the 1960s and 1970s visual artists, some of whom had followed Cage's lectures, began taking experiments with sound into visual art galleries and museums or into the streets in the form of public installations.[17] An early example of the latter is Max Neuhaus's *Times Square* (1977–1992, 2002–present), which consisted of speakers located beneath a metal vent on Broadway between Forty-Fifth and Forty-Sixth Streets, emitting a soothing drone just audible beneath the cacophony of Times Square. As LaBelle remarks, "With sound installation and the works of Neuhaus and others, sound art finds definition, demarcating itself from the legacy of experimental music and entering into a more thorough conversation with the visual arts."[18] Of course, not everyone accepts the idea that what has emerged in the course of this history is an independent art form that should be called "sound art." They complain that the many things called sound art are too heterogeneous and are often hybrids with other art forms, an objection one might also make against a category of scent or olfactory art.[19]

Yet sound art theorists like LaBelle, Christophe Cox, Carmen Pardo, and Alan Licht find neither heterogeneity nor hybridity a sufficient objection to the usefulness of the category of sound art, although all admit that sound art is a vast field, difficult to define, with unfixed boundaries.[20] Cox proposes that we think of it as "works of art that focus attention on the materiality and transmission of sound . . . presented in galleries, museums and public spaces," a definition parallel to Drobnick's definition of olfactory art and what I am calling olfactory/scent art in the singular.[21] Moreover Cox suggests, as I have, that we not think of "sound art" as "exhaustive or . . . precise, but merely heuristic."[22] Both "sound art" and "scent art" can be useful ways of grouping certain kinds of artworks for purposes of reflection and discussion, always keeping in mind that there are innumerable other ways of sorting things into categories depending on one's purpose.

The History of Olfactory/Scent Art

There are also a couple of pragmatic criteria for judging the usefulness of thinking of scent or olfactory art as an art category along the lines of "sound art." First, both sound and scent art share parallel histories; second, there are artists involved in both tendencies who self-identify as "sound artists" or as "olfactory/scent artists." Art historians have already begun to construct a history for scent or olfactory art. Both contemporary scent/olfactory art and sound art can trace their recent rise to the 1960s, which saw a turn toward a pluralization of materials and techniques in contemporary art. Moreover, just as historians of sound art have found precedents going all the way back to Futurism, so the historian Caro Verbeek has found forerunners of olfactory art in both Futurism and in the work of Marcel Duchamp. In addition to Carra's 1911 manifesto, "The Painting of Sounds, Noises and Smells," that we discussed earlier, Verbeek has called attention to the recently discovered manifesto "The Art of Scent" (c. 1916) by the Italian Futurist Ennio Valentinelli, who declares:

> We must refine our nostrils. We have to start conquering the senses that have been elusive till now, we have to assimilate a new world and create a new lyricism. We have to force the reluctance of the senses when it comes to stench, we have to imprison, withhold, and enjoy, overcome nausea.[23]

He sounds like Sissel Tolaas a hundred years ago!

Duchamp's most important contribution to the history of olfactory art was his design of the 1938 International Surrealist Exposition in Paris that involved odors. Duchamp kept the room semidark, carpeted the floor with oak leaves, ferns, and grasses, and had a water-filled pond with water lilies and reeds installed as part of it. In addition to the odors that came from these elements, odiferous bags of coal hung from the ceiling and a coffee-roasting machine was going that, as Duchamp remarked in a later interview, gave a "marvelous odor in the room. And that was part of the exposition, it was surrealist after all."[24]

Although there were a few works involving odors exhibited in the 1940s and 1950s, the next notable works that foreshadowed today's proliferation of olfactory art did not appear until the 1960s and 1970s, by which time the so-called postmedium or postdisciplinary turn in the art world was in full swing. Edward Kienholz's *The State Hospital* (1966), for example, consisted of a prison-like door with bars through which one looks into a spare room where two naked figures lie in bunk beds, their arms strapped to the railings, their heads fishbowls. Crucially, Kienholz infused the room with a disagreeable hospital disinfectant smell that greeted any visitor who peered through the bars.[25] A 1975 video installation by Bill Viola had a boiling cauldron of eucalyptus leaves in front of a video of a woman dropping leaves into a similar boiling pot, an ingeniously simple provocation to think about the different communicative powers of vision and smell. Cildo Meireles's 1980 *Volátil* made visitors enter barefoot into a darkened room with a single burning candle on a table, but what gave the installation its anxiety-inducing power was the smell of natural gas that intimated things might blow up at any moment. By the late 1980s and early 1990s a number of installation artists were using materials that gave off distinct odors, such as Jana Sterbak's *Meat Dress for an Albino Anorexic* (1987). The rotting cow's head crawling with flies that was part of Damien Hirst's *A Thousand Years* (1990) released a memorable stench. By the end of the 1990s there were enough such works, in which smell was prominent and seemed intentional, that the term "olfactory art" came to be a focus of discussion by curators and critics like Jim Drobnick in his pioneering article "Reveries, Assaults, and Evaporating Presences."[26]

Thus, although what I am calling olfactory or scent art has only gained a modest place in today's art world, it clearly has a history, and one that runs parallel to the history of sound art, which combines sound with sculpture, installation, performance, and other kinds of conceptual art. Although both sound art and scent art have emerged out of visual art's postmedium turn, sound art is by comparison relatively well established and accepted by the art world, whereas olfactory art is still making its way. Among recent recognitions of the place of smell in the arts, apart from the 2015 *Belle Haleine: The Scent of Art* exhibit, is the inclusion of smell in exhibitions at the Tate Britain (*Sensorium*, 2015), the

Albright-Knox (*Out of Sight! Art of the Senses*, 2017–2018), and the Cooper-Hewitt (*The Senses: Design beyond Vision*, 2018), as well as the award of the prestigious Hugo Boss Prize to Annick Yi in 2016, followed by a show of her work at the Guggenheim in 2017.

Some Self-Identified Scent or Olfactory Artists

A second indicator that olfactory or scent art is becoming a new art category, similar to sound art alongside other pragmatic groupings of the arts, is the fact that there are several artists who not only consistently use smell in their work but who sometimes refer to themselves as olfactory or scent artists, such as Clara Ursitti, Peter de Cupere, Oswaldo Maciá, Brian Goeltzenleuchter, and, of course, Christophe Laudamiel and Wolfgang Georgsdorf.

Clara Ursitti, whose *Self-Portraits in Scent* and *Pheromone Link* we have already encountered, is one of the pioneers of olfactory art and has been combining odors with photography, installation, performance, and video work since the 1990s. Like so many other scent artists, she sees her work as subverting the Western disdain for smell and challenging some of the social prejudices associated with it. An example of her kind of sharp and humorous commentary is *Poison Ladies* (2013). She asked twenty-eight women over sixty to attend an art opening wearing Dior's powerful perfume *Poison*, which reviewers have called "extreme," "titanic," "punk rock," or "a camphor-and-tuberose beast." As the women gradually entered the room unannounced, the average age rose substantially as the air filled with *Poison*.[27]

Then there is Peter de Cupere, who has been as prolific and varied as any artist working with smell. As we have seen, his works have ranged from inventing and playing a scent piano, to scent "paintings" and sculptures, to performance works, to installation and participatory works. Although de Cupere has expressed some discomfort with the label "scent artist" as too limiting, he has been one of the most visible and vocal advocates for olfactory art.[28] In 2014, he issued an "Olfactory Art Manifest" that closes, as all good manifestoes should, with a call to action:

> This manifest calls all artists to enter into the smell experiment.
> This manifest calls all curators, museum directors . . . to show more olfactory art.
> This manifest calls EVERYONE to smell harder![29]

Oswaldo Maciá has been making works involving scents for over twenty years. In his 2013 "Manifesto for Olfactory-Acoustic Sculpture (A Guide to Creating an Uncomfortable Question)" he says that his works seek to downplay the visual by making it merely a "plinth" for a kind of sculpture that "fills space with volumes of sound and smell."[30] One of his most complex and suggestive works is *The Library of Cynicism* (2013), which draws its title from the Greek philosophical movement associated with Diogenes. A series of aquarium-like tanks emit the smells, one of them labeled "dog" ("Cynic" in Greek is literally "doglike") at the same time that the sounds coming from a series of small speakers are derived from animal species on the brink of extinction.[31]

Brian Goeltzenleuchter characterizes his current artwork as focusing on "the way in which personal and cultural narratives can be expressed through the sense of smell."[32] He has created collaborative works with writers (*Olfactory Memoirs*) and musicians (*Odophonics*) and participatory works such as (*Sillage*). His *Institutional Wellbeing* projects use scents to comment on and hopefully improve the functioning of institutions, usually art museums. His *Institutional Wellbeing: An Olfactory Plan for Oceanside Museum of Art* in San Diego (2009) included a scent installation dispensing a custom-designed smell that was intended to "show how fragrance could function as both olfactory brand and gentle critique."[33]

Finally, we should take note of Christophe Laudamiel and Wolfgang Georgsdorf. So far, almost all of Georgsdorf's olfactory artwork has centered on developing and demonstrating *Smeller 2.0*. Laudamiel is unique among contemporary artists who focus their careers on smell since he is by profession a perfumer who has his own scent design company and in the past has designed scents for major fashion and cosmetic houses like Abercrombie & Fitch and Estée Lauder. Over the last few years, Laudamiel has increasingly turned his attention to creating scent artworks for the galleries that represent him in Berlin and New York. Among his many devices for exhibiting his scent creations is "scent squares" (2014), open white rectangles that emit scents when visitors press a button along the bottom. They were originally designed for a Berlin gallery to be set a few feet in front of Jakob Kupfer's slowly moving digital prints so that the visitor could view Kupfer's work through the open rectangles while smelling a fragrance. Like de Cupere and Macia, Laudamiel has also published an eloquent manifesto, *Liberté, Egalité, Fragrancité* (2016), whose fifty demands range from calling for scent education in the schools to establishing copyright protection for perfumes.[34]

Exhibition, Conservation, and Ontology

Given odors' volatility and evanescence on the one hand and the human sense of smell's tendency to habituation and idiosyncratic responses on the other, scent artworks in the gallery or museum present a number of problems for exhibition, conservation, and collecting, practical problems that can also raise theoretical issues. In the case of exhibition and conservation, the curators of the Tinguely Museum's 2015 survey point out that the odors of many works threatened to drift into each other so that, just as with sound art exhibitions, the curators ended up presenting many of the olfactory works in closed spaces.[35] A more specific problem with exhibiting odorous works is that they may require constant replenishment, either because the sources may dissipate too quickly, or the qualities may change with time. In the case of Wolfgang Laib's *Milkstone*, for example, a work that involves milk poured into the depressed surface of a rectangular stone slab, the work will gradually turn from emitting the faint, sweet smell of fresh milk to a sour stench—unless the milk is removed and replaced each day.

If scent artworks present complex challenges for museum curators and audiences, they can be even more challenging for collectors (and for artists who are trying to make a living from their art). It is one thing to buy a painting or sculpture, or even an installation work that demands a good deal of space; but how does one show off one's collection of olfactory or scent artworks, especially when many are deliberately designed either to challenge the normal range of olfactory tolerance or to defy collecting itself by evaporating? Christophe Laudamiel has come up with an ingenious solution to the permanence problem by selling vials of his constructed scents through a gallery along with copies of one or another of the devices he originally used to present them in the gallery. One of these devices is the scent squares mentioned already; another is the porcelain "parabol," made to his design. The parabols look like large, shallow serving dishes with covers. The collector is given a bottle of the scent liquid with a dropper to impregnate four pieces of chalk that are placed in the bowl and covered with the lid. The artwork is "experienced" simply by lifting the lid of the bowl and smelling. Of course, after a period of time the pieces of chalk need replenishing, and when the vial of liquid runs out, a replacement can be ordered from the artist. What I find theoretically interesting about Laudamiel's scent squares and parabols is the way they reflect an effort to "normalize" scent art by giving it something resembling a traditional "collectable" form. This represents the opposite pole from those artists who see the use of scent in their artworks as a way of defying collection and the machinations of the art market. Of course, something of this span of attitudes toward the economics of the art world has marked contemporary art for over a century and is not peculiar to olfactory art.

It might seem that these issues in the area of exhibition, conservation, and collecting are largely practical and of little theoretical or philosophical significance, but these pragmatic matters contain serious ontological and aesthetic problems concerning both the concept of art and issues of identity. After all, the kind of objections raised by Hegel, Beardsley, or Scruton, namely that odors are ephemeral and cannot be organized into coherently structured works, assume, as Jaquet says, a certain traditional concept of the permanence of the artwork that no longer holds. But the volatility and ephemerality of odors and the need to keep replenishing certain works are also philosophically significant insofar as curators are trying to ensure that the work on exhibit or in storage remains ontologically the "same" work each day or on each occasion of its subsequent exhibition. These kinds of philosophical issues about "sameness" go back to the famous debate about whether the ship of Theseus that gradually had every plank and element replaced over time was the same ship. The philosopher Sherri Irvin has brought that debate up to date in a valuable discussion of the ontological issues raised by many works of installation and performance art, and most of her points would apply to olfactory art as well.[36]

The Interpretation of Olfactory Art

As the category of olfactory art becomes more accepted, another challenge facing curators, critics, and art theorists will be to establish interpretive principles. Obviously, general aesthetic theories provide an important framework for formulating critical approaches, although the relationship is a dialectical one since critical practices are part of the pragmatic constraint on aesthetic theorizing, even as theories attempt to provide accounts of the nature of art and aesthetic judgment that may in turn affect critical and curatorial practices. Unfortunately, whereas we can trace a long history of aesthetic theories and critical approaches tailored to the dramatic, visual, and auditory arts, we lack such a framework for thinking about olfactory art that is typically presented in art galleries and museums. I will suggest a couple of current avenues toward developing such a critical matrix.

The first approach comes from the Japanese philosopher Yoko Iwasaki's suggestion that a key to understanding olfactory or scent art is the Japanese concept of space in the arts.[37] For Iwasaki, what makes the Japanese understanding of space helpful for interpreting olfactory artworks is the now familiar observation that in Japanese painting and sculpture what Westerners often see as "empty" space is equally noticed and valued as something positive. Similarly, in Japanese poetry, she suggests, the words and the objects called to mind are not the exclusive focus of attention, but what she calls "the atmosphere that arises between the words," for which she cites the famous haiku by Bashō:

> Peace of the old pond
> A frog jumps in
> Sound of water[38]

Here, what is important, she says, is not simply the objects, the pond, the frog jumping, or even the sound itself, but the peaceful silence in which the sound resonates. One is also reminded of the Zen garden at Ryoangi, where the rocks are not arranged primarily for disinterested aesthetic contemplation in the Western sense, but for deeply engaged meditation. I think Iwasaki is trying to get at a similar phenomenon in the case of scent art. The very intangibility, evanescence, and pervasiveness of smells and our difficulty in objectifying them give them unique potentialities as vehicles for atmospheric artworks.[39]

Iwasaki's use of the atmospheric implications of Japanese spatial conceptions offers an appealing perspective on many kinds of scent artworks: one thinks of Maki Ueda's scent-infused room *Invisible White*, or Meg Webster's *Moss Bed, Queen*, whose gentle moss odor filled a large gallery, or Ernesto Neto's aroma-rich *Mother Body Densities*, which pervades most of a museum. Perhaps that would include some of Oswaldo Maciá's sound-and-scent sculpture installations such as *Transition*, where the visitor sits on a bench inside a tunnel filled with plant scents and animal cries that together create the kind of atmospheric experience Iwasaki describes. But her approach does not work as well with other kinds of scent art such as perfume-like works made from pungent body odors (Ursitti's *Self-Portraits in Scent*) or installation works with a sharper political edge (Yi's *You Can Call Me F*) or that use acrid odors (de Cupere's *Tree Virus*) or outright stenches (Delvouye's *Cloaca Professional*). These kinds of works seem obviously less amenable to the serene meditative experiences Iwasaki describes. Yet stenches and offensive smells of many kinds abound in contemporary olfactory art, and they are often used with a political edge.

These rougher kinds of works would seem to require a different interpretive approach, one with a more historical and political framework. Such an approach could be derived from Caroline Jones's historical study of Greenbergian formalism, *Eyesight Alone: Clement Greenberg's Modernism and the Bureaucratization of the Senses*, which argues that the triumph of formalism in the 1950s was due in part to the way the senses, and especially smell, were regulated, segmented, and controlled "under eyesight alone."[40] Emblematic of this "modernist sensorium," of course, was the emergence of the "white cube" gallery in all its featureless and odorless austerity. But at the very height of this reign of austerity, figures like Alan Kaprow with his happenings and Carolee Schneemann with her raunchy, feminist, body-focused performances broke through the immaculate purity of formalism. As Jones notes, these messy, and sometimes smelly, avant-garde works of the 1950s and 1960s opened the way toward "the installation/performance art axis that has burgeoned in the new millennium."[41]

The cogency of such a historical framework for interpreting works of contemporary scent art becomes apparent in a dialogue between Jones and Anicka Yi in connection with Yi's trilogy of exhibitions from 2013 and 2014, called *Denial, Divorce,* and *Death*. Yi remarks that she knew little about Greenberg and formalism until reading Jones's book, but now sees more clearly that what she has been trying to do is, in fact, "to shift perception through the other senses."[42] The scent aspects of *You Can Call Me F* with which we opened this chapter, for example, are "a form of parasitism between the [female] bacteria's smell and the white patriarchal 'non-smell' of the Gagosian Gallery."[43] By using "all that . . . stink," Yi says, she wants to tell gallery visitors: "Get uncomfortable, get aroused, get in your pathetic body. *This isn't an abstract painting.*"[44] Asked about an earlier, perfume-like scent artwork called *Shigenobu Twilight*, honoring the founder of the Japanese Red Army (Fusako Shigenobu), Yi responded, "Smell is . . . like a form of cannibalism, which is why the portrait format is so appropriate. To take these molecules into your nasal passages is to take these women into your body—to smell them is to eat them."[45] Perhaps unknowingly, Yi is taking direct aim at the Kantian fear of the smell of others invading us. Viewed, more broadly, Yi's program of reordering the senses is consciously aimed at upending the traditional contemplative relation between viewers and artworks by using the most visceral means she can find.

Taken together, Jones's critique of the modernist ocular-centric sensorium and Yi's reflections on that critique offer ample material for developing a cogent interpretive framework apposite to a large segment of scent art. Given the hybridity and enormous variety of contemporary olfactory or scent art, there is not likely to be a single interpretive approach that will be appropriate to all types. There will be a place for Iwasaki's mindfulness perspective as well as the historical/political approach implicit in the Jones-Yi dialogue. Of course, a good interpretation of olfactory or scent art will not only employ a historical perspective like Jones's, but will also put individual smell works in the context of similar olfactory experiments, as well as in the broader context of conceptual and installation art generally. And good criticism will also need to be undergirded by some knowledge of both the psychology and the cultural history of smell. This is the approach taken by the leading olfactory art critic Jim Drobnick, who is unique among today's art curators and critics in his broad knowledge of the historical and social science literature on smell as well as having followed the development of olfactory art closely over the last three decades. Since the background of many art critics has typically been in the history of the visual arts, those lacking Drobnick's knowledge of olfactory science will have to make some effort to understand the contextual, material, and technical challenges scent artists face as well as the psychological characteristics of the sense of smell that visitors to art exhibitions bring with them. Critics lacking this background knowledge are likely to end up focusing on everything about a smell art hybrid except its olfactory qualities. In this early stage of the development of scent or olfactory art, one of the chief tasks of critical interpretation will be to examine not only what each artist has been able to achieve through the use of odors, but at the same time to educate the public in how to approach such works.

Olfactory Art and Aesthetic Appreciation

By using "sublime stenches" in the title of this chapter I was engaging in a bit of hyperbole, since only some scent artworks actually employ stenches, and, as Emily Brady has pointed out, powerful stenches do not fit classic definitions of the sublime.[46] Kant did not even consider smell in relation to the sublime, whereas Burke gave "intolerable stenches" a minor role, and then only when represented in a "*description or narrative.*" According to Burke, actual taste or smell experiences that are not deeply threatening ("terrible") but only disagreeable, are not sublime but "merely odious, as toads and spiders" are.[47] My aim in using the term "sublime," whose underlying concept has gone through many permutations since the eighteenth century, has been to call attention to the central role in so much olfactory art of unaccustomed, offensive, and sometimes intensely discomforting smells. Traditional aesthetic theory tended to focus on issues of beauty and pleasure, leaving aside the disturbing and the threatening, which the idea of the sublime was partly intended to cover. Moreover, as Carolyn Korsmeyer has shown with respect to gustatory taste in *Savouring Disgust*, there is a need for theoretical reflection on the aesthetic value of works that use repellant means.[48] No doubt, there are limits to the sorts of smells that can bear significance despite their unpleasantness, as we saw in the case of fecal odors on stage, since outright, visceral disgust and nausea seem to have a biological basis, even if it is culturally shaped.

We noted a similar set of issues with respect to "antiaesthetic" conceptual art that stresses challenge and social critique. A good deal of olfactory art, like Haug's *U-deur* and Annicka Yi's *You Can Call Me F*, shows little direct concern for what one might think of as purely aesthetic experiences, that is, experiences focused primarily on the pleasures derived from formal and sensory qualities. Yet, as we said of Haug's *U-deur*, conceptually oriented works often make their conceptual points in such a way that the sensory aspects of the work whether pleasant or unpleasant are inseparable from the conceptual aspects leading to reflection. Of course, there are an enormous variety of olfactory or scent artworks running the gamut from the highly conceptual and disturbing like those of Yi to the almost purely perceptual and sensually pleasant, like Ernesto Neto's aromatic sculptures.

Another way of thinking about the aesthetic experience of olfactory art is that much contemporary olfactory art, whatever its other conceptual and social aims, also intends to make us more aware of the odors around us and of the importance of our sense of smell. These works enlarge our range of experience by inviting us to pay attention via our sense of smell to things we may seldom notice.[49] Enlarging the range of our olfactory experience is especially apparent in the work of Sissel Tolaas, who is as much a social activist on behalf of smell as an artist using smells as a medium. Those who visited a work like *The Fear of Smell and the Smell of Fear* were not simply exposed to pungent body odors in the unaccustomed setting of an art gallery; they were asked to become agents who released the odors by touching and breaking the tiny paint capsules. In this way visitors were not only drawn out of the traditional distanced relation to artworks shown in a "do not touch" gallery setting, but also drawn out of their usual relation to the smell of the other and invited to think about both the nature of art and the nature of smell.

I want to close this chapter with a striking work that enlarges the range of one's aesthetic experience at the same time it uses odors to intensify the artist's political message, Otobong Nkanda's work *Anamnesis*, which I mentioned in the introduction. When I first encountered *Anamnesis* as part of an exhibition of her works at the Chicago Museum of Contemporary Art in 2018, I found it a particularly effective way to engage its audience and bring alive its postcolonial message. As one entered the large gallery containing a variety of her works, one couldn't miss *Anamnesis*, a tall, thick, free-standing white wall over half the length of the gallery with a jagged dark brown slash of varying width running around it. From a distance the wall could be taken for an abstract sculpture and the undulating slash as perhaps representing a meandering river. When I drew closer and read the notation in front of the work, I discovered that the brown slash was made up coffee beans, fragments of tobacco leaves, cloves, and other spices of the kind that were part of the Western colonial exploitation of Africa. Visitors were invited to come close and smell this stream of trade goods. When I had first walked into the big gallery and visually registered the wall and its slash from a distance, my relation to it was relatively detached, but as I came close and followed the river of scent, inhaling its shifting odors, these familiar yet exotic smells literally entered my body and the wall came alive. Nkanda says in a video interview accompanying the exhibition that she intentionally used materials whose smells would not only be familiar, but would "trigger memories that are linked to experiences, that are linked to family histories, that are linked to colonial histories."[50] Nkanda's comment applies to the memories of both those who labored to produce the trade spices, beans, and plants and to those who consumed them. For some Westerners who visit *Anamnesis*, the olfactory aspect of the experience not only makes the reality of colonial exploitation palpable, but also reminds us that the senses have played an important role in our desire to exploit the riches of other lands. Moreover, like many of the other works we have met in this chapter, *Anamnesis*, by breaking the occular-centric spell of the odorless white cube, is a powerful reminder of our embodiment and our connection to others through the proximal senses.

Some aesthetic theorists who would accept the kinds of hybrids we have discussed in this chapter as legitimate artworks worthy of serious aesthetic engagement might still have doubts as to whether such "pure" olfactory works as incense or perfume should be considered part of the fine arts. In the case of incense, which we traditionally think of only in religious or household and commercial atmospheric contexts, there is at least one striking exception: the Japanese practice of kodo, a unique olfactory form combining aspects of social ritual and literary appreciation. Kodo will be the subject of the next interlude and a fitting transition to the discussion of whether perfumes should be considered part of the fine arts.

Notes

1. Chris Sharp, "Anicka Yi's Allegorical Bouquets," *Cura* 22 (2016): 103.
2. Wendy Vogel, "What's That Smell in the Kitchen? Art's Olfactory Turn," *Art in America*, April 8, 2015, https://www.artinamericamagazine.com/.../whats-that-smell-in-the-kitchen-arts-olfacto.
3. Karen Rosenberg, "Scent of 100 Women: Artist Anicka Yi on Her New Viral Feminism Campaign at the Kitchen," *Artspace*, March 12, 2015, https://www.artspace.com.
4. Karen Rosenberg, "Anicka Yi: 'Divorce,'" *New York Times*, May 22, 2014, https://www.nytimes.com/2014/05/23/arts/design/anicka-yi-divorce.html
5. Drobnick, "Smell: The Hybrid Art," 173.
6. Jim Drobnick, "Reveries, Assaults and Evaporating Presences: Olfactory Dimensions in Contemporary Art," *Parachute* 89 (Winter 1998): 10–19.
7. Jim Drobnick has also written insightfully on this theme in "Inhaling Passions: Art, Sex and Scent," *Sexuality and Culture* 4, no. 3 (2000): 37–56.
8. Brian Goeltzenleuchter, "Scenting the Antiseptic Institution," in Henshaw et al., *Designing with Smell*, 2–18, 248.
9. Among the surveys one might consult are Barbara Pollack's "Scents & Sensibility," ARTnews 110, no. 3 (2011): 88–95; Wetzel et al., *Belle Haleine*; Marcel van Brakel, Wander Eikelboom, and Frederik Duerinck, eds., *Sense of Smell* (Breda: Eriskay Connection, 2014).
10. Bacci, "Scent-ific Art in Context," in Wetzel et al., *Belle Haleine*, 129–30.
11. Gwenn-Aël Lynn, "On Olfactory Space," in Henshaw et al., *Designing with Smell*, 26–28.
12. "Atmospheres" as I am using it here is a pragmatic category for sorting olfactory or scent artworks into broad types, to be distinguished from the more expansive use of the term "atmospheres" by Gernot Böhme to describe "the full range of aesthetic work." He speaks of "autonomous art" shown in galleries and museums as only one kind of atmospheric work alongside the atmospheres of staging in politics, theater, or advertising. Gernot Böhme, *Aesthetics of Atmospheres*, ed. Jean-Paul Thibaud (New York: Routledge, 2017), 13–14. I discuss Böhme's work in Part IV.

13. The exhibition, called *Let the Fancy*, was cocurated by Jeff Robinson and Allison Lacher at the University of Illinois at Springfield and ran from September 27 to October 18, 2018. Morris creates his own perfumes but sometimes simply uses commercial ones. See "Matt Morris on His Process," *School of the Art Institute of Chicago: A Biannual Magazine*, Spring 2019, 43.
14. Drobnick, "Smell: A Hybrid Art," 175.
15. Brandon LaBelle, *Background Noise: Perspectives on Sound Art* (New York: Continuum, 2006), ix.
16. Peter Köster, "The Specific Characteristics of the Sense of Smell," in Rouby, *Olfaction*, 31.
17. LaBelle, *Background Noise*, xii–xiv.
18. LaBelle, *Background Noise*, xiv.
19. Interestingly, one of the naysayers has been Max Neuhaus himself, who wrote that one might as well propose a category of "steel art" to encompass everything that is made of steel. Max Niehaus, "Sound Art?," in *Sound*, ed. Caleb Kelly (Cambridge, MA: MIT Press), 72-73.
20. Pardo views sound art simply as "a hybrid form that establishes various relationships between the visual and the aural." Carmen Pardo, "The Emergence of Sound Art: Opening the Cages of Sound," *Journal of Aesthetics and Art Criticism* 75, no. 1 (2017): 36. Alan Licht considers sound art's distinctness to be that it belongs to an "exhibition situation rather than a performance situation." Licht, *Sound Art: Beyond Music, between Categories* (New York: Rizzoli International, 2007), 14.
21. Christophe Cox, "From Music to Sound: Being as Time in the Sonic Arts," in Kelly, *Sound*, 86.
22. Cox, "From Music to Sound," 186.
23. Caro Verbeek, "In Search of Lost Scents: The Role of Olfaction in Futurism," in van Brakel et al., *Sense of Smell*, 61. See also Verbeek's " 'Inhaling Futurism': On the Use of Olfaction in Futurism and Olfactory (Re)constructions," in Henshaw et al., *Designing with Smell*, 201–3.
24. Caro Verbeek, "Surreal Aromas—(Re)constructing the Volatile Heritage of Marcel Duchamp," in Wetzel et al., *Belle Haleine*, 118. Marcel Duchamp's first work using scent was an assisted readymade in 1921, *Belle Haleine: Eau de Voilette* (Beautiful Breath: Veiling Water), although it was a work *about* smell and perfume intended to be looked at and thought about, rather than smelled.
25. Kevin Sweeney, who saw the Kienholz work in the 1960s, remembered it smelling like the disinfectant Lysol. Personal communication.
26. Drobnick, "Reveries, Assaults, and Evaporating Presences, 10–19.
27. Jim Drobnick has done an interesting profile of Ursitti in "Clara Ursitti: Scents of a Woman," *Tessera*, June 2002, 85–97.
28. Peter de Cupere, *Scent in Context: Olfactory Art* (Antwerp: Stockmans Art Publishers, 2016), 9.
29. Peter de Cupere, "Olfactory Art Manifest," July 1, 2014, https://www.olfactoryartmanifest.com ›. The manifesto can also be found in Peter de Cupere, *Scent in Context: Olfactory Art* (Antwerp: Stockmans Art Publishers, 2016), 101–104.

30. Oswaldo Maciá, *Manifesto for Olfactory: Acoustic Sculpture Compositions,* https://www.oswaldomacia.com.
31. See Drobnick's brief profile of Maciá in "Smell: The Hybrid Art," 176-79.
32. See Goeltzenleuchter's website at https://www.bgprojects.com/.
33. Goeltzenleuchter, "Scenting the Antiseptic Institution," 253.
34. The manifesto can be found at the website of Laudamiel's company Dream Air, https://www.dreamair.mobi?christophe-laudamiel.
35. Roland Wetzel, "Introduction to and Review of the Exhibition," in Wetzel et al., *Belle Haleine,* 11. An insightful recent discussion of several issues raised by olfactory art for museums is Hsuan L. Hsu, "Olfactory Art, Transcorporeality, and the Museum Environment, *Resilience: A Journal of the Environmental Humanities,* 4, no.1 (2016), 1-24.
36. Sherri Irvin, "Installation Art and Performance: A Shared Ontology," in *Art and Abstract Objects,* ed. Christy Mag Uidhir (Oxford: Oxford University Press, 2013), 242-62.
37. Yoko Iwasaki, "La reconnaissance de l'espace et l'art olfactif Japonais," in Jaquet, *L'Art olfactif,* 223.
38. Iwasaki, "La reconnaissance de l'espace," 226.
39. Iwasaki, "La reconnaissance de l'espace," 227.
40. Caroline A. Jones, *Eyesight Alone: Clement Greenberg's Modernism and the Bureaucratization of the Senses* (Chicago: University of Chicago Press, 2005), 391.
41. Jones, *Eyesight Alone,* 390.
42. From "Quasi/Verbatim: An Exchange between Caroline A. Jones and Anicka Yi," in *Anicka Yi, 6, 070, 430 K of Digital* Spit, ed. Alise Upitis (Cambridge, MA: MIT List Visual Arts Center, 2015), 10.
43. Yi, "Quasi/Verbatim," 16.
44. Yi, "Quasi/Verbatim," 13.
45. Yi, "Quasi/Verbatim," 11.
46. Emily Brady, "The Negative Aesthetics of Odor," paper presented at the workshop "Scent, Science and Aesthetic: Understanding Smell and Anosmia," Barcelona, May 22-24, 2013. I am grateful to Professor Brady for sharing the text of her paper.
47. Edmund Burke, *A Philosophical Enquiry into the Origin of our Ideas of the Sublime and Beautiful* (Notre Dame: University of Notre Dame Press, 1968), 85-86.
48. Carolyn Korsmeyer, *Savoring Disgust: The Foul and the Fair in Aesthetics* (Oxford: Oxford University Press, 2011).
49. In the case of vision, Dominic Lopes speaks of images as "perceptual prostheses, expanding the powers of vision." *Aesthetics on the Edge,* 14.
50. The quotation is from a video interview playing continuously outside the exhibition. The exhibition as a whole was called *To Dig a Hole and Fill It Up Again,* Museum of Contemporary Art, Chicago, March 31-September 2, 2018. I am grateful to Jeff Robinson for calling my attention to this exhibition.

Interlude

Kodo, an Art of Incense

As Montaigne reminded us, incense has been used for religious purposes in almost every part of the world for millennia with the intent to "arouse and purify our senses" and "make us more fit for contemplation."[1] Although the religious use of incense is still global, as are many secular uses, only in Japan has incense also been part of a practice whose more sophisticated versions have been claimed to be a unique art form. This interlude will introduce *kodo* and examine the claims of three contemporary aesthetic theorists regarding its art and aesthetic status.

What Kodo Is

Although writers often refer to kodo as a "ceremony" or "game," the more serious versions of kodo far exceed such categories since they involve a small group of people contemplating the scents of rare incense woods in a meditative and poetic context. The room chosen for a traditional kodo event will typically hold eight to twelve people seated in a square and will be relatively sober except for an alcove with a painting or calligraphy scroll. Participants enter silently and once seated Japanese style, the kodo master enters carrying a wooden box with the incense materials, accompanied by a recorder, who distributes paper, ink, and brushes to the participants. The master will have prepared several miniscule pieces of aromatic wood and have chosen a particular *kumikoh*, or form, that prescribes a certain combination of woods with a short poem or other literary piece. After ritually cleansing the instruments, the master places a sliver of wood on a small mica plate that sits atop a mound of hot ash in a cup. As the heat gently releases the aroma, the master inhales the scent while holding one hand in front of the cup, then tells the group the scent's name and the cup is passed from person to person to inhale and consider. Once all the aromas have been sampled, the master circulates them again in a different order and each person writes an identification of the scents. After the individual sheets are handed in and recorded, the results are announced in the terms of the poem or story that has informed that particular session and the record is passed around for all to see how they fared. The master asks if everyone has enjoyed the session, and the person who has identified the most scents is presented with the official record. Then everyone bows and thanks each other for sharing the time.

Although kodo's roots go back to medieval aristocratic activities described in the *Tale of Genji*, kodo as a formalized practice only began to emerge in the 1400s under the emperor Ashikaga, who also promoted two other classic Japanese arts: *chado*, the tea ceremony, and *kado*, flower arranging (also known as *ikebana*). In each case, the *do*, derived from the Chinese *tao*, can be variously translated as "way," "art," "appreciation," "truth." Although kodo, as the way or art of scent, flourished until the late nineteenth century, it went into decline with the Meiji Restoration and modernization, and, despite its modest revival since the 1960s, is still much less known and practiced in Japan and abroad than *chado* and *kado*. Yet for several centuries, it held a rough parity with the other two classic Japanese arts, and it merits our attention because its more sophisticated forms suggest a unique aesthetic practice that raises questions about the Western concept of art.[2]

Poems and stories became an essential part of kodo as its practice spread gradually from aristocrats to the samurai class and on to the urban middle classes, finally reaching a large swath of the educated population by the Edo period (1603–1867). A defining aspect of kodo is the *kumikoh*, or form, combining of a set of woods with a literary source, chosen by the master from hundreds of *kumikohs* developed over the years. Kiyoko Morita of Tufts University gives this example of a *kumikoh* called "Shirakawa Border Station," based on a short poem about a monk's journey on foot over several months from the capital, Kyoto, to Shirakawa, some 370 miles away.

> I left the capital
> Veiled in Spring mist.
> An Autumn wind blows here,
> At Shirakawa Border Station.

In this *kumikoh* there are only three woods (other *kumikoh* may have up to twenty-five), called Mist in the Capital, Autumn Wind, and Shirakawa Border Station. In the initial trial round, only two woods are announced and sent around; the challenge to the participants is to remember them and distinguish them from the third. When the recorder announces the results, he or she does not give out a score but says for those who correctly identified all three, "Crossed the Border," for those who missed all three, "Stopped at the Border," for those who got only Spring in the Capital, "Spring Wind," for those who got only Autumn

Wind, "Fallen Leaves," for those who got only Shirakawa Border Station, "Travel Garments."[3]

Given kodo's ritualized context, slow pace, and literary references, it is a much more meditative and reflective aesthetic activity than the frequent use of the terms "game" or even "ceremony" to translate kumikoh suggests. One can see the meditative and reflective aspect in the general practice of speaking of "listening" to the woods rather than simply "smelling" them. One version of the origin of this practice attributes it to a Mahayana sutra that speaks of listening to the Buddha's words in the form of incense.[4] The idea of listening also corresponds to the focus on the spiritual resonance of the scents and their names. The mindfulness aspect of kodo is reflected in the first four items on a sixteenth-century list of the "Ten Virtues of Incense":

1. Sharpens the senses
2. Purifies mind and body
3. Removes spiritual pollutants
4. Promotes alertness[5]

We should also note that kodo developed two main "schools" over time: the Oie school derived from aristocratic practices that emphasized elegance and the literary allusions, and the more sober Shino school, developed under samurai and Zen influence that stressed simplicity and meditative attitudes.[6] Although the basic structure of kodo has remained largely unchanged, its aims gradually diversified as it gained a wider audience thanks to the availability of less expensive incense sticks to replace rare and costly imported woods. From an emphasis on meditative and aesthetic concentration along with knowledge of literary sources many people began to focus on success in naming and the use of elaborate paraphernalia.[7] Today, for example, the website of the Shoyeido Incense Company, which sells kodo boxes, cups, and incense sticks, mentions only the "fun" of the "game" of kodo.[8] But the recent revival of kodo has also involved training masters who observe the more reflective aesthetic approach that once made kodo an equal partner with the art of tea and the art of flower arranging. Outside Japan some intellectuals have adapted kodo's core elements to fit their own cultures. In the United States, for example, Kiyoko Morita experimented with *kumikohs* that use the work of Western poets such as Emily Dickinson and encouraged discussion at the end of sessions.[9] But of most interest for philosophical aesthetics are the claims of three theorists that certain aspects of kodo are exemplary of things essential to any kind of refined aesthetic practice or fine art form.

Art Form or Aesthetic Practice?

Yoko Iwasaki mentions several exemplary aspects of kodo that she believes qualify it as an art form. First she notes that the notion of "listening" to incense and the traditional practice of classifying incenses partly in terms of the five tastes, plus the obvious tactile and visual aspects of the ceremony, point to kodo's multisensory character. "In Kodo all of the senses are triggered and alert," but this, Iwasaki boldly claims, "is the essence of every art."[10] A second aesthetic aspect of kodo is what Iwasaki calls the "atmosphere" that pervades the kodo event. The crucial aesthetic dimension of kodo is made up not just of the objects and people gathered in the room or those mentioned in the accompanying poem, but the feeling pervading them all. Similarly, in smelling the incense pieces, it is not the degree of success in identification that matters so much as the imaginative and meditative experience of the activity as a whole.[11] In the third place, Iwasaki stresses that despite people's personal associations with particular smells, kodo joins its participants in a "common sensory experience," something Iwasaki believes is also typical of our most significant experiences of several other arts.[12] Although one may question whether multisensory experience is the "essence of every art," Iwasaki's stress on the way kodo's atmosphere is favorable to imagination and reflection connects with mainstream Western thinking about the nature of aesthetic experiences prompted by fine art.

Hiroshi Yamagata, like Iwasaki, also emphasizes the intersubjective nature of the kodo experience, but his case for treating the reflective practices of kodo as an art form draws on a different kind of argument. He appeals primarily to the Western philosophical notion of the *sensus communis* found from Aristotle to Kant, the idea that there is a common faculty, a sort of sixth sense, that unites all the other senses.[13] Kant used a version of it to buttress the final stage of his argument for the universality of aesthetic judgments. Yamagata's own version of the *sensus communis* is closer to Aristotle's. Yamagata believes that each of us possesses something like an internal "sense" that underlies and unites all the senses and is intermediate between body and mind. Because this internal sense is common to all people, it allows for "the formation of a community and communication through the senses."[14]

Yamagata suggests that kodo's use of traditional metaphoric names for scents such as "Spring Light" and "Dreams of a Sleepless Night," and its frequent invocation of poetry, show that kodo's original focus was not on the competitive guessing of names, but that kodo was in fact "an elevated and refined art" connected to literature.[15] His central argument is that such kodo sessions create "a kind of spiritual community" by means of both "synesthesia and the common sense." "By using several sorts of woods, one 'hears' the scents that arise, distinguishes their subtle differences, and creates a symbolic atmosphere with the help of literary materials, in collaboration or competition with other participants."[16]

Unlike Iwasaki and Yamagata, the French philosopher Chantal Jaquet leans toward emphasizing the "way" aspect of the *do* in kodo rather than the "art" aspect, arguing that kodo does not fit such traditional Western fine art notions as "work," "originality," or "creation." She believes we will gain a better understanding of kodo by placing it in the context of the Japanese aesthetic emphasis on the beauty of the impermanent that one finds in other traditional Japanese arts such as *chado* and *kado*. Like *chado*, kodo has both a spiritual/aesthetic aspect and a moral aspect involving simplicity and respect.[17] Jaquet also points to several practical parallels between kodo and *chado*: a master who sets the themes and chooses the kind of tea or wood used, a simple room, a ritual cleansing of utensils, and a passing of the tea bowl or cup of ash among a small group of participants.[18]

Where kodo differs from *chado* is the central role of the *kumikoh*, the particular combination of different scents and their connection to poetic references, whether explicit or implicit. Kodo's uniqueness, Jaquet stresses, lies in this interplay of scent and sound, wood and word as orchestrated by the master. And although the master's guidance is based in traditional *kumikohs*, there is room for innovation, even improvisation, by varying the number and sequence of scents and poetic sources. It is an "artistic approach," she remarks, "similar to free jazz" in that it takes a given source and varies and recombines its elements.[19] Yet kodo is not only a "combinatory art," for Jaquet; it is also a "collective art" in which "the harmonious encounter of the participants . . . supersedes individual genius."[20] Like Iwasaki and Yamagata, Jaquet concludes that kodo in its more serious forms is less about identifying scents than letting one's mind and spirit be opened, in the company of others, "to a veritable olfactory poetics."[21]

What strikes me about the interpretations of Iwasaki, Yamagata, and Jaquet is that, despite their differences, all emphasize the atmosphere of a shared experience in a unique aesthetic practice or art form that seems to escape many of our Western categories. Although one might simply attribute this to a supposedly "Asian" tendency toward the communal versus a "Western" tendency toward individualism, I think their arguments cut deeper. Yamagata, by connecting the shared aesthetic experience of kodo to the tradition of the common sense, shows that the human sense of smell can be a vehicle for what he calls "a third aesthetic" alongside traditional Western aesthetics based on sight and hearing.[22] And Jaquet's claim that kodo is a kind of "collective performance," in which participants share a fleeting moment of olfactory beauty, suggests a profound supplement to the Western model of an audience moved by the expression of individual genius.

At this point a skeptic might object that although kodo may, in some loose sense, be an *aesthetic* practice, it is not an *art* practice and not only for the reasons Jaquet lists—the absence of a work, of originality, of creativity. The skeptic would point out that even if we update the standard Western fine art categories to include recent conceptually oriented modes of art that involve appropriation, improvisation, or ephemerality, kodo still does not fit. One cannot claim kodo as an example of "performance art," since in kodo the "performers" are not artists; there is no "artist" but only a master and participants who follow traditional roles and rules. Even the recent Western category of "participatory" or "relational" art, which requires that spectators take some action to complete the work, seems inappropriate to an activity like kodo that does not have an artist setting up its conditions. Instead of a true artist-creator, the "master" mostly follows long-established precedents, even if there is room for innovations.[23]

How should we respond to these objections? We could retort that these complaints against art status for kodo simply reveal the inadequacies of Western concepts and categories of art to do justice to non-Western cultures, even when the concepts are updated to include conceptual and performance modes; therefore, we need a broader concept of art. Another answer would be to embrace the skeptic's claim that kodo is an aesthetic rather than an art practice, yet deny that as an aesthetic practice, kodo is of a lesser aesthetic value than Western fine art practices.[24] My own suggestion would combine aspects of both answers. If we are to break out of the narrow confines that I believe still handicap a good deal of Western aesthetic thinking, we need to embrace pluralistic concepts of both art and the aesthetic, something I will propose in more detail later.[25]

But could the arguments in favor of considering kodo an art form, or at least an aesthetic practice in no way inferior to the Western fine arts, also be applied to the claims of some perfumers and perfume enthusiasts that we should recognize perfume as a fine art form? Chantal Jaquet is convinced they can. For Jaquet, kodo not only shows that "a pure olfactory contemplation is possible," but it also suggests that Westerners' rejection of fine art status for perfume "rests on an ethnocentrism that shows the poverty of the Western imagination in the domain of olfaction."[26] Jaquet has even organized a state-funded interdisciplinary research program in France called Project KÔDÔ with the triple aim of examining (1) the philosophical, historical, and sociological conditions of the emergence of kodo, (2) kodo's neurophysiological implications, and (3) kodo and the place of olfactory creativity in contemporary art.[27] I agree with Jaquet that the existence of kodo in its more sophisticated forms shows that in principle a fine art of perfume is *possible*. After all, the differences between incense and perfume at a technical level seem relatively insignificant: incense typically comes in solid form and its molecules are released by heating or burning, whereas perfume is typically in liquid form and its molecules released by evaporation. But there remain at least two serious barriers to the elevation of perfume to fine art status that are based on cultural differences between the kodo experience and Western perfume use and appreciation. The first, and most obvious, is that the focus of perfume use and appreciation in the modern West is the *individual* and his or her self-presentation, not a shared aesthetic experience—although the rise of internet perfume blogs might eventuate in some sort of social practice of contemplating perfumes in a way similar to traditional kodo. A second, and more difficult, barrier to fine art status for perfumes is that whereas incense has deep religious and spiritual associations, even in some of its secular uses such as kodo, perfume use in the West has associations with vanity, deception, and seduction. Even a philosopher like Frank Sibley, as we saw earlier, who was willing to grant that perfumes could be objects of aesthetic appreciation, considered them trivial compared to the established arts. The next two chapters will tackle this nest of problems.

Notes

1. Montaigne, *Complete Essays*, 221.
2. Yoko Iwasaki, "Art and the Sense of Smell: The Traditional Japanese Art of Scents (Ko)," *Aesthetics* (Japanese society for Aesthetics) 11 (2004): 63.
3. Kiyoko Morita, *The Book of Incense: Enjoying the Traditional Art of Japanese Scents* (Tokyo: Kodansha International, 1992), 74–78.
4. Morita, *Book of Incense*, 41–43. It is worth noting that the word s*entire* in Italian means both to hear and to smell.
5. David Pybus, *Kodo: The Way of Incense* (Boston: Tuttle Publishing, 2001), 7.
6. Chantal Jaquet, *Philosophie du Kodo: L'esthétique japonaise des fragrances* (Paris: Presses Universitaires de France, 2018), 75–95.
7. Morita, *Book of Incense*, 40–7.
8. "the main idea is to have fun with fragrance," https://www.shoyeido.co.jp › english › culture › ceremony.
9. A group led by Morita presented a *kumikoh* centered on Dickinson's poem *May-Flower* at the Japan Festival in Boston in 1992. Jaquet, *Philosophie du Kodo*, 114–16.
10. Iwasaki, "Art and the Sense of Smell, 63.
11. Iwasaki, "La reconnaissance de l'espace," 224–25.
12. Iwasaki, "Art and the Sense of Smell, 66.
13. For a discussion of the history and implications of the concept of the common sense see David Summers, *The Judgment of Sense: Renaissance Naturalism and the Rise of Aesthetics* (Cambridge: Cambridge University Press, 1990).
14. Hiroshi Yamagata, "Pour une esthétique du troisième sens: L'odorat," *Revue d'esthétique* 21 (1992): 84.
15. Yamagata, "Pour une esthétique," 86.
16. Yamagata, "Pour une esthétique," 86.

17. The aesthetic of both *chado* and kodo expresses "the magnificence of impermanence" and the "plenitude of...silence." Jaquet, *Philosophie du Kodo,* 202.
18. Jaquet, *Philosophie du Kodo,* 78, 192–201.
19. Jaquet, *Philosophie du Kodo,* 231.
20. Jaquet, *Philosophie du Kodo,* 202.
21. Jaquet, *Philosophie du Kodo,* 15.
22. Yamagata, "Pour une esthétique," 87.
23. Similar criticisms of the claim that *chado* should be considered a fine art form have been developed in Daniel Wilson's article "The Japanese Tea Ceremony and Pancultural Definitions of Art," *Journal of Aesthetics and Art Criticism* 76, no. 1 (2018): 33–42.
24. For a theoretical account of aesthetic value that could be used to justify the latter direction see Dominic McIver Lopes, *Being for Beauty: Aesthetic Agency and Value* (Oxford: Oxford University Press), 2018).
25. I indicate the direction of such a proposal for the concept of art in the postlude to Part III, "Free Art versus Design Art," and I do something similar for the concept of the aesthetic in the introduction to Part IV.
26. Jaquet, *Philosophie de l'odorat,* 291.
27. See the kodo project website at http://kodo.univ-paris1.fr/en/anr-kodo/research-programme/.

11
Beautiful Fragrances
Is Perfume a Fine Art?

Baudelaire understood the peculiar aesthetic power of incense and perfumes as well as anyone. Consider these lines from "The Perfume":

> Reader, have you ever inhaled
> With intoxication and slow gourmandize,
> The grain of incense that fills a church,
> Or the enduring musk of a sachet?
> The profound, magical spell by which the
> Restored past makes us inebriate in the present![1]

Surely the phrase "intoxication and slow gourmandize" describes an important aspect of some of our finest aesthetic experiences. And the "magical spell" by which a scent may restore the past and "make us inebriate in the present" anticipates Proust's notion of the way sensory experiences of taste and smell can become moments of illuminating remembrance. To the extent that the power to induce such experiences is typical of works of fine art, one might argue that the capacity of some of the finer perfumes to provoke those experiences is a reason to treat perfume as fine art. Of course, many other things induce intoxication or invite slow gourmandize—Emily Dickinson was even "inebriate of air."[2] If we want to show that perfume can be a fine art, we will need to join the argument from intoxicating effects to an argument that shows perfumes can have the kind of complex expressive structures that other works of fine art do. If that is true, perfumes may not only heighten our sensibilities but may at the same time lead us to carefully attend to their formal and expressive characteristics, one prominent way of defining the aesthetic appreciation of art.

We find just such an argument on behalf of perfumes as fine art already articulated by Duc Jean des Esseintes, the protagonist of Huysmans's novel *Against Nature*, published in 1884.[3] When des Esseintes decides to create his own perfume to banish an olfactory hallucination, we are told that he had always been convinced the human sense of smell could "experience pleasures equal to those of hearing and sight, each sense being capable, by natural disposition or expert cultivation of composing them into that whole which constitutes a work of art."

Art Scents. Larry Shiner, Oxford University Press (2020) © Oxford University Press.
DOI: 10.1093/oso/9780190089818.003.0024

This belief leads des Esseintes to draw an explicit parallel between a sensitive response to the most sophisticated perfumes and our response to works of fine art such as painting or music:

> Just as no one without an intuitive ability expressly developed by study could distinguish a painting by an old master from a daub . . . no more could anyone, without prior instruction, at first help confusing a bouquet created by a real artist, with a potpourri fabricated by some industrial manufacturer to be sold in grocery shops and cheap bazaars.[4]

Here, in 1884, Huysmans has his protagonist make a case that perfumes can prompt an aesthetic experience similar to painting and music that would not have been out of place in 1984 or 2014. In fact, when the Museum of Arts and Design mounted its exhibition of perfumes as *The Art of Scent: 1889–2012*, the justification given by the curator, Chandler Burr, was strikingly parallel to that of des Esseintes. Burr, you will remember, presented a dozen classic scents, spritzed from shallow indentations in the gallery walls with the aim of convincing visitors that perfumes are "actually works of art, beautiful and aesthetically important . . . equal . . . to painting, sculpture, music, architecture, and film."[5] Unfortunately, Burr offered little in the way of explanation to support his claim, relying mostly on various presentational conceits.[6] But Burr's description of perfumes as "beautiful and aesthetically important" could have been developed into an argument. Many professional perfumers such as Robert Calkin, J. Stephen Jellineck, and Jean Claude Ellena have cited the beauty and harmony of the finest perfumes as a reason to consider perfume a fine art form.[7] And at least one widely admired perfumer, Edmund Roudnitska, devoted an entire book to arguing for perfume's fine art status by drawing on Kant's contrast of the merely "agreeable arts," which aim at ordinary sensory satisfaction, with the "beautiful" or "fine" arts (*schöne künste*), which aim at the properly aesthetic pleasures of reflection.[8]

Yet there is a problem with relying solely on an argument from beauty, since "beauty" as the name for consummate artistic achievement hasn't been central to art criticism and theory since at least the 1950s. The attempts at restoring beauty's place within criticism and philosophical aesthetics that began in the 1990s probably owed something to the very conceptual and postdisciplinary turn that has exploded the limits of what can count as fine art. Whereas figurative painting or tonal compositions were dismissed as retrograde in the formalist criticism of the 1950s, in today's art world one can be a figurative painter or compose classical symphonies, and can even pursue "beauty"—and have still one's art respected for its achievement. That's one meaning of the much-talked-of "pluralism" in the arts today. But in a pluralistic situation, beauty is no longer top dog, and although a case for perfumes as fine art can still appeal to beauty, it sounds a bit old-fashioned.[9] In order for perfumes to be taken seriously as *contemporary* art, we need to show that they are capable of manifesting some of the qualities besides beauty that have been at the forefront of contemporary art criticism since the postmedium turn: the challenging, the subversive, the transgressive, and so on. Such qualities are obviously at the forefront of the criticism of gallery and museum works of olfactory or scent art, but does it make any sense to think of perfumes in that way?

One of the two major philosophical approaches to the postdisciplinary situation in the arts defines a work of fine art as something that prompts aesthetic experiences of many kinds, although a few who champion an aesthetic theory of art claim that we should still use "beauty" as the term for consummate aesthetic excellence.[10] But aesthetic definitions of art, of course, are only one of the ways art theorists and philosophers have responded to the conceptual and postdisciplinary turn. The second major approach is made up of various contextual and historical theories of art that define something as fine art depending on the context and/or history from which it has emerged, giving special importance to established art practices and institutions. Although there are many varieties of both the aesthetic and contextual approaches to defining art (and of attempts to combine them), I think the issues at stake in the debate over perfume's fine art status will stand out more clearly if we explore first one approach and then the other. I will first build an aesthetic case that perfumes can be used to make objects of high aesthetic regard and meet one set of criteria for inclusion in the fine arts. Then I will develop a contextual and historical case against considering perfumes fine art.

An Aesthetic Case for Perfume as Fine Art

Aesthetic theories of fine art claim that for something to be a work of art it must have the capacity to prompt an appropriate aesthetic response in qualified observers.[11] Of course, the trick here is to define what an appropriate aesthetic experience is. Despite considerable disagreement on details, most contemporary definitions include such features as focused attention, with understanding, on formal, expressive, and other aesthetic qualities for their own sake.[12] Finally, some theories of aesthetic experience stress that genuine aesthetic experience is itself a state of mind enjoyed for its own sake.[13]

What are the characteristics an aesthetic object must have if it is to prompt or invite this special kind of experience? First, we should note that almost any kind of object can and has been found to have some of the characteristics that provoke a degree of aesthetic response: natural scenery, mathematical proofs, eloquent oratory, well-designed cars, furniture, clothing, and so on. But for many aesthetic theorists, what is most relevant to the issue of whether some object or practice can be considered part of a fine art form is that it can elicit a particularly high level of aesthetic response.[14] Among the most important of those characteristics are formal devices (arrangements of colors, shapes, materials, tones, literary tropes) that successfully embody a work's aims and which, however numerous the devices may be, achieve a satisfying interplay or even unity. Equally important to formal complexity and interaction are expressive qualities that serve the work's aims and are able to suggest a full range of emotional meanings: sadness, joy, terror, hope, love, hate, yearning, satisfaction, and so forth. The list of other more specific aesthetic characteristic is almost endless, but those we have indicated are enough for us to ask: How do perfumes fare in the light of these formal and expressive criteria? Can perfumes prompt the kinds of elevated and refined aesthetic experiences that would lead us to consider them works of fine art?

We have already briefly examined the negative answer implied by the tradition running from Kant to Scruton that held that the sense of smell in general lacks the cognitive powers to be a basis for either the creation or appreciation with understanding of works of fine art. Parts I and II of this book argued that the sense of smell does have a cognitive dimension that informs its emotional and hedonic aspects and that it is capable of being developed to a high degree among professional perfumers and presumptively among many laypeople who have made the requisite effort. Now we need to move beyond those general arguments to show that perfumes in particular can become genuine aesthetic objects worthy of the status of fine art. To do that, the first thing we need to do is show how claims like Monroe Beardsley's assertion, embraced by Scruton and others, that smells lack the "balance, climax, development, or pattern" required to "construct aesthetic objects" does not apply to perfumes any more than such arguments apply to the hybrid scent arts or to works like *Green Aria* examined previously.

In reply to the Beardsley-type argument, we may note that because perfumes are made up of substances possessing differing volatilities, they have not only a formal structure, but also a temporal sequence. Moreover, perfumers generate the structure of a perfume in much the same way that artists in other media create their works. Creating a perfume is not simply a matter of blending a set of odorous substances, but of imagining a complex form, a composition involving a variety of odor molecules called "notes," differing in quality and intensity as well as in volatility. Some notes will actually be part of preconstructed harmonious units called "accords." Accords are the memorable aesthetic gestalts that form the core of a perfume, giving it an underlying identity.[15] The perfumer Annick Menardo argues that a fragrance is good to the extent its accord(s) is good: when "I see the pattern, I feel the melody."[16] Menardo's musical reference is apt since a perfume's formal structure is constantly developing over time. Although all the notes and accords are present from the beginning, the perfumer can order the differing volatilities so that one can experience a temporal sequence and even a certain rhythm as the notes and accords evaporate. It's as if we were at a play and when the curtain went up all the characters would be revealed on stage, but not every one of them would immediately speak and attract our attention; then over time singly or in groups, they would move, speak, interact, and then exit one after another. The traditional way of describing the temporality of a perfume has been to speak of top or head notes, middle or heart notes, and bottom or base notes, identified roughly in terms of descending volatility. The head notes, such as most citruses, will evaporate in a few minutes, the heart notes, such as florals, in a few hours, and the base notes, woody or musky, might last a day or more. Of course, this tripartite division is really just a schema, a heuristic for ordering the enormous variety of temporal possibilities. Moreover, many perfumes since the 1980s, especially those aimed at the mass market, do not follow this traditional sequence, often making their dominant note apparent from the beginning and remaining more or less constant through dry down. The more interesting and respected perfumes often have complex structures and temporal patterns.[17] Part of what gives an outstanding perfume its intellectual as well as sensory interest is the way this complex compositional structure results in an experience of variety in unity over time, a classic criterion of aesthetic quality since the eighteenth century. Of course, as in the case of a fine wine, the average person may not be able to identify a complex and carefully designed perfume's individual elements or its sequences without considerable experience and training.

If what we have said about the structural complexity and temporality of the best perfumes is true, they are clearly capable of embodying aims far higher than immediate sensory gratification. As Menardo says of accords, "It's the idea that is precious."[18] One need only read a few of the more respected perfumers or perfume critics to see that when they discuss perfumes they describe both scent qualities and temporally unfolding structures. This is also reflected in the fact that most perfumers will tell you that an unusually acute sense of smell is less important for perfume creation than the mental discipline to develop one's powers of discrimination and memory.[19] Despite the habit of referring to a professional perfumer as "a nose," perfume creation is not primarily a sensory craft, but demands a kind of intelligence and imagination similar to what goes into other creative arts. As Edmund Roudnitska memorably put it, "I do not compose my perfumes with my nose, but with my brain and even if I were to lose my power of smell I could still invent and compose perfumes."[20] (One cannot help but think of Beethoven, who composed his Ninth Symphony after he had lost his hearing.) Such capacities are also attested by the neuroscience studies we reviewed earlier that show continuing brain plasticity and creative ability in perfumers to generate images and patterns as the perfumer ages.

Perfumers have long seen the intellectual and aesthetic demands of perfume creation as similar to music composition: it is only, writes Jean-Claude Ellena, "by seeking a pattern, a melody" that "I create an olfactory form."[21] But there is an even more specific parallel between perfume design and music composition than the superficial similarities suggested by terms like "composition," "note," "accord" or "melody." The perfumers' goal is not a single olfactory artifact, but a formula that can be reproduced, much as the composer of a musical work typically does not create a single sound artifact but a score. Here we have an olfactory version of what philosophers of music call the type/token problem. Is the work of art the type that is, the musical score or perfume formula, or is the work of art the token, the tones of this particular performance or the odors of this perfume? There are similar type/token parallels in standard fashion or product design. Although a fashion or product designer might occasionally make her own prototype, or even produce small production runs, the usual aim of a designer is to produce a pattern or set of coordinates that will be turned into multiple instances by others. The sleek modernist chair that we admire in the Museum of Modern Art in New York or in the Victoria and Albert in London is a token of the design created by Marcel Breuer, just as the vapor from a bottle of *Trésor* that we test at a perfume counter is a token of the formula created by Sophia Grosjman.

If these remarks suggest that perfumes do have a formal and temporal structure that appeals to our cognitive as well as our sensory capacities, we then need to ask: is a perfume's structure also capable of representation and expression, two other traditional aesthetic characteristics of works of fine art? Perfumes seem capable of representation in certain senses of a term that is highly contested in philosophy. Certainly, perfumes may sometimes resemble, exemplify, and/or symbolize. Thus citrus odors have a brightness and lightness that has made them useful in symbolizing cleanliness and alertness, so they frequently show up not only in detergents or hand soaps, but also as notes in certain perfumes and colognes aimed at men. The perfumer Jean-Claude Ellena has created perfumes that use the more obvious kinds of association and exemplification to suggest a place, such as using green tea odors to evoke the idea of Japan, or mango aromas to suggest Egypt.[22] Another perfumer, Dominic Ropion, has written fondly of certain delicate notes frequently associated with babies and small children, orange blossom, bergamot, vanilla, molecules that are often known as "white musks" and that can be used to represent innocence. Their sweetness, he says,

> evokes all the most agreeable notes of the skin. They illustrate the fascination of innocence . . . they remind me of my children when they were babies. I loved to smell their hair, their tummies, their feet. . . . We make copious use of them in perfumery to suggest virginal flesh, a consensual idea of purity.

But, of course, there are also contrasting skin notes that say, "I'm highly sexual," such as civet, cumin, and indole. Cumin by itself, Ropion points out, hints at body sweat, but when joined with certain other notes "it is a wonderful vehicle for sensuality."[23] Reading these passages from Ropion's memoir, one can't help thinking of Baudelaire's *Correspondences*: "perfumes fresh as the flesh of infants, / Mellow as oboes, green as meadows, / And others, corrupted, rich and triumphant."

Although Roger Scruton, as we noted earlier, is willing to admit that smells can suggest meaning by association, he denies they are capable of expressing meanings directly in the way he thinks paintings or sculptures do.[24] And even Frank Sibley, who was more sympathetic to the aesthetic possibilities of odors, suggested that "perfumes and flavors . . . unlike the major arts . . . have no expressive connection with emotions, love or hate, death, grief, joy," a position taken up by Dennis Dutton.[25] But many perfumers, such as Ropion or Ellena, have attempted something more complex and ambitious than mere association, namely, to express the feelings evoked by a person or a place. Ellena gives this example of how he composed a perfume called *Un jardin en méditerranée* that was released by Hermès in 2003. On a visit to an aromatic garden in Tunisia one day, he watched a young woman tear a fig leaf and sniff it with pleasure, and on his return to France he attempted to create an olfactory equivalent of his experience. Given current headspace technology, he could have gone back and sampled the garden and run it through a gas chromatograph and mass spectrometer to get a formula he could attempt to reproduce.[26] But that kind of literalism, he remarks, would be like a tourist snapshot that missed "the emotional tone of the place."[27] Instead of attempting to replicate the odors, he tried to compose a perfume that would express a "poetic memory" of the garden.[28]

In reading Ellena's account of how he composed *Un jardin en méditerranée*, I was struck by two things. First, he sought to achieve what we might call cognitive expressivity by using artificially constructed elements that would retain sufficient identity to provide both relevant associations as well as interest in the way they worked together. Second, his comment about the failure of a snapshot or the literal reproduction of an odor to capture "the emotional tone" of a place suggests an overarching creative intention not unlike that of many writers, composers, and painters who seek an equivalent not an imitation or copy.[29] Whether or not Ellena's perfume succeeds in expressing and communicating to everyone the particular emotional tone he intended, his account illustrates the fact not only that some perfumes may be intentionally structured to be expressive of both feelings and ideas, but that those who appreciate them may experience and imaginatively judge them *as* attempts at expression, something that is clear from perfume reviews.[30]

There are also obvious parallels between this level of aesthetic appreciation of form and expression in perfumes and the aesthetic appreciation of form and expression in a fine wine. Kevin Sweeney has drawn attention to the movement from orthonasal to retronasal smell in tasting wine, pointing out that "because of the temporally extended nature of a complex wine, one is apt imaginatively to recall and structurally integrate the wine's qualities."[31] This is one reason that the better wine reviews, like the better perfume reviews, focus on describing the sequencing and imaginative interpretation of sensory experience. As for the question of association versus expressivity raised by Scruton, Cain Todd has argued that the expressive properties of fine wines "are not reducible to mere association" since "the recognition of intention in wine can suffice to turn mere association into expression."[32]

Moreover, the expressive content of perfumes, like that of wines, ranges widely. Although perfumers like Ellena and Roudnitska have tended to stress notions of beauty, balance, and harmony, there are also some "niche" perfumes that play on the edge of offense, just as there are wines that push the envelope of our normal taste expectations.[33] Many of these small niche companies have become commercially viable thanks to the internet, which has also spawned several perfume blogs by connoisseurs, a few of which show an informed interest in the structural and expressive complexity of scents that clearly goes beyond matters of either liking/disliking or wearability.[34] Among niche perfume producers the aesthetic range is wide indeed. At one end of the spectrum are companies like Etat Libre d'Orange, which tells us its *Secretions Magnifiques* is "like blood, sweat, sperm, saliva . . . as real as an olfactory coitus that sends one into raptures."[35] At the other end of the aesthetic spectrum, there are little companies like Juniper Ridge, whose sustainably "wildharvested" fragrance called *Big Sur Backpacker*, we are told, allows us to follow the scent of "deep redwood canyons giving way to endless wildflower meadows under Junipero Serra Peak." Some of the niche perfumes aspire to associations with profounder emotions; Serge Lutens's company even has a soothing one called *De Profundis*, a memorial bouquet of chrysanthemum notes in honor of Oscar Wilde. Whether many niche products will approach the kind of confrontational odors used by the creators of contemporary olfactory or scent art remains to be seen (after all, most niche companies are small and need to connect with consumers who want to wear the perfumes). Certainly, there are already individual perfumers who go beyond merely using contrasting elements and attempt to achieve something like genuine dissonance. Dominique Ropion's *Olfactory Hommage to Francis Poulenc* (2011), for example, attempts to "transcribe the notion of dissonance into perfume," by finding notes that would simulate the clash of transparency and sensual depth in Poulenc's music.[36]

Whether we are thinking of mainstream or niche perfumes, it is obvious that an informed and experienced public is crucial if aesthetically complex perfumes are to be appreciated as a fine art form. Here is the olfactory neuroscientist Andre

Holley describing how knowledgeable persons experience a fine perfume: they "add to their sensory delight ... a cognitive jubilation when they recognize the subtle syntax, the daring joining of notes thought to be incompatible, the elegance of a sober but expressive style."[37] Here we have everything needed to satisfy even a Kantian demand for a play of imagination and understanding resulting in a reflective rather than purely sensory pleasure.

If our arguments about the formal structure, temporality, representation, symbolism, and expressive range of perfume scents, along with the idea of a relatively established discourse for discussing them, are correct, then an aesthetic empiricist or formalist ought to conclude that perfumes can afford aesthetic experiences worthy of fine art status. Of course, that does not mean, as the fictional des Esseintes noted, that every run-of-the-mill perfume "sold in grocery shops and cheap bazars" is a work of fine art any more than every painting, poem, or piece of music we encounter is worthy of fine art status. Cain Todd has argued on similar grounds that some wines may justifiably be considered works of fine art. And several contemporary aesthetic theories of the nature of art, such as those of Gary Iseminger and Nick Zangwill, not only embrace the traditional list of fine arts, but explicitly include such things as industrial design, advertising, and weaving (or in Zangwill's case, even whistling, cake-decorating, arranging rooms, religious rituals, and fireworks displays); their theories would surely have room for perfumes.[38]

A Contextual Case against Perfumes as Fine Art

So far everything looks good for the claim that perfumes can be a fine art form. But there's a problem: purely aesthetic theories of the fine arts have difficulty accommodating many avant-garde artworks, especially those that deliberately aim at resisting traditional aesthetic responses. The main examples usually cited are Marcel Duchamp's *Fountain* (1917), the men's urinal he signed and put in an art exhibition, and John Cage's equally notorious *4′ 33″* (1952), the piano work in which the pianist sits at the keyboard without playing and the music consists of the ambient noses of coughing, rustling programs, passing cars, and so on. Similar artworks have multiplied over the last sixty years as conceptual, installation, performance, and participatory art have come to dominate cutting-edge art making. A good recent example is Cuban artist Tanya Bruguera's *Tatlin's Whisper #6 (Havana Version)* (2009 and 2015), in which audience members were invited to come to a microphone on stage and speak uncensored for one minute in defiance of the Cuban authorities. Aesthetic criteria seem irrelevant to works like these, yet these kinds of works are widely accepted as fine art by art critics.

Faced with such cases, many philosophers have rejected aesthetic definitions of art in favor of theories that define fine art in relation to the context or history within which an artwork is made. Approaches to defining art that emphasize the historical aspect take two forms. One form gives a central place to the *narratives* that establish something as belonging to the fine arts (Noël Carroll); the other form makes the criterion of art the fact that someone *intends* their work to be regarded in the way that past fine artworks have been regarded (Jerrold Levinson).[39] Approaches to defining art that emphasize the contextual aspect, on the other hand, tend to emphasize that what makes something art is its relation to a context of artistic *practices* made up of shared media, norms, roles, and institutions.

Although most versions of the historical/contextual understanding of the category of fine art would be open in principle to admitting new art forms—after all, these approaches originated in an attempt to embrace the new modalities like installation and performance art that emerged from the postmedium turn—this might not apply to perfumes. Given the commercial and functional context of standard perfume creation and consumption, it might seem obvious that both the history and context from which most perfumes emerge is more like the practice of product design than like the practices of the contemporary fine arts. Even so we, need to offer reasons to think perfumery does not fit the historical/contextual demands of typical contemporary fine art practices. In what follows I will briefly describe a contextual model of my own for identifying a fine art practice and use it to compare the practices involved in creating a perfume-like work as part of a contemporary art practice, to the practices involved in creating a contemporary luxury perfume for the commercial market. The theoretical model I am proposing draws together several of the most frequently encountered elements for comparing art and nonart practices, each of which is always open to modification and challenge from within the various practices themselves.[40] As networks of shared assumptions, histories, and activities, the practices involving the fine arts can be analyzed on several levels, but I will focus on the macro level of general characteristics distinguishing the fine arts as a whole from the design arts.[41]

Among the most general features of art practices are roles, intentions, media, norms, and institutions. A great deal could be said about each of these features, but for present purposes I will comment on only three. Given our interest in smell, it is important to keep in mind that an art *medium* does not simply consist of a material, but, as David Davis and Dominic Lopes remind us, requires a set of techniques that turn some resource or vehicle (paint, found objects, body movements, musical tones, ideas, odors, etc.) into a medium for an artwork.[42] Among the many *roles* that go to make up an art practice—our concern is to compare the typical practice of commercial perfume creation by professional perfumers and the typical practice of perfume creation by professional artists—it is important to specify that the role of artist in fine art practices often involves an *intention* to make a statement to an audience, *sanctioned* by presenting it within the context of an art *institution*.[43] Putting these aspects together, we get this contextualist model of a (fine) art practice: someone assuming the role of artist intends to make a statement to others in their role as audience within the context of an art practice by transforming a set of resources into a medium for a work that is *typically* sanctioned as art by presenting through an art institution.[44] On this model, musical works like *Green Aria*, installations like Tolaas's *Smell of Fear and the Fear of Smell*, and perfume-like, smell artworks such as Clara Ursitti's *Self-Portraits in Scent,* discussed previously, are clearly works of contemporary fine art. All involve artistic intentions and are hybrids combining scents with other recognized art media within various normative practices, and their art status is further confirmed by their typically being presented in the context of established art institutions.

But, one might ask at this point, doesn't a contextual understanding of fine art mean that the curator, Chandler Burr, actually turned the classic perfumes exhibited at the Museum of Arts and Design into fine art by presenting them in the *context* of an art *institution*? The answer would have to be yes if the contextual model I am proposing consisted only of an institutional component and the role played by curators or impresarios. In such a narrow version of a contextual view, one could simply make something into fine art by fiat.[45] The purely institutional understanding of context misses the crucial role of both publicly acknowledged normative practices and of artistic intentions. That is why the heuristic model I have proposed includes the notions of historically shaped practice norms, artist's intentions, and media traditions, in addition to the idea of recognition by art institutions.

Let's use this multifaceted model to analyze the practices involved in two artists' generation of perfumes. In 2010, the well-known artist Kiki Smith asked a perfumer to create a perfume from various smells she liked: patchouli, sandalwood, musk, and boxwood, accented with notes of chamomile, fig, and black currant. She issued it in a "limited edition" under the name *Kiki*, and it is sold through the shop of the New Museum in New York. According our method for identifying art kinds, in what ways is the perfume *Kiki* an instance of a contemporary fine art practice? The fact that Kiki Smith is an artist and the perfume *Kiki* is sold in a limited edition by an art museum shop does not by itself mean that *Kiki* is a work of contemporary (fine) art. Smith's having a perfume created under her name did not necessarily express an artistic intention in her role as an artist, nor did the commissioning of the perfume explicitly occur within the context of some currently recognized type of fine art practice, especially since she apparently made no effort to sanction it as art by *exhibiting* it in the museum, but simply had it sold through the New Museum shop. Absent any such explicit indications of her perfume's art status, it seems plausible that the *practice* involved in Smith's commissioning the perfume is only marginally different from the typical practice of other celebrities, whether Lady Gaga or Michael Jordan, who have commissioned perfumes or cologne's under their name.

The second artistic act of perfume generation that I want to analyze is Lisa Kirk's creation of *Revolution Pipe Bomb* in 2007. Stimulated by a chance encounter with Ulrich Lang, the head of an eponymous niche perfume company, Kirk, whose works often deal with political and social issues, decided to commission a perfume on the theme of revolution and use it as part of an installation. As she said in an interview, "If we can't start a revolution at least we can create a fragrance that symbolizes rebellion."[46] She first conducted an informal survey of a number of journalists and political radicals, asking them to tell her what a revolution smells like and got these replies: smoke, tear gas, burned rubber, gasoline, and decaying flesh. She then asked a perfumer at the Symrise scent production company to design a perfume expressing these odors. The resulting perfume has notes of birch tar, ambergris, leather, musk, vetiver, wood, and civet that together give off a dominantly metallic, smoky odor.

Kirk first released *Revolution Pipe Bomb* in a limited edition of twenty-eight in 2008, contained in vessels shaped like a pipe bomb that were designed by the jeweler Jelena Behrend and made of precious metals (silver, gold, platinum). An exemplar of the vessels was initially exhibited at the Museum of Modern Art's PS1 gallery from October, 2007 to January 2008 as part of an installation work that included a simulated lab with peeling wallpaper and Molotov cocktails in evidence, along with materials to suggest the lab was the site of the perfume's production. In March of that same year, *Revolution Pipe Bomb* was released for sale to the public through Participant Inc., an art organization that coproduces limited edition artworks by offering collectors a tax deduction for part of their purchase. There were twenty silver versions at $3,750 each, five gold versions at $27,350, and three in platinum at $47,750. At the Participant Inc. opening, models in ski masks sprayed the perfume on visitors. One can't help wondering, were the bourgeoisie being asked to pay handsomely for the privilege of owing a work of art symbolizing their eventual overthrow? Or did Kirk's work also mock the pretensions of some self-styled revolutionaries and anarchists who play with the external trappings of revolution?

Kirk's *Revolution Pipe Bomb* project is obviously more clearly the product of a fine art intention and sanction than is Smith's *Kiki*, since Kirk's decision to have a perfume composed for her was subsidiary to a larger intention to create a work of art within the context of the wider practices of installation and participatory art, approaches that have characterized much of her other work. One could characterize Kirk's own art practice as oriented toward a critique of consumerism and raising political issues. Of equal importance to this general orientation, Kirk's specific procedure in executing the *Revolution Pipe Bomb* project involved her turning the commissioned perfume from Symrise into a medium for a work of conceptual art that she then used as part of installation and performance works at two art institutions, PS1 and Participant Inc. respectively. We could also apply our practice model to the reception side of Kirk's *Revolution Pipe Bomb* project. According to the norms of most installation and performance art, the audience's role is to complete the work by physically entering the installation or performance site, usually in the context of an art institution, thereby becoming participants in an art project whose approach is rooted in the historical norms of broadly conceptual, postdisciplinary art practices since the 1960s.[47]

Now that we have unpacked some of the reasons that make the *Revolution Pipe Bomb* project a work of contemporary (fine) art practice, let's see what a comparison of it with standard perfumery practices can tell us about the "Perfume is art" claim. In comparison to the practices that generated *Revolution Pipe Bomb*, the practices involved in the production and consumption of even the most complex, imaginative, and aesthetically compelling luxury perfumes often lack the requisite intentions, norms, and institutional contexts of circulation to make them fit comfortably into the wider practices of contemporary art. Among these differences in creative practices, the norm of artistic freedom is perhaps the most important. In the practice of most mainstream perfume creation, perfumers are seldom free to create a perfume for no other purpose than to make an artistic statement for an audience. Rather, the perfume creation process typically begins with what is called a "perfume brief," similar to the design brief in many areas of industrial production. The "brief" guides the perfumer in developing a formula for a product that will satisfy the company initiating the brief and be appropriate for use by the purchaser. Briefs are typically written by major fashion and cosmetic houses, only a few of which actually have perfumers on their staff; instead, the fashion and cosmetic firms commission perfumes through fragrance- and flavor-manufacturing companies that provide the firms with a concentrate made to satisfy the fashion or cosmetic houses' briefs.[48]

A perfume brief typically specifies the general aims and marketing constraints within which the perfumer is to work, although sometimes the brief will even suggest the odor families that should be used in the work. Some briefs are extremely brief, specifying only the general theme, and assuming the perfumer is aware of the kinds of consumers the company typically serves. Other briefs may include such things as how the proposed theme will fit in with the firm's other scents and products, how it will relate to competitor's perfumes, what the characteristics of the target audience are, and, of course, the cost limits of ingredients.

Most perfume briefs also assume a set of other constraining factors that are simply part of the general practice of professional perfumery, such as attention to skin and environmental safety (for which there are legal regulations), general hedonic acceptability (no viscerally offensive or disgusting odors), and chemical and marketing issues such as color, stability, and so on.

Even smaller niche firms that may not be as concerned with guiding potential perfume designers in terms of competitive market position or using the results of consumer panels may still give potential perfume designers a brief that specifies themes and consumer profiles. Thus, when the creative director of the relatively small German firm of Humiecki & Graef, which has a line of eight perfumes, decided that the next perfume should be called *Trust*, he wrote up a brief consisting of three short descriptive statements of ideas he associated with "trust" and accompanied them with three photos he thought exemplified different types of trust relation. The brief even suggested some scent components that seemed to him appropriate. The brief was discussed at length with the two perfumers he had commissioned to produce the scent, Christoph Hornetz and Christophe Laudamiel, who have their own scent design firm, Dream Air.

We actually posses a fairly detailed account of the interaction of the two perfumers with the creative director at Humiecki, thanks to two social scientists studying organizational behavior, who were given access to every stage of the genesis of this particular perfume.[49] According to these researchers, in addition to an initial forty-five-minute discussion between the perfumers and the company's creative director, the two perfumers kept the written brief on their desks and constantly referred back to it as they attempted to come up with a perfume exemplifying "trust." Although the two perfumers amount of freedom was greater than that of many perfumers who work within large corporate settings, they were not free to interpret the "trust" theme without consideration of the creative director's opinions. Interestingly, the social scientists following the development of the perfume referred to their study as illuminating the "emerging field of artistic perfumery . . . the growing niche market characterized by conceptually advanced and experimental fragrances that serve highly symbolic functions."[50] But the constraints the two *Trust* designers and other perfumers experience in their normal contract work are more typical of the constraints and practices of the designers of other commercial products than they are like the situation of most artists, even of artists who decide to create a perfume-like scent. And if such artists commission a fragrance as part of a hybrid work of smell art, as in Lisa Kirk's case, the artists will be the ones issuing a constraining brief to some perfumer.

Moreover, in the perfume industry as a whole, the constraining hand of those who commission a perfume extends far beyond the contents of a brief.

Whether we are talking about large fashion and cosmetics brands like Dior and Yves Saint Laurent or about small niche perfume firms like Annick Goutal or Serge Lutens, there will be a considerable amount of discussion and even guiding and monitoring of the perfumer's work (which may take months or even years) during the development process. The artistic director of a big company or sometimes the CEO of a smaller one is likely to take an active part at almost every stage of a perfume's development. The best of these directors do not see themselves as imposing something, but they do try to make sure the perfumer sticks to the theme and builds toward a product that will be commercially viable for the firm. Even in the case of the respected niche leader Frédéric Malle, who is celebrated for putting his perfumers' names along with his own on his labels, this involvement can be close and intense. Dominic Ropion describes how he and Malle constantly exchanged ideas during the development of the perfume *Portrait of a Lady*, including Malle's suggesting certain notes, a fact Ropion not only accepted but also found stimulating and useful. Ropion says he even values the group conversations when he is working for a larger firm. "If I am an 'artist,' he writes, "it is only in the context of other people, because my creations have no existence without a strong link to their environment. I need to be surrounded by talk, different ideas and opinions, a hubbub that fills my brain and nostrils."[51]

But what about Jean-Claude Ellena's process in creating *Un jardin en méditerranée*, which he describes so eloquently, making it sound like a completely free, poetic, and artistic activity? The fact is that Ellena came up with this perfume partly in response to a thematic brief. In 2002 the director of Hermès's perfume division, who was acquainted with a Tunisian woman in charge of designing Hermès's window displays (the woman also had a garden at her home in Tunis), sent out a brief to a number of perfumers that included the statement: "Make me a perfume that smells of the scents found in this Tunisian garden."[52] Ellena visited the garden, sent in a preliminary scent and won the brief and as it turned out, the success of *Un jardin en méditerranée* led to his becoming the in-house perfumer for Hermès. As in-house perfumer, Ellena is still subject to following the theme set each year by the company's president, which is then given greater specification by the director of the perfume division. Certainly Ellena is much freer in his choice of materials and creative practice than many other perfumers who work directly for one of the big concentrate manufacturing firms like Givaudan or International Flavors and Fragrances, but for all his creativity and artistic flair, he is still constrained by the contextual fact that he is designing not only perfumes to express a theme chosen for him, but also constrained by the fact that the perfumes are primarily intended to be marketed and worn (and thus subject to safety and hedonic acceptability rules). Moreover, the perfumes are presented to consumers through standard commercial institutions, not through art institutions, and they are evaluated by perfume critics and consumers in relation to the norms by which past and present perfumes intended for wear are evaluated. By contrast Kirk decided on her own to create *Revolution Pipe Bomb*, intending it to be experienced primarily as an artistic/political statement in the context of the contemporary art world. Her artistic aim was central, whereas any practical use that the resulting perfume might be put to by its audience, including wearing the scent (which smells pretty gritty) was secondary at best. Moreover, her work was circulated through standard art world institutions and evaluated by art critics and art audiences in relation to the norms for past and present works of conceptual and installation art. An equally important difference in the two practices concerns the place of each perfume in the two creators' individual bodies of work. Whereas Kirk's works before and after the *Revolution* project have nearly all been mixtures of conceptual, installation, or performance practices, such as *House of Cards* (2008), which mocked real estate speculation by offering visitors the chance to buy time shares in a small shanty she constructed, Ellena has devoted nearly all of his distinguished career to designing perfumes intended for wear and to be circulated through standard perfume channels. In calling attention to the various ways in which Ellena's process, like that of most perfumers, is normally subject to constraints typical of the practice of design, I wish in no way to downplay the creativity and artistry demanded of the best perfumers, only to call attention to many differences in the practice contexts of contemporary art compared to contemporary design professions like perfumery.

An Impasse

Is perfume a fine art form? From the perspective of an aesthetic understanding of art, many commercial perfumes clearly have all the formal, cognitive, and expressive potential it takes to be (fine) art and the most complex and imaginative perfumes are worthy of the kind of careful attention paid to works of the established (fine) arts and their creators deserve to be called "artists." From the perspective of a contextual understanding of fine art, most standard and even many niche perfumes, however much creativity they show, still lack crucial elements of the practices that characterize the contemporary art world and should continue to be treated as part of design and their creators considered "designers."

We seem to be at something of an impasse. The next chapter will attempt to break the deadlock between the aesthetic and contextual approaches to the nature of art and to the question of whether perfume should be considered one of the fine arts.

Notes

1. Baudelaire, *Les Fleurs du Mal* (Paris: Calmann Levy, 1866), 59. "Perfume" is the second part of a poem called "Un Fantome" that was not included in the first edition of 1857. The translation is my own. I am grateful to the Knox College library for giving me access to the volume.
2. From the poem beginning "I taste a liquor never brewed" in Emily Dickinson, *Complete Poems*, ed. Thomas H. Johnson, (Boston: Little Brown and Company, 1960), 98–99.
3. Huysmans is thought to have modeled des Esseintes On the notorious aesthete and dandy of the late nineteenth century, Count Robert de Montesquiou.
4. J.-K. Huysmans, *Against Nature*, trans. Brendan King (Sawtry, UK: Daedalus, 2008), 136.
5. Quoted in Pollack, "Scents & Sensibility," 92.
6. The catalog text of the exhibition never discusses the issue of fine art status, but simply offers rhetorical tropes from art history such as calling *L'Interdit*, released in 1957, "Abstract Expressionism" or *Eau d'Issy*, of 1992, "Minimalism." Burr, *Art of Scent, 1889–2012* (pages unnumbered).
7. Robert R. Calkin and J. Stephan Jellinek, *Perfumery: Practice and Principle* (New York: John Wiley & Sons, 1994).
8. Edmund Roudnitska, *L'esthétique en question: Introduction à une esthétique de l'odorat* (Paris: Presses Universitaires de France, 1977).
9. That may be why the olfactory psychologist Avery Gilbert called Burr's strategy in the "Art of Scent" exhibition "reactionary." Gilbert, *First Nerve*, December 7, 2012, http:// www.firstnerve.com. For a stimulating collection of essays on multiple aspects of beauty in aesthetic discourse see Peg Zeglin Brand, ed. *Beauty Unlimited* (Bloomington: Indiana University Press, 2012.
10. Nick Zangwill, *The Metaphysics of Beauty* (Ithaca: Cornell University Press, 2001).
11. Alan Goldman, "The Broad View of Aesthetic Experience," *Journal of Aesthetics and Art Criticism* 71, no. 4 (2013): 330. The definition of the aesthetic is, of course, a deeply contested issue in philosophy. For an overview see James Shelley, "The Concept of the Aesthetic," *Stanford Encyclopedia of Philosophy*, ed. Edward N. Zalta (Winter 2017 ed.), https://www.plato.stanford.edu/archives/win2017/entries/aesthetic-concept/.
12. Traditionally, the focused attention aspect of aesthetic experience was called "disinterestedness," but many theorists now prefer to follow Arnold Berleant and speak of "engagement" in order to stress that aesthetic experience is a matter not only of careful attention but also of affective involvement. Arnold Berleant, *Art and Engagement* (Philadelphia: Temple University Press, 1991). The inclusion of "other aesthetic" properties in my list reflects the widespread acceptance in philosophical aesthetics of Frank Sibley's demonstration that there are innumerable aesthetic properties beyond the eighteenth-century triumvirate of the beautiful, the sublime, and the picturesque. Frank Sibley, "Aesthetic Concepts," in *Approaches to Aesthetics*, 1–23.
13. Gary Iseminger, *The Aesthetic Function of Art* (Ithaca: Cornell University Press, 2004); Jerrold Levinson, *Aesthetic Pursuits: Essays in Philosophy of Art* (Oxford: Oxford University Press, 2016).

14. Goldman, "The Broad View of Aesthetic Experience," 332.
15. Marina Jung Allégret, *Aux Sources du Parfum: Un Essai pour tout comprendre sur cet art merveilleux* (Paris: Vérone editions, 2016), 25–9. As Allégret points out, there is a certain ambiguity in the distinction between a "note" and an "accord," since a perfumer might use more than one accord (or established unit of several notes) in which case the accord becomes in effect a note in the overall structure (*Aux Sources du Parfum*, 60–62).
16. Annick Menardo, in an interview with Denys Beaulieu, *Nez: The Olfactory Magazine* 4 (Autumn–Winter 2017), 63.
17. Jean Claude Ellena, *Perfume: The Alchemy of Scent* (New York: Arcade Publishing, 2011), 66–68.
18. Menardo, *Nez*, 4, 63.
19. Calkin and Jellinek, *Perfumery*, 3–4.
20. Roudnitska, *L'esthétique en question*, 70.
21. Ellena, *Perfume*, 54.
22. Ellena, *Perfume*, 46.
23. Dominique Ropion, *Aphorisms of a Perfumer*, trans. Annie Tate-Harte (Paris: NEZlittérature, 2018).
24. Scruton, *I Drink Therefore I Am*, 163.
25. Sibley, *Approach to Aesthetics*, 248; Dutton, *Art Instinct*, 212.
26. Headspace technology was first developed in the 1980s and uses an airtight dome placed over an object. Inert gases are then pumped in to remove odor compounds, which are trapped and then analyzed by gas chromatography and mass-spectrometry.
27. Ellena, *Perfume*, 53.
28. Ellena, *Perfume*, 54.
29. Ernst Gombrich, *Art and Illusion: A Study in the Psychology of Pictorial Representation* (Princeton: Princeton University Press, 1961).
30. For examples of knowledgeable perfume reviews see Luca Turin and Tania Sanchez, *Perfumes: The Guide* (New York: Viking Penguin, 2008).
31. Sweeney, "Can Olfactory Sensing Lead to Imaginative Aesthetic Experience?," 7.
32. Todd, *Philosophy of Wine*, 143.
33. Carolyn Korsmeyer has offered a penetrating account of the way aversive elements play a necessary role in intensifying and deepening aesthetic absorption and meaning in *Savoring Disgust: The Foul and the Fair in Aesthetics* (Oxford: Oxford University Press, 2011). I am indebted to Barry Lee of the University of York for reminding me of the "darker" perfume experiments when I was in the early stages of researching this book.
34. Ellena, *Perfume*, 64. One such niche perfume company is Ulrich Lang New York, whose head, Ulrich Lang, put Lisa Kirk in touch with Patricia Choux of Symtech and later helped Kirk market the 2010 version of *Revolution*.
35. Although I don't think that the sample I tried lives up to the hype, it certainly could not be accused of being conventionally pleasant.

36. For a discussion of Ropion's homage to Poulenc and similar works by other perfumers see Marie-Anouch Sarkissian, "Trois oeuvres d'art olfactiives mettant en pratique le concept d'interface Parfum-Musique," in Jaquet, *L'Art olfactif*, 155–69.
37. André Holley, "Les trois piliers de l'art du parfum," in Jaquet, *L'Art olfactif*, 55–61.
38. Iseminger, *The Aesthetic Function of Art*, 100–101; Nick Zangwill, "Are There Counterexamples to Aesthetic Theories of Art?," *Journal of Aesthetics and Art Criticism* 60 (2002): 116.
39. Noël Carroll, "Historical Narratives and the Philosophy of Art," *Journal of Aesthetics and Art Criticism* 51, no. 3 (1993): 313–26. Similar arguments can be found in Carroll's *Art in Three Dimensions* (Oxford: Oxford University Press, 2010), 19–52. Jerrold Levinson, "The Irreducible Historicality of the Concept of Art," *British Journal of Aesthetics* 42, no. 4 (2002): 367–79.
40. For a more extended discussion see my "Art Scents: Perfume, Design and Olfactory Art," *British Journal of Aesthetics* 55, no. 3 (2015): 382–83. Two works that initially influenced my model were David Davies's *Art as Performance* (Oxford: Blackwell, 2004), and Peter Larmarque's "Wittgenstein, Literature, and the Idea of a Practice," *British Journal of Aesthetics* 50, no. 4 (2010): 375–388. The model's emphasis on the historical aspect of practices is indebted to Alasdair MacIntyre's *After Virtue* (Notre Dame: University of Notre Dame Press, 1981), 62, 190–95, as well as to Noel Carroll's narrative approach. Among the first to work out this type of approach was George Dickie in the second version of his "institutional" theory of art. See *The Art Circle: A Theory of Art* (New York: Haven, 1984). I first presented the model used here in a paper at the ASA Pacific Division meeting at Asilomar in 2014 and am grateful to Gary Iseminger for helpful comments. That same year Dominic Lopes presented a somewhat different model in his *Beyond Art* (Oxford: Oxford University Press, 2014). Finally, one may consult with profit Nicholas Wolterstorff's adaptation and modification of MacIntyre's approach to social practices in *Art Rethought: The Social Practices of Art* (Oxford: Oxford University Press, 2015), 83–106.
41. At the next level down one can distinguish within the fine arts among conceptual, installation or performance modes, etc., or, in the case of design, among fashion, product, and graphic practices. And within each of these more specific levels of practice one can identify the features of an individual artist's or designer's specific practice.
42. On the relationship between vehicle and medium, see Lopes, *Beyond Art*, 133–44, and Davies, *Art as Performance*, 56–62.
43. In giving a central role to intention, I am not only drawing on Levinson's historical definition, but also modulating it with the help of Sherri Irvin's important notion of "artistic sanction," which shows some of the ways in which an intention need not be verbally expressed. See Sherri Irvin, "The Artist's Sanction in Contemporary Art," *Journal of Aesthetics and Art Criticism* 63, no. 4 (2005): 315–26.
44. Someone will immediately object that there are "loner" artists or "outsider" artists who aren't grasped by this model. But this is a heuristic model, not an attempt at an essentialist definition using necessary and sufficient conditions.

45. Philosophers familiar with the debates of the 1970s and 1980s surrounding the "institutional" theories of art will recognize in these comments the classic objection made by Arthur Danto and Richard Wollheim to George Dickie's initial version of his "institutional theory."
46. Pollock, "Scents and Sensibility," 94. Much of Kirk's practice is "conceptual" in the broader sense critics now give the category "conceptual art." For an excellent philosophical airing of these issues see Peter Goldie and Elisabeth Schellekens, eds., *Whose Afraid of Conceptual Art?* (London: Routledge, 2009).
47. I should note that each of the pipe bombs that were the key element in Kirk's work could be opened and did contain a wearable vial of the scent, although none of the purchasers has contacted her for a refill. Personal communication from the artist. For a brief historical and theoretical analysis of installation art see Claire Bishop, *Installation Art: A Critical History* (London: Tate Publishing, 2005). Participation art (sometimes called "relational aesthetics") is a newer and less settled category. See Claire Bishop, ed., *Participation* (London: Whitechapel Gallery, 2006).
48. For a general discussion of perfume briefs see Charles S. Sell and David H. Pybus, *The Chemistry of Fragrances: From Perfumer to Consumer* (Cambridge: Royal Chemical Society Publishing, 2006), 138–42.
49. Nada Endrissat and Claus Noppeney, "Materializing the Immaterial: Relational Movements in a Perfume's Becoming," in *How Matter Matters: Objects, Artifacts, and Materiality in organization Studies*, ed. Paul R. Carlisle et al. (Oxford: Oxford University Press, 2013), 58–91.
50. Endrissat and Noppeney, "Materializing the Immaterial," 59.
51. Dominique Ropion, *Aphorisms of a Perfumer* (Paris: NEZlittérature, 2017), 88.
52. Chandler Burr, *The Perfect Scent: A Year inside the Perfume Industry in Paris and New York* (New York: Henry Holt, 2007), 6–7.

12
Perfume between Art and Design

From "Art" to Art

On a visit to the modern wing of Chicago's Art Institute some time ago, I overheard this exchange between a boy, who looked to be ten or eleven, and his father. They were walking around a huge tree trunk lying on its side that was the sole object in the gallery, taking up most of the floor. I had visited the gallery many times and always liked this work, especially after I stopped to read the explanatory panel by the artist, Charles Ray. *Hinoki* ("cypress" in Japanese) is the replica of a tree he had come across driving through the California woods; he was so taken by it that he had it carted to his studio and sent it in pieces to Japan, where expert carvers reproduced it as exactly as possible. I have enjoyed watching people's reactions to it. Many don't stop to read the panel, but simple stare at the tree a moment and move on. But that day, as the father and son paused briefly, I heard the boy ask, "What's that?" Answer: "A tree." "Why are we looking at it?" "It's art." As a philosophy teacher, I held my breath, hoping the next question would be "*Why* is it art?" but the two fell silent and moved on.

Although some philosophers are often interested in listening in on ordinary talk about art, many are suspicious of what they call "folk concepts." After all, the thinking embedded in our common usages is often highly ambiguous if not contradictory. And surely the little word "art" is a particularly unruly one. On the one hand we use it to mean the visual arts in contrast to the musical arts. But we also use it to mean the (fine) arts as a group, which includes music, dance, theater, and literature as well as photography, film, and so on. But even more pervasive than our use of "art" to refer to fine or high, or capital-*A* Art, is the use of "art" in the small-*a* sense. The small-*a* sense covers an enormous span of things and activities from pottery, archery, and cooking to teaching, politics, and medicine—as when we say medicine is "an art" as well as a science, or when we speak of the "the art of" motorcycle maintenance or the art of exciting a crowd at a political rally. In ordinary usage, then, art with a small *a* can refer to almost any human activity done with some skill, intelligence, and grace. But there is at least a fourth major way we use "art" that is especially germane to our interest in whether we should call things like perfume or fashion "art," namely, when we use "art" with a modifier to name various groupings of arts that seem to stand closer to the so-called fine arts than the everyday arts of cookery, carpentry, or

Art Scents. Larry Shiner, Oxford University Press (2020) © Oxford University Press.
DOI: 10.1093/oso/9780190089818.003.0025

motorcycle maintenance, yet are often contrasted to the fine arts as categorically different and of lower standing: applied arts, craft arts, decorative arts, commercial arts, design arts. Moreover, some of the confusion over what is and is not Art (fine art) stems from the fact that so many writers, including philosophers, use "art" or "arts" without any modifier, leaving it wholly to context to determine what is meant, whether fine art or some other kind of art.

Many philosophers find these overlapping usages of ordinary talk about art altogether too messy. After all, our everyday linguistic uses, even though we seem to get by pretty well with them, do at times appear inconsistent if not contradictory. For example, in ordinary talk we often use the term "art" in a way that suggests we believe all the human arts form a vast continuum with no sharp internal boundaries or status divisions, but at other times we seem to accept the traditional hierarchical divisions of the arts into capital *A* versus small *a*, high versus low, major versus minor, or the fine versus the many applied, design, and everyday arts—as if those divisions did have relatively clear boundaries. Philosophy's responsibility, as some see it, is one of correcting, radically if need be, our ordinary concepts and linguistic usages by developing logically consistent theories.[1] On the other hand, there is also a stream of philosophy that tends to think everyday language and practices often get things about right, and philosophy's job is to help clarify or reconcile contradictions in ordinary usage rather than generate alternate theories. As Noel Carroll has pointed out, although the highly variable and sometimes inconsistent criteria many of us use in talking about the arts may not be reducible to a classic *definition*, nevertheless, our grab bag of mixed criteria does allow us to *detect* fine art in many situations.[2] For example, when we come across something unexpected, we can often, like the father and son at the Art Institute, quickly classify it and move on. The father passing the tree replica may have assumed that since it was in an art museum, it must be (fine) art, or at least is considered so by experts in the matter, even if he did not know their reasons.

Although some philosophers of art do acknowledge the vast realm of art with a small *a*, many are so focused on developing theories to explain what is or is not *fine* art that they simply contrast "art" to "nonart." Unfortunately, that practice often leaves the enormous number of both everyday and intermediate arts in a kind of theoretical limbo. Yet when confronted with controversial issues such as the art status of food, wine, fashion, or perfume, even philosophers and art critics as well as laypeople too often fall back on the traditional hierarchies of high versus low, major versus minor. In his book *What Art Is*, for example, Arthur Danto contrasts Warhol's *Brillo Box* made of painted plywood as "fine art" to actual Brillo box cartons, whose design he calls "commercial art." The Chicago Art Institute curator, Zoë Ryan, on the other hand, has applied the same polarity in the opposite direction, claiming that the work of certain fashion designers is not "purely commercial" because their conceptual approach "elevates" their designs "to the status of fine art."[3] In *Making Sense of Taste*, Carolyn Korsmeyer invoked both the fine art versus applied art and major versus minor contrasts in arguing that fine cuisine and wine should not be considered fine arts, whereas Cain Todd, in *Philosophy of Wine*, took the opposite tack, wanting to "elevate fine wine on to the pedestal of art."[4] Given the persistence of such hierarchies and the frequent talk of "elevation," it is understandable that many perfumers and perfume lovers, as well as the curator of the *Art of Scent* exhibition at the Museum of Arts and Design, have been eager to "elevate" perfumes onto the "pedestal" of the fine arts.

In the previous chapter we considered the claim that perfume is fine art in the light of the two most prevalent contemporary philosophical approaches to bringing our unruly concept of art to heel, the aesthetic and the contextual/historical approaches. But these ended in a deadlock. The aesthetic approach seemed to be quite comfortable with including perfume among the fine arts, but the contextual approach led to strong reasons to suggest it belonged to the intermediate realm of the design arts instead. Despite this conflict, each approach did afford many insights into the nature of perfumery whether one calls it an art practice or a design practice. Nevertheless, it is time to see if there is a way out of the impasse.

Ways Out of the Impasse between Fine Art and Design

Given the plethora of competing definitions of art among contemporary philosophers, it would be tedious to survey all the ways of overcoming the impasse between the aesthetic and contextual approaches to the question "Can perfumes be fine art?" Instead, I will examine two general directions. The first direction takes the obvious route of attempting some combination, often in the form of a disjunction, of the aesthetic and the contextual/historical definitions. The second direction attempts to get beyond the aesthetic/contextual divide by turning away from any attempt to define (fine) art and adopting a "local analogies" strategy.

Combining Aesthetic and Contextual Theories

There are several prominent ways of combining aesthetic and contextual criteria to produce a concept of (fine) art. At one end of the spectrum of composite theories are the highly pluralistic "cluster" views, according to which something can be an art form if it possesses or promotes any combination of art-making characteristics (how many is left vague). The characteristics often include promoting aesthetic responses, manifesting formal complexity or skill, expressing emotion, being original, being presented in an art context, and so on. Cluster theories are appealing because they tend to follow closely our everyday or folkways of detecting (fine) art. By using such an open-ended set of criteria, perfume, as well as almost any other art practice in the small-*a* sense, could be claimed as a *fine* art form. But cluster theories seem more like ways of redescribing ordinary usage than trying to clarify it, let alone resolve conflicting views.

More theoretically satisfying are several of the more structured composite or disjunctive theories. One kind of combinatory theory subordinates one side to the other. The most informative example is Gary Iseminger's theory of art that reinterprets such contextual criteria of fine art as intentions, practices, or institutions (the art world) as themselves primarily serving an aesthetic function. Specifically, "The function of the artworld and the practice of art is to promote aesthetic communication."[5] In this way, Iseminger is able to bring both the (fine) arts and the decorative and design arts under the wing of aesthetic functionalism and to include in (fine) art "some parts of the practices of furniture design, cooking, gardening, glassblowing, wood-turning, and the like."[6] Certainly, if "some parts" of furniture design, cooking, and gardening can be accepted as fine arts, so could some perfumes.

Several other combinatory theories of art attempt a more even balance between the aesthetic and the contextual/historical aspects of their definitions, such as the theories of Robert Stecker or Stephen Davies. Each of them tries to construct a definition broad enough to embrace both the traditional fine arts as well as avant-garde works like Duchamp's *Fountain* or Cage's *4′ 33″*. And Davies also wants a definition that can include the arts of non-Western cultures along with prehistoric cave paintings and carvings. Since there was nothing like what we call an art world context with its specialized institutions fifty thousand years ago, Davies claims that these works must be considered "art" for aesthetic reasons. Thus, he comes up with this combination definition: something is art if it "(a) shows excellence of skill and achievement in realizing significant aesthetic goals . . . *or* (b) it falls under an art genre or art form established and publicly recognized within an art tradition . . . *or* (c) if it is intended by its maker/presenter to be art."[7] What is interesting about Davies's definition in relation to perfume is that, although the aesthetic (a) part of his definition is motivated by finding a way to embrace Paleolithic art, Davies himself is ready to apply the aesthetic criterion to cars, furniture, and clothes, although specifying that "only the most superb examples . . . should be accorded art status."[8] Similarly, Robert Stecker grants that disjunctive definitions like his own and Davies's imply that "almost anything can be art," but adds the codicil that whatever falls outside the currently accepted "central art forms" will have to "meet a higher standard."[9] Given the definitions of Davies and Stecker, it seems likely that the best perfumes, like those of Ellena or

Grosjman, would meet Davies's criteria of "superb examples" or Stecker's "higher standards."

But not all philosophers who construct composite theories are as optimistic about including "almost anything" in (fine) art. Although another Davies, David Davies of McGill University, has also suggested we should define fine art by combining contextual and aesthetic elements, he believes any such combinations, including his own, cannot draw "a clear boundary between art and non-art." Should we, he asks, include in (fine) art "such practices as carpet-weaving, figure-skating, cake design, and pottery?" Such activities "have always occupied an uncertain position on the fringes of the artworld, and this suggests we should not be looking for a sharp division between art and non-art at this level of analysis." Yet David Davies also thinks that "within an acknowledged system of the artworld" such as ours, the habit of not including such things in the fine arts is "about right," which would suggest that he might let something like perfume linger on the "fringes of the artworld."[10]

Apart from David Davies's objection to including "fringe" arts within fine art, the other composite approaches we have examined, whether Stecker's and Stephen Davies's disjunctive definitions or Iseminger's "aesthetic institutionalism," seem open to admitting the best examples of hitherto excluded practices like fashion design or perfumery into the (fine) arts. Accordingly, should we consider the impasse resolved? It seems that all we need to do in order for perfume to become a fine art form is accept any one of these attractive and carefully constructed theories. Yet, with the exception of Iseminger's subordination of contextual practices to their aesthetic function, none of the disjunctive accounts truly integrates the aesthetic and contextual approaches, but simply allows us to alternate between one and the other. Thus, we could use aesthetic criteria to justify making a perfume like *Un jardin en méditerranée* a work of fine art, and switch to contextual criteria for understanding *Revolution Pipe Bomb* or other avant-garde artworks whose main point is not to provoke a traditional aesthetic response. But the permissiveness of disjunctive theories is so broad that they indeed let in "almost anything." Although it might seem good for perfumery that so many different theories of the nature of (fine) art would enfranchise perfume as Art with a capital *A*, if hardly anything is excluded by these theories, the victory seems a bit hollow. Moreover, allowing almost anything into fine art demands a more radical revision of our ordinary way of thinking and talking about and interacting with the long-established arts than many people would be willing to accept, since it would imply abolishing the difference between art and entertainment, or art and craft, or art and design. I believe there is, indeed, a sense in which those distinctions no longer hold in their traditional hierarchical form, certainly not when accompanied by the invidious implications of superiority/inferiority or by the gender, race, and class biases they once had. In the postlude to Part III, which follows this chapter, I will suggest a radical revision of the high/low, major/minor, fine/applied distinctions. But first I want to explore a different kind of solution to the aesthetic versus contextual impasse that I find appealing because it would preserve the distinction between (fine) art and design, but without the invidious status connotations still clinging to the concept of "fine art."

Local Analogies and the Possibility of "Art Perfumes"

Dominic Lopes argues in *Beyond Art* that we ought to set aside the attempt to find a global definition of fine art or the fine arts (actually he seldom uses the term "fine art," but just contrasts "art" with "nonart"). Instead of trying to define the category of art as a whole, he argues that we should pursue the more modest project of considering each of the arts individually. In the case of arts knocking at the door of what he calls the "art club," he suggests that one way to check their qualifications for entry is to compare them to other practices that were also once outside but are now almost universally accepted as members (photography, jazz, quilts, computer art, etc.). Such comparisons may uncover some analogies that would help us answer the contextual objections to perfume's entry that we encountered in the previous chapter. After all, even arts like photography had to prove themselves through decades of debate about their qualifications to be considered fine art before they were allowed into the club, and even now not all photography is considered fine art photography.

One thing that should encourage us to follow a local analogies approach is that the category of the fine arts itself is hardly something eternally fixed but has been continually evolving. From the first uses of the term "fine art" in the eighteenth century, the list of arts has included a core of four or five arts (painting, sculpture, poetry, and music plus architecture) to which various theorists, even at the time, added one or two other arts, such as engraving, dance, garden design, or rhetoric. And membership kept changing. By the late nineteenth century, garden design and rhetoric had dropped off many lists, while engraving, ballet, and the novel were generally accepted, and photography (invented in 1839) was knocking at the door. By the 1950s not only photography, but film and jazz had joined most lists of the fine arts and, as mentioned earlier, from the 1960s on there has been a veritable explosion in the number of art forms that most art theorists and philosophers are willing to include in the fine arts. Lopes himself mentions the broad acceptance today of such things as earthworks, installation art, happenings, popular music, quilts, street art, computer art, videogames, music videos, even comic books.[11] In the light of this history, it is not surprising that Lopes has referred to the fine art category as a "clique," a term that conveys something of the class, gender, and racial baggage the fine art tradition has often dragged with it until recently.[12]

Another virtue of Lopes's local analogies approach is that once we give up trying to construct a single, unified theory of (fine) art or of the (fine) arts as a group, we are free to use different theoretical approaches for theorizing different individual arts and/or to combine aspects of hitherto opposed theories, but on a strictly local level.[13] By treating each case locally and independently, we would avoid the implication of some of the composite general definitions that we can admit perfumes into fine art on an aesthetic basis, and simply ignore the contextual arguments that have been made against their fine art status. That sort of composite strategy would win the case too cheaply and in the long run would probably not support a broad acceptance of perfumes as a legitimate fine art form. What I propose instead is that we see where Lopes's local analogies approach might lead in the case of perfume. That means considering the arguments and strategies used by proponents of other arts that were once outside the fine art club, but that, unlike perfume, have now been widely admitted.

As a first step, I suggest we evaluate the "perfume is art" claim in the light of a distinction often employed by art critics, curators, and historians with respect to photography and quilts. Let's begin with photography. Many curators and art historians distinguish between the exhibition and study of photography *as art* (selected travel, scientific, and journalistic photographs originally made for practical purposes) and the exhibition and study of *art photography* (a now recognized [fine] art practice that explores the aesthetic possibilities of the photographic medium and has its own history and canon of masters and masterpieces).

But there is a second sense in which critics and curators sometimes speak of "art photography," namely, to describe the use of photographic processes by contemporary artists who may be less interested in exploring and advancing photography as an independent art medium than in exploiting it as part of conceptual or installation works. The theorist Lucy Souter calls the first kind of art photography "modernist" or "fine art photography" and the second "postmodernist," or "contemporary art photography." Whereas art photography in the modernist, fine art vein has stressed formal and expressive values along with technical excellence (Edward Weston), the postmodernist or contemporary art tendency in photography often rejects such values as originality, craftsmanship, and personal expression in favor ideas and social commentary (Andreas Serrano). Moreover, the two approaches differ in practical ways: the modernists often call themselves "photographers" and show their work in specialized fine art photography galleries, whereas the postmodernists typically call themselves "artists" who happen to use photography and show their work in contemporary art galleries. Yet as Souter points out, there has always been considerable overlap between the two worlds, particularly in recent years. Thus, Jeff Wall, who has been critical of modernist "fine art" photography and typically produces huge, arranged tableaux in light boxes as a way of exploring the nature of representation, nevertheless, attends carefully to both the formal considerations and high production values that were and are part of modernist photography.[14]

Although the distinctions between appreciating a photograph made for practical purposes "as art" and appreciating either of the two kinds of "art photography" leave plenty of room for borderline cases, I believe the distinctions suggest a useful way of thinking about the "perfume is art" claim. The Museum of Arts and Design perfume exhibition was clearly engaged in presenting "perfume *as art*," even if its curator talked as if there were already a tradition of *art perfume*, by declaring a canon of masterpieces and applying fanciful historical parallels with the visual arts. But, of course, no such art historical tradition or practice exists in the sense that one exists for art photography. Here again a comparison with Kirk's *Revolution Pipe Bomb* is revealing. Kirk's *Revolution Pipe Bomb* could only be considered an *art perfume* in the second sense of "appreciating *art photography*," for example, in the way we might speak of Andres Serrano's works as instances of art photography. In much of his work, Serrano seems less interested in exploring and expanding the medium of photography than in using photographs to make what are primarily conceptual art pieces, just as the perfume Kirk commissioned for *Revolution Pipe Bomb* was a vehicle for a work combining conceptual and installation approaches to art making, rather than an attempt to explore the possibilities of perfume as an independent art form.

So what would it take to move from exhibiting commercial perfumes *as* art to a theoretically justified practice of creating and appreciating *art perfumes*? Finding analogies and paths to fine art acceptance for perfumes is going to be a lot more difficult than it was to find paths for photography or quilts to gain admittance, since there has never been a widely recognized art form that is addressed to the sense of smell in the way that art photography or art quilts could be compared to painting, which is typically addressed to vision. Even what we have called "scent art" or "olfactory art" cannot as yet provide the kind of analogies Lopes proposes since, as we have seen, most of these works are hybrids and the category of "olfactory art" is not that widely recognized. Consequently, we may have to draw on partial analogies with several existing arts and seek several "paths" that perfume might take toward fine art status.

Among the arts forming the traditional core of the fine arts that might provide a path toward fine art status, music immediately comes to mind. Not only does the practice of perfumery already use a number of musical concepts (composition, notes, accords) but, as we have seen, perfume also raises some parallel critical issues (multiple instances, relations of type and token). It also seems that there are possible general analogies between sounds and odors (temporality, evanescence, invisibility). And then there are the parallels between scent/olfactory art intended for galleries and museums and sound art, which is also typically intended for galleries or museums. Yet despite these rather general parallels, "music" may be too broad a category for comparison. In any case, music has always been one of the fine arts, so that it cannot truly be an instructive model for how something gets added to an existing set of the fine arts. Hence, I will turn to analogies with arts that were once clearly outside the fine arts club but now have a relatively secure place within it.

Let's go back to photography. The principal reasons for excluding it from fine art were based on the following claims. First, it was said, machine technology does most of the work, so that photography is primarily a technical craft lacking the scope for the kind of intelligence and imagination one finds in the best painting and sculpture. Alfred Stieglitz replied that in the hands of a mere craftsman, photography may be a matter of mechanical reproduction, but in the hands of a genuine artist, its point becomes *"what you have to say and how you say it."*[15] In the previous chapter we made similar aesthetic arguments for the cognitive and expressive aspects of perfume creation. Two other objections to photography were similar to the contextual arguments we made against perfumes as fine art, namely that photographs are produced as multiples and are often made and sold for instrumental purposes. Advocates of photography's art status answered these argument by pointing out that graphic works by great artists like Rembrandt were often produced in large runs, and that art photographs, like prints, can be made in numbered, limited editions and intended for contemplation in art galleries or museums rather than used for some practical purpose. In the case of perfumes, a similar reply to the multiples/instrumental charge would be that perfumes, like photographs, can also be issued in numbered, limited editions and can sometimes be created with the intention that they be presented in the context of an art gallery or museum, and not simply exhibited as part of campaign to gain fine art status.

In fact, as we mentioned in the previous chapter, at least one prominent professional perfumer has already begun to create perfumes that he intends to be experienced as artworks and has exhibited in fine art venues, Christophe Laudamiel's art perfumes presented in "scent parabols," which he showed in Berlin and New York art galleries. These works, I believe, are a harbinger of something like a category of art perfumery similar to art photography. The gallery visitor or the collector raises the lid of the parabols to inhale and contemplate the scent coming from the knobs of chalk they contain. With the scents presented in these parabols we have works that are clearly an extension of perfumery practice, but are also intended as artworks to be presented in an institutional art context. Moreover, the formulas for the scents embedded in the nodules inside the bowls are constructed in a way similar to standard perfumes and given titles, yet these named scents are not meant for wear but purely for the enjoyment of their formal and expressive character. The galleries representing Laudamiel do not sell bottles of the scents separately (many of which would not be suitable for wear anyway); rather Laudamiel offers purchasers of his work a guarantee of replenishment. Each vial of scent comes with a signed and dated parabol and a certificate of authenticity. By selling his olfactory art works to art collectors in limited, numbered editions in this way, Laudamiel is following the standard procedures used for fine art prints or fine art photographs.

Let's turn now to some different analogies, those suggested by the history of the art quilt. In 1971, the Whitney Museum of American Art presented a collection of traditional Amish quilts, pointing out that their designs bore a strong formal resemblance to the geometric abstraction of modernist painting. At that time quilting was still a female-identified craft or decorative art, yet the idea of quilts as an art form quickly caught on. This rapid acceptance was helped along, of course, by the feminist movement of the 1970s, with its critiques of the exclusion of women painters and sculptors from art histories and of women's underrepresentation in art museums. By the late 1970s many critics and curators had begun to distinguish between exhibiting traditional quilts *as* art, as the Whitney had done, and *art quilts* deliberately created by artists to hang on the wall and be circulated through art galleries and museums (although selected traditional quilts might be incorporated into art quilt histories and exhibitions).[16]

Importantly, the case in favor of the newly emerged art quilt was not only based on critiques of patriarchy, but also on aesthetic arguments celebrating the quilts' formal and expressive originality. But frequently there were also contextual differences between art quilts and traditional quilts, such as many artists making their quilts too small to be mistaken for bed covers and facilitating audiences viewing them in the way paintings are typically viewed by hanging them on gallery walls. This made it easier to distinguish "art quilts" from the majority of quilts made by nonprofessional artists who typically took inspiration from traditional patterns and intended their quilts for use in the home or display in traditional quilt venues where they would be judged partly on craftsmanship.

Obviously, an art perfume practice that developed along the lines of the art quilt would involve a similar contextual freedom from traditional forms and functions. Instead of norms related to wearability or a focus on pleasure, harmony, and beauty, art perfumes might favor scents that were intended to be appreciated for their combination of innovative structural complexity with an expressivity and symbolism that challenged the receiver's expectations of what a perfume should smell like. In the previous chapter we already saw instances of some niche perfumes bordering on such ambitions and noted that there are perfume enthusiasts who collect and enjoy scents for their complexity and edginess and not just to wear. Another series of works by Christophe Laudamiel indicate what the next step toward an art perfume on analogy with the emergence of the art quilt might be.

In the spring of 2017, I was able to visit Laudamiel's show called *Over 21* that I mentioned in the introduction. The main room in the Dillion and Lee gallery held a long dinner table with twelve black place mats, but instead of plates on them there were large aluminum canisters of scent, each with a small hole in the top, and instead of cutlery, there were perfume blotters one could dip in the canisters. "Menus" were provided that had the title of each work along with bits of information about the scent ingredients, including the availability and price of each scent in a particular presentational form (as scent squares, parabols, etc.). Among the more interesting scents were *Leather Kings and Queens* (2013) (described as "true ambergris home-made infusion"), *Elephant in Musth* (2007) ("the molecules are somewhat animalic but several are rather grassy, honey, lavender or mimosa like"), and *Green Fairy in Chelsea* (2017) ("Absinth, also called Green Fairy . . . from French recipes mixing Wormwood, Anise, Fennel and a touch of coriander"). Over the years, I have spent enormous amounts of time looking at paintings, photographs, sculptures, and installations in galleries and museums, but I have to say that I personally found this visit a deeply engaging and thought-provoking aesthetic experience. On a purely experiential level these works of Laudamiel's certainly felt as much like contemporary artworks as anything I have seen or heard in other art venues.

If enough perfume creators were to follow in Laudamiel's footsteps and a wider circle of galleries, critics, and members of the art public began to frequent such work, something like a distinct practice of the art perfume similar to art photography or the art quilt might eventually emerge. Once established as a distinct practice apart from the practices of commercial perfumery, the exhibition and history of such art perfumes might, as exhibitions and histories of art photography and art quilts have done, incorporate selected instances of traditional perfumes as forerunners because of their remarkable formal and expressive qualities (the sort of thing the *Art of Scent: 1889–2012* exhibition was hoping to do). The emergence of such institutions as the perfume museum in Grasse, France, and the Osmotèque in Versailles are beginning to create archives that could be the basis for such a history. At the same time, educational institutions have sprung up to teach a broader public how perfumes are made, ranging from the small Institute for Art and Olfaction in Los Angeles to the Grand Musée du Parfum in Paris that opened in 2016.[17]

Of course, the emergence of something like an established practice of creating and appreciating art perfumes as distinct from ordinary perfumes might take a long time. In the case of photography, wide public recognition took over fifty years, although in the case of the art quilt it was less than a decade. Perfume, of course, faces not only a partially undeserved image of superficiality and association with seduction, but also suffers from widespread ignorance and prejudice with respect to odors in general. Even so, one can even imagine an eventual social ritual developing among connoisseurs of perfume similar to the Japanese kodo ceremony in which varieties of niche perfumes are passed around for pleasure and identification.

Art Perfumes, Scent Art, and Standard Perfumes

Let's take stock of how far the local analogies approach has led us in our attempt to resolve the impasse between the aesthetic case for and the contextual case against classifying finer, more complex perfumes as fine art. Recall that we examined the analogies with art photography and art quilts because the composite definitions of fine art seemed not only to admit almost any type of practice into the fine art world but also to ignore the specific contextual arguments in favor of treating perfumes as part of design rather than fine art. Although the local analogies approach of Lopes has allowed us to answer several of the contextual arguments against fine art status for perfumes, it appears to have led to a drawback of its own. For it solves the impasse by dividing perfume into two kinds: a tiny minority of potential art perfumes and a vast swath of standard perfumes, which would seem to remain identified with design. But before concluding that such a division would be a setback for those who seek greater cultural respectability for the best perfumes, we need to ask how such a new art form as art perfumes, if it did become a reality, would relate to olfactory or scent art hybrids on the one hand, and to standard design art perfumes on the other.

Art perfumes as we have imagined them on the model of art photography or the art quilt would not, like the vast majority of olfactory artworks, be hybrids in which odors play an adjunct role, but would be artworks in which scent itself played the dominant role. This is clearly the case with Laudamiel's recent works presented in his scent squares or scent parabols. Second, to be considered art *perfumes*, art perfumes would also have to retain some connection to the ancient art of perfumery, just as the art quilt remains connected to the ancient art of quilting, and art photography remains connected to, even blends into, the many documentary, commercial, and recreational uses of photography. One might think of a possible art perfume art form as an extension of some of the more experimental niche perfumes that perfume fanciers buy and enjoy primarily for their scent with only a secondary concern for wearing them.

Thus, there would not be a sharp line separating works in a possible art perfumes genre and such scent art hybrids as Ursitti's *Self-Portraits in Scent* or Kirk's *Revolution Pipe Bomb*, especially since all would circulate through contemporary art institutions. This kind of overlapping would be like that between the first and second kinds of art photography, where the difference sometimes concerns the degree to which the technical innovations and historical references of a photograph are subordinated to other artistic aims, such as conceptual ones. In terms of perfumery technique, for example, Ursitti's and Kirk's works involved perfumes that showed little in the way of technical innovation or historical reference to other perfumes. The same is not true of Laudamiel's works like the scents shown in his scent squares or scent parabols.

But what about standard perfumes intended for wear? The idea of a possible new art form, the art perfume, would restrict the rest of standard perfumery to the status of a design art along with fine fashion, furniture, cars, and much else. But perfumers and perfume enthusiasts should not despair that this is a negative outcome for the cultural status of the finest perfumes—unless, of course, one accepts the traditional hierarchical polarities that set the "fine" arts not only apart from but intellectually and imaginatively above all other arts, including design. One solution to the question of the art status of the best standard perfumes for wear would be to adopt one of the composite definitions of (fine) art that I discussed at the beginning of this chapter: Iseminger's aesthetic institutionalism or the disjunctive definitions of Stephen Davies or Robert Stecker, who admit into fine art any "superb" or "higher" instances of fashion, furniture, or cars and presumably could admit certain perfumes. Although I am sympathetic to these composite approaches, they could easily reinforce the invidious traditional polarities of high/low, major/minor by suggesting that we simply "elevate" a handful of individual items from the category of design to the status of fine art, leaving everything else in the nether regions. Thus, I would be ready to embrace a composite approach only if it were understood in the context of a nonhierarchical view of the whole spectrum of art categories, especially with respect to the relation of fine art and design art. Although it would be beyond the scope of this book to work out such a nonhierarchical view in detail, the following postlude, "Free Art versus Design Art," will suggest one way to begin.

Notes

1. Aaron Meskin, "From Defining Art to Defining the Individual Arts: The Role of Theory in Philosophies of Arts," in *New Waves in Aesthetics*, ed. Kathleen Stock and Katherine Thomson-Jones (New York: Palgrave Macmillan, 2008), 125–49.
2. I take Noel Carroll's "nondefinition" of art to be along these lines. See *Art in Three Dimensions*, 27–30. See also Lopes, *Beyond Art*, 68–69.
3. Arthur C. Danto, *What Art Is* (New Haven: Yale University Press, 2012), 44; Zoë Ryan, *Bless, Boudicca and Sandra Backlund* (Chicago: Art Institute of Chicago, 2012), 11, 14.

4. Korsmeyer, *Making Sense of Taste*, 144; Cain Todd, *Philosophy of Wine: A Case of Truth, Beauty and Intoxication* (Montreal: McGill-Queens University Press, 2010), 145.
5. Iseminger, *Aesthetic Function of Art*, 23.
6. Iseminger, *Aesthetic Function of Art*, 101.
7. Stephen Davies, "Defining Art and Artworlds," *Journal of Aesthetics and Art Criticism* 73, no. 4 (2015): 377–78. Davies third criterion thus includes historical definitions such as that of Levinson.
8. Davies, "Defining Art," 379.
9. Robert Stecker, "Defining Art," in *The Oxford Handbook of Aesthetics*, ed. Jarrold Levinson (Oxford: Oxford University Press, 2003), 151.
10. David Davies, *Art as Performance* (Oxford: Oxford University Press, 2004), 253.
11. Lopes, *Beyond Art*, 116.
12. For a discussion of this history and its gender, social class, and institutional aspects, see Larry Shiner, *The Invention of Art: A Cultural History* (Chicago: University of Chicago Press, 2001).
13. Lopes, *Beyond Art*, 62, 126.
14. Lucy Soutter, *Why Art Photography?* (London: Routledge, 2013), 1–12. See also Susan Bright, *Art Photography Now* (London: Thames & Hudson, 2011), 7–17.
15. Alfred Stieglitz, "Pictorial Photography," in *Photography: Essays and Images*, ed. Beaumont Newhall (New York: Museum of Modern Art, 1980), 164.
16. Carolyn Ducey and Amy Shell, *Perspectives: Art, Craft, Design & the Studio Quilt* (Lincoln, NE: International Quilt Study Center & Museum, 2009), 6–7; Kate Lenkowsky, *Contemporary Quilt Art: An Introduction and Guide* (Bloomington: Indiana University Press, 2008), ix–xi. As both of these works point out, the terms "art quilt," "studio quilt" and "quilt art" are often used interchangeably.
17. As of this writing the museum has been closed due to financial difficulties. This is most unfortunate since its exhibits, which I had the pleasure of visiting in the summer of 2018, were excellently designed interactive educational opportunities.

Postlude
Free Art versus Design Art

The particular nonhierarchical way of thinking about the relation of fine art and design within an overall scheme of the arts that I am about to propose should not only be of interest to aestheticians and philosophers of art, but also go some way to assuaging the concern of those who admire the most complex and aesthetically interesting perfumes, but regret that I have not given them the elevated position they believe the epithet "fine art" or "Art" with a capital *A* would bestow. In sketching a position on the differences between fine art and design art, I will not be giving yet another "definition" of either (fine) art or of design art in terms of necessary and sufficient conditions, but will draw attention to certain characteristics that both distinguish the two yet allow for a continuity that fosters fruitful combinations.

Although perfumery, like fashion, may produce works as aesthetically interesting and deserving of cultural respect as many works in established fine art forms, I have argued that the typical practices of perfume creation make it a different kind of practice than those typical of contemporary olfactory or scent art that involve hybrids of odors with conceptual, performance, or installation art forms shown in galleries and museums. The primary reason standard perfumery practice is different from the practices of creating olfactory art (which would include possible art perfumes) is that standard perfumery shares with other product design practices a simple but central normative assumption: standard perfumes must serve the practical and symbolic functions of adorning the body. As Glenn Parsons and Jane Forsey have argued in their respective books on philosophy of design, practical function is the key norm among the several norms that distinguish the practice of design art from fine art.[1] Although Parsons has carefully worked out a definition of design ("the intentional solution of a problem by the creation of [plans for] a new sort of thing)," what is most interesting for our purposes is the way he cashes out this definition though a series of comparisons of design practice with several adjacent practices, including (fine) art.[2]

To begin with, he contrasts a capitalized "Design," as something carried out by trained professionals, with a lowercase "design" as a general cognitive activity engaged in by all sorts of people and professions, from legislators who can be said to "design" laws to scientists who sometimes speak of "designing" experiments.

This is an important move and parallels my stress on the difference between "Art" and "art." Among the innumerable small-*d* design practices in modern society, the one closest to a capitalized Design is engineering. Generally, engineers in the modern industrial world design the "insides" of machines or the structural elements of bridges and buildings, whereas professional designers including architects focus on aesthetic "surfaces," "shapes," "volumes," and "user interfaces," the latter often concerned with the safety and ease of the way things function. With respect to fine art, Parsons stresses two crucial differences between (Fine) Art and Design so conceived. First of all, artists working in the context of the art world are primarily concerned with helping us understand and interpret the world in ways that may or may not lead us to change it, but artists need not seek to change the world, whereas designers are engaged in directly changing the world by planning new things or new arrangements of existing things. Second, we think of artists as primarily expressing *themselves* in their work, even if they do so on our behalf, but for the designer, self-expression is usually secondary to creating something that will appropriately serve its function—which may be decorative or symbolic as well as practical in the narrower sense. Hence, Parsons speaks of the designer as "shackled" in comparison to the artist's expansive freedom.[3] "Shackled" is perhaps too strong a term for the constraints that the designer must embrace compared to the relative freedom of artists. In fact, even the term "constraint" can be misleading since it too suggests something externally imposed, whereas the best designers ideally take a proactive concern for the safety and well-being of users as well as aiming at users' aesthetic satisfaction. Hence, I would prefer to speak of design art as a "responsive" art, that is, as responding to the various needs of users as well as to social and economic conditions. Thus, designers have a responsibility or obligation to society that artists may choose to assume, but normatively need not.

Keeping these characteristics of design in mind, let's consider standard perfume as a design product. In standard perfume design, whether mainstream or niche, although the aesthetic effects of the design are of central importance, any standard or niche perfume must also meet the practical norms associated with wearability. Hence in addition to their aesthetic preoccupations perfumers must also be concerned with achieving the right balance between volatility and persistence and avoiding scents that are harmful or viscerally disgusting.[4] I speak of "visceral disgust" here to distinguish it from what Carolyn Korsmeyer calls "aesthetic disgust," namely the way a certain kind of negative emotional reaction can play an essential role in positive aesthetic experiences.[5] Although many niche perfumes play on the edge of offense and thus employ something like "aesthetic disgust," odors that are immediately and deeply repellant are obviously not marketable enough to be part of either standard mainstream or niche perfumery. Accordingly, most perfumers who design either mainstream or niche perfumes today are more like fashion or product designers, whose practice norms direct them to consider the needs and desires of clients and users, than they are like contemporary artists, whose practice norms leave them free to make or use anything, in any way, for any reason, although that freedom even includes permission to make or do something that serves some practical or political function(s).[6] From this perspective it would be more descriptive to call most contemporary (fine) art practices "free" art rather than "fine" or "high" or "major" art given the misleading status implications of the latter terms, which also reflect outmoded social and gender hierarchies. Of course, in most of the other European languages besides English, we would be speaking of "free arts" instead of various expressions that signify "beautiful arts," for example, *beaux arts, schöne künste, belle arte, bellas artes*, etc.

Yet, because the fine or "free" arts and the design or "responsive" arts are distinguished from each other by *norms* of practice rather than rules, there is no fixed boundary between the two, but a transitional border area. Thus, an individual designer such as Philippe Starck, with his famous *Juicy Salif* (a sculptural but rather impractical citrus squeezer), may not only minimize the normative constraints of practicality but also occasionally shift into the role of contemporary artist, making expressivity and aesthetic form trump function. Conversely, an individual artist like Andrea Zittel (with her *A-Z* living units, furniture, and clothing) may shift into the role of designer, creating hybrid works that the art critic Alex Coles calls "DesignArt."[7] Zittel's works look like functional design objects and could actually be used (though many lack the durability one expects from good design). Yet artists like Zittel intend their works to be considered art and typically circulate them through art galleries and museums, and the works are reviewed in art periodicals like *Art Forum*.[8]

Lisa Kirk's *Revolution Pipe Bomb* was a kind of "DesignArt" hybrid in Coles sense, since it looked like a work of perfume design and, in fact, consisted partly of a perfume in a luxury bottle, although the work as a whole was intended for presentation in the world of contemporary art and was shown as part of a (fine) art installation at MoMA's PS1 gallery. But Kirk subsequently went even farther in playing with the role of designer. In 2010, just two years after *Revolution Pipe Bomb*, she released a perfume called simply *Revolution*. It contained essentially the same liquid as *Revolution Pipe Bomb* but was presented in a generic little bottle an inch high, with a simple pasted-on label, and at this writing it sells for $90 over the internet. Kirk developed *Revolution* at the suggestion of Ulrich Lang, head of the niche perfume company Ulrich Lang, New York, whose chance conversation with her several years before had led to *Revolution Pipe Bomb*. Moreover, the little bottle of *Revolution* is not only sold over the internet through several standard perfume sites in addition to her own web site, but also through Ulrich Lang's European outlets. Thus, both the medium and the institutional context for distributing *Revolution* seem to be almost identical to that of many niche perfumes sold by specialist boutiques. As a result, although both the earlier *Revolution Pipe Bomb* and the more recent *Revolution* contain a similar elixir, unlike *Revolution Pipe Bomb*, which was clearly a work of contemporary art, the art status of the little bottle of *Revolution* stands ambiguously on the border between fine art practices and design art practices. When you buy a bottle of *Revolution* from an online site or one of Ulrich Lang's boutiques, are you buying a niche perfume or a piece of contemporary art? Of course, one could argue that the implied intention here seems more about Kirk letting a wider audience participate in her art project of creating "a perfume to symbolize rebellion" than it is to compete in the niche perfume market. Yet since the ordinary buyer is given no explicit indication that the little bottle called *Revolution* is part of a larger art project meant to "symbolize rebellion," one could with equal justice argue that *Revolution* is in fact just another niche perfume.

If Kirk's selling *Revolution* in perfume boutiques and over the internet is an example of an artist taking on the role of a designer (at least indirectly by commissioning a certain type of scent), Christophe Laudamiel's scents to be used with scent squares or scent parabols for exhibition in art galleries are obvious examples of a perfume designer turned artist. These kinds of border crossings from Art to Design and Design to Art have been occurring even longer in the realm of fashion. A quick look at the dynamics of the relation between what we might call "art fashion" and "standard high fashion" can help us see more clearly the options that are being opened up to the world of perfumery by someone like Laudamiel.

Since the 1980s some haute couture designers, called the "fashion avant-garde" by fashion historians, have occasionally designed garments that verge on the purely artistic and unwearable: Martin Margiella's gilet made from broken crockery, Viktor and Rolf's evening gowns with gaping circular holes cut through them, Rei Kawakubo's dresses with crude protrusions, and Hussein Chalayan's dresses that mechanically change shape. Although less radical, Alexander McQueen's outlandish, one-off designs that were meant primarily for multiarts runway spectacles were as dazzling as the costumes and sets of some smaller operas.[9] Yet, by themselves, these periodic crossovers between haute couture and fine art, even if enthusiastically embraced by art world institutions (McQueen, Pierre Cardin, Giorgio Armani, Jean Paul Gaultier, and Christian Dior have all had shows in major art museums), do not automatically make the general practice of fashion design a part of contemporary (fine) art. As the Belgian designer Walter Van Beirendonck has remarked: "The way I work on a fashion project is completely different [from] how I approach an art project . . . it's very important to me that the fashion collections can be bought and worn and have a link to reality and the consumer."[10]

Similarly, the normative constraints on standard perfumery deriving from the need to be responsive to the practical interests of the consumer who wants to wear the perfume (as well as the economic and legal constraints on ingredients) mean that most perfume creation is still best understood as part of design art practices rather than fine art practices. The art and design practices of Laudamiel, like those of Walter Van Beirendonck, are exemplary in this respect. Laudamiel has moved comfortably back and forth from his role as an artist creating artworks for the two galleries that represent him to his role as a designer within the firm Dream Air, of which he is a founding partner. In his capacity as designer, as we saw in the case of Humiecki & Graef's *Trust*, he is ready to follow a perfume brief in order to meet the practical needs of both the perfume distributor and the eventual users.

Professional perfumers seem to be divided on both the "art" status of perfume and whether they should think of themselves as "artists" with all the romantic baggage that term often carries with it.[11] Some, like Michael Roudnitska, insist on what they call "auteur perfumery," the idea that the perfumer, as in a certain idea of the film director, is the artist solely responsible for the perfume as a work of art.[12] Although Jean-Claude Ellena is also quite insistent on the labels "fine art" and "artist" for perfumes and perfumers, he also recognizes that there is both a collaborative and an artisanal aspect to perfume creation.[13] Moreover, Dominique Ropion rightly points out that there is also a scientific aspect to perfume creation, since both scientific knowledge and technical precision are important in working out a formula.[14] As he says of the perfumer's rigorous training, "Once you are no longer taking your first steps beyond raspberry waffles and the smell of tires, you have to swap the poetry of fragrances for the science of molecules; once you have perfected the latter, you are better able to create the former."[15] I think both Ropion and the independent perfumer Francis Kurkdjian get the balance of science, craftsmanship, and artistry in perfume design about right. Kurkdjian says he was taught in perfumery school that perfume is art, but he now thinks not; rather, it's "like clothing; its purpose is to be worn, in contrast to a work of art which is hung on the wall and is sufficient unto itself." But Kurkdjian like Laudamiel is also quite clear that he can use his perfumery skills to take on the role of an artist when he wants to by developing installations or collaborations whereby he can show "routes towards a purely artistic . . . olfactory composition."[16] More recently, the perfumer Philip Kraft has distinguished between "olfactory art" and "scent design," emphasizing that the former is perceived differently by everyone, whereas "a functional scent design should send the same lifestyle message to everyone."[17]

But surely these differences between fine art practices and design art practices in perfumery do not mean that standard perfume designs for wear are intrinsically less worthy of serious aesthetic or intellectual engagement than are "art perfumes" like those that Laudamiel or Kirkdjian create for exhibitions. Although artists and designers differ in the degrees of freedom that are available to them in their respective roles, the greater freedom of the artist does not make the role of artist aesthetically or intellectually superior to the role of the designer. In some ways, as Parsons points out, the creative role of the designer, who must take into consideration so many factors—aesthetic, symbolic, functional, economic, and legal—is far more arduous and demanding than that of the artist. And this remains true despite the emergence of artificial intelligence algorithms such as Philyra that can juggle together millions of existing perfume formulas with marketing data to generate perfumes for particular demographics.[18] I believe that the anxiety of some perfumers and perfume connoisseurs to have perfume listed among the fine arts and perfumers called artists reflects a lingering tendency to overvalue fine art and to undervalue design art. One is not intrinsically superior to the other; they are simply different kinds of practices despite the many areas of overlap.

Moreover, as the philosophers Carolyn Korsmeyer and Yuriko Saito have argued in other contexts, trying to shoehorn everything worthy of serious aesthetic appreciation into the category of fine art ends up distorting our normal experience.[19] This is also true of perfume. Even if a separate "art perfume" practice were to eventually be recognized as part of the fine or free arts, its very freedom to explore uncharted regions and ignore the norms of wearability, safety, and so on means that the resulting art perfumes are likely to *lack* some of the deeper meanings and values that standard perfumes gain from being worn on the body and experienced in everyday situations (we will consider some of those meanings in Chapter 13). What the philosopher Llewellyn Negrin says of "treating fashion as a visual art" is also true of treating standard perfume as art rather than as something designed to be worn: it is "done at the cost of severing its links with the body and lived experience."[20]

No doubt the lingering prestige of the epithet "fine art" is such that to deny that label to something people admire may seem to diminish it, especially given the invidious polarities of capital *A* versus small *a*, high versus low, major versus minor, and so on, which continue to distort thinking about human creativity. To insist that we treat perfume as (fine) art in order to honor it misprizes both art and design. On the one hand, the talk of "elevation" and "pedestals," while reflecting a certain traditional social reality, overlooks the historical contingency of what has been included in the fine arts. On the other hand, the elevation talk overlooks the fact that the arts as a whole form a vast continuum, which Western culture has historically divided up in a variety of ways for a variety of purposes. I think Nicholas Wolterstorff is getting at something similar in his book *Art Rethought* through his eloquent interpretations of memorial art, Orthodox icons, social protest art, and work songs, examples of the many different kinds of art practices and the many different ways of engaging with them that have emerged throughout history and that do not fit comfortably into the traditional Western hierarchical polarities. Given the historical contingency of the fine art category, I believe we need to quit thinking of the arts primarily in vertical or hierarchical terms and begin to think of them "horizontally" as practices differing in media, aims, and scope, within each of which there are works of greater and lesser aesthetic value.

Working out a general theory to support such a radical revision of the concept of (fine) art could draw on several contemporary approaches, but is probably closest to Christy Mag Uidhir and P. D. Marcus's proposal that we replace "monistic" concepts of art (which, in their view, include even disjunctive or combinatorial attempts) with what they call "art concept pluralism." Drawing on the way biology is able to get along with three different major concepts of "species," each of which is adequate to a certain range of data and is useful for organizing different research programs, Mag Uidhir and Marcus suggest that the philosophy of art might adopt a similar approach. This would not collapse into conceptual anarchy since each major type of art concept would have to show its credentials for organizing information about a particular range of practices and its usefulness for a coherent research program. Although my proposal for a nonhierarchical approach to the arts in general might carry pluralism farther than Mag Uidhir and Marcus would be willing to endorse, a pluralistic theory along these lines would anchor the claim that the fine or "free" arts are different from but not superior to the design or "responsive" arts.[21] Such an art concept pluralism would also open the way to developing categories of art that could do justice to non-Western collaborative practices like kodo that we discussed earlier.

I conclude that the best works of perfume and fashion have as much intrinsic importance and dignity when understood as works of design as they would have by labeling them "fine art" or "contemporary art." In fact, I contend that labeling them "fine art" or treating them as contemporary artworks could be highly misleading for how we understand and treat them. By calling standard perfumes "fine art," we suggest that their practical and symbolic functions in daily life are less important and valuable than engaging them in a specialized art setting, whereas, in fact, their practical and symbolic functions are crucial to appreciating them as the kind of arts they are. As Lopes asks of graphic novels or comics whose recognition as (fine) art forms has recently been debated: is one "likely to gain much insight into that genre by campaigning for its art status, once its character and value have been fully appreciated?"[22] Similarly, once we learn to value perfumes as formally complex and emotionally expressive design works that can enhance people's lives, adding the epithet "fine art" may only be misleading, honorific froth.

Notes

1. Glenn Parsons, *The Philosophy of Design* (Oxford: Polity Press, 2015), 11–12. Jane Forsey also has an interesting comparison of design to both art and craft in *The Aesthetics of Design* (Oxford: Oxford University Press, 2013), 23–66.
2. Parsons, *Philosophy of Design*, 11–12, 21–23.
3. Parsons, *Philosophy of Design*, 107.
4. Ellena, *Perfume*, 41.
5. Korsmeyer, *Savouring Disgust*, 3–4, 30–31, 68–70.
6. Of course, the kinds of practices usually grouped under fine art and design respectively have also been historically intertwined through design's close relation to architecture.
7. Alex Coles, *DesignArt* (London: Tate Publishing, 2005). Although some designers relish the idea of their works being considered fine art, others such as Hella Jongerius are put off by the term "design art," and remain content to think of themselves simply as designers. See Louise Schouwenberg, ed., *Hella Jongerius: Misfit* (London: Phaidon Press, 2010), 189–90.
8. See Andrea Zittel's website, https://www.zittel.org.
9. Sonia Delaunay, Alexander Rodchenko, and Giacomo Balla were three early twentieth-century artists who designed ready-to-wear. For curators like Andrew Bolton of the Metropolitan Museum in New York, the fact that high-fashion designs are a vehicle for expressing ideas and are highly crafted is enough to make fashion (fine) art. Of course, if that is all it takes, philosophical and anthropological writings are fine artworks too. For Bolton's views, see *Art/Fashion in the 21st Century*, ed. Michael Oakley Smith, Alison Kubler, and Daphne Guiness (London: Thames & Hudson, 2013), 160.
10. Interviewed by Mitchell Oakley Smith in *Art/Fashion in the 21st Century*, 60.
11. Sarah Bouasse, "Is the Perfumer an Artist?," *NEZ: The Olfactory Magazine* 4 (Autumn–Winter 2017): 87–92. Bouasse quotes the perfumer Calice Becker to the effect that the romantic view of the lone, suffering artist is hardly applicable to the perfumer who works for clients and often with a team.
12. Quoted in *NEZ: The Olfactory Magazine* 4 (Autumn–Winter 2017): 91–92.
13. Quoted in *NEZ: The Olfactory Magazine* 4 (Autumn–Winter 2017): 88.
14. The same is true, of course, of ceramics, where knowledge of the physics and chemistry of clay and of glazes and their transformation under intense heat is an important part of being able to achieve intended aesthetic effects.
15. Ropion, *Aphorisms of a Perfumer*, 21.
16. Quoted in *NEZ: The Olfactory Magazine* 4 (Autumn–Winter, 2017): 90–91.
17. Kraft's wide-ranging essay develops a method of analysis he applies to both functional perfumes and artists' perfumes. Philip Kraft, "The Odor Value Concept in the Formal Analysis of Olfactory Art," *Helvetica* 102, no. 1 (January, 2019), https://doi.org/10.1002/hlca.201800185.
18. The big fragrance and flavor corporation Symrise supplied IBM with 1.7 million formulas, and once IBM developed Philyra, Symrise used it to produce a successful perfume aimed at Brazilian millennials (although one of their master perfumers tweaked the final product a bit). No doubt the big fragrance and flavor producers will

embrace things like Philyra to cut time and costs, but the implications for the status of design of the use of AI in developing perfumes is no different than previous uses of AI to play chess or generate music, visual images, or clothing designs. It certainly does not show that the practice of perfume design is intellectually or creatively inferior to the creation of olfactory art. See Chavie Liever, "Is AI the Future of Perfume? IBM Is Betting on It," *Vox*, October 24, 2018, https:// www.vox.com /the-goods/2018/10/24/ 18019918/ibm-artificial.

19. Korsmeyer, *Making Sense of Taste*, 141; Yuriko Saito, *The Aesthetics of Everyday Life* (Oxford: Oxford University, 2009), 13–18.
20. Llewellyn Negrin, "Fashion and Aesthetics: A Fraught Relationship," in *Fashion and Art*, ed. Adam Geczy and Vicki Karaminas (London: Berg, 2012), 43.
21. Christy Mag Uidhir and P. D. Marcus, "Art Concept Pluralism," *Metaphilosophy* 42, nos. 1–2 (2011): 83–97.
22. Lopes, *Beyond Art*, 204. On the comics as art question see *The Art of Comics: A Philosophical Approach*, ed. Aaron Meskin and Roy T. Clark (Malden, MA: Wiley-Blackwell, 2012).

PART IV
THE AESTHETICS AND ETHICS OF SCENTING

Overview

Varieties of Aesthetic Experience

In this final part of the book we turn to three areas of aesthetic practice that raise unavoidable ethical as well as aesthetic issues. If we are to do justice to both the aesthetics and ethics of scenting bodies, places, and foods, we will need an understanding of aesthetic experience and judgment that goes beyond views of aesthetics based primarily on the appreciation of the fine arts. On the one hand, not even all fine artworks have been meant to be experienced purely aesthetically, but also to engage us morally, religiously, or politically. On the other hand, aesthetic experience itself has always been concerned with nature, design, and everyday life in addition to the arts. Although Kant's aesthetic was framed with nature as well as the arts in mind, from Hegel down into the late twentieth century philosophical aesthetics focused most of its attention on the fine arts. But thanks to the pioneering work of Ronald Hepburn, Arnold Berleant, Allen Carlson, and others, the aesthetics of nature has received increased attention in recent decades. We will consider some of this work in a later interlude on smell in nature. In the case of design and everyday life, which will be the main concern of Part IV, we can now call on at least three contemporary approaches to aesthetics that focus on design and everyday practices in ways that can help us articulate the aesthetic and ethical issues raised by deliberate scenting: "functional beauty," "everyday aesthetics," and the "aesthetics of atmospheres."

Glenn Parsons and Allen Carlson's *Functional Beauty* expands aesthetics from its focus on the fine arts to include the entire designed world and argues for the place of function and purpose in aesthetic appreciation. They group manifestations of functional beauty into several types, including the idea of "looking fit for function," the perception of simplicity and unity as in modernist design, and what they call "a pleasing dissonance," as in some postmodernist architecture.[1] Although we noted earlier that Parsons and Carlson are among those contemporary philosophers who deny that proximal senses such as smell can be involved in properly aesthetic perception, much of their analysis of functional beauty could actually be adapted to account for the role of function in our aesthetic response to olfactory artifacts such as perfumes.[2] Thus, one could argue that the temporal profile of most standard perfumes for wear makes them smell "fit for function" and that many niche perfumes featuring discordant notes set up a "pleasing dissonance."

A second major contemporary attempt to broaden the practice of aesthetics is "everyday aesthetics," which focuses on such ordinary experiences as the smell of baking bread, the sound of rain on a tin roof, the feel a silk scarf, the visual patterns of hanging laundry, or the multisensory pleasures of walking through a city. Although there are several approaches to everyday aesthetics, the main directions are nicely captured in the titles of two works by leading theorists.[3] Tom Leddy's *The Extraordinary in the Ordinary* (2012) stresses those moments of revelation in our everyday lives when what he calls the "aura" of some everyday object or activity emerges into consciousness.[4] Yuriko Saito's *The Aesthetics of the Familiar* (2017), on the other hand, focuses on what she calls "the aesthetic texture of ordinary life," those moments when we savor our experiences in their very familiarity.[5]

An example of what Saito is getting at that involves smell is her discussion of laundry hanging.[6] Laundry hanging is a "lost" art in much of the United States, especially in middle-class suburban communities, many of which forbid hanging laundry outdoors as aesthetically "unsightly" (a classist subtext may also be at work here). Yet the aesthetic pleasures some people take in laundry hanging go well beyond visual satisfaction at the artful arrangements of clothes on a line. The multiple pleasures of laundry hanging also include the rhythmic movements of the act of hanging up and taking down, the tactile feel of both wet and the dry clothes and linens against the hands and face, and, once the laundry has dried in the open air on a summer day, "the smell of sun-soaked clothes."[7] Saito's comment about the smell of sun-dried laundry brought vividly to mind one of my fondest childhood experiences; I would be sent out to the backyard in the summer to bring in the laundry that had been drying all afternoon in the Kansas sun. I remember not only the warm caress of the brilliant white sheets on my face but especially the wonderfully fresh smell as the sheets enveloped me as I pulled them down. By the time I finished college, the electric dryer was already becoming ubiquitous in middle-class suburbia and eventually people began throwing scented tissues into their dryers to give their clothes the artificial smell of fresh air (or so advertisers claimed).

Some readers already familiar with both the "everyday aesthetics" and "functional beauty" approaches might be surprised at my pairing them in this methodological discussion since the most severe criticism of everyday aesthetics has come from Parsons and Carlson. Their main charge in *Functional Beauty* is that the "the flight into sensuality characteristic . . . of everyday aesthetics" will end in subjectivism and relativity.[8] Given the reasons developed in the first two parts of this book on behalf of the cognitive capacities of smell, such critiques based on assumptions that the proximal senses are irredeemably subjective are highly questionable. In addition, such complaints, as we suggested earlier, overlook the actual connection between the proximal senses and the aesthetic appreciation of many functional objects, whether the scent profile of a perfume, the flavor profile of a dinner, or the tactile feel of a dress material.

A third contemporary way of broadening aesthetics that is useful for thinking about design and the everyday is Gernot Böhme's "aesthetics of atmospheres." Böhme's starting point is our ordinary experience of an atmosphere pervading a place, a social gathering, or emanating from a person, group, or thing, for example, such qualities as cheerful, gloomy, uplifting, menacing, chic, business like, or middle class. The experience of atmospheres, in Böhme's view, is a quintessentially *intermediate* phenomenon, linking subject and object through bodily feelings, thereby creating a "common reality of the perceiver and the perceived."[9] On the object side, atmospheres can be consciously produced through what Böhme calls "staging," which he views as part of the aesthetic economy of late capitalism. Staging takes place not only in theater, architecture, and installation art, but also in advertising, educational meetings, festivals, and political rallies. On the subject side, these arrangements are "experienced in a state of affective resonance," corresponding to the way situations can be "tuned" to produce a certain effect.[10] Finally, because atmospheres are experienced in a multisensory fashion through feeling the environment around us, Böhme explicitly credits smell with an important role in our experience of them.[11]

Given the differences among these three efforts to enlarge aesthetics beyond the fine arts, could the three approaches be united within a single definition of the aesthetic? Although there are many contemporary attempts to provide a unified concept of the aesthetic, in my view we should treat "aesthetic" in a way parallel to what I suggested for "art," that is, as the name of a nonhierarchical continuum, in this case, a continuum of types of experience and judgment. Such a continuum would reach from the traditional "for itself" appreciation associated with the fine arts, through the functional beauty approach to the design arts, to the everyday aesthetics approach to objects and activities experienced as "extraordinary," and on to savoring the aesthetic texture of everyday life in its familiarity, including "atmospheres." There should also be a place on such a continuum for the kind of meditative and poetic experiences of non-Western aesthetic practices like kodo.[12] The profile of the different kinds of aesthetic experiences and judgments appropriate to these different regions of objects and activities would shift as we move from one to another, in a way not unlike Mag Uidhir and Marcus's idea of art concept pluralism.

This is not the place to work out the details of a general theory of "aesthetic concept pluralism," but one pluralistic theory, focused on *aesthetic value*, has recently been developed by Dominic Lopes in *Being for Beauty: Aesthetic Agency and Value*.[13] Lopes contrasts his pluralistic understanding of aesthetic value with the monism of those contemporary theories of aesthetic value that still make the appreciation of (fine) art their paradigm and that typically link such appreciation to a standard of taste exemplified by the joint verdict of ideal or "true judges," those Humean paragons who reliably identify artistic masterworks.[14] Instead, Lopes focuses on the innumerable flesh-and-blood "aesthetic experts" who come from all social classes and walks of life and exercise their special skills in the context of specific collaborative practices such as maintaining a public garden, conserving classic video games, recruiting and teaching in a dance school, the kinds of activities to which we might add leading sessions of kodo. Lopes's focus on ordinary "aesthetic experts" suggests that since we are "all cut of different cloth, a diversity of aesthetic opportunity is better than aesthetic monoculture."[15]

So far I have not addressed the question of how any of the three particular ways of enlarging our understanding of aesthetics I have mentioned relates to ethics. One way they are open to moral considerations is that all three approaches reject views of the aesthetic that take as their paradigm the contemplation of art "for itself" apart from any practical or moral purpose. Although Parsons and Carlson's *Functional Beauty* does not specifically discuss ethics, Parson's follow-up book on design does, arguing that design intrinsically involves not only safety issues, but also such things as respect for user freedom, economic availability, and environmental sustainability, issues that, as we will see, have some counterparts in the practices of scenting bodies, places, and foods.[16] The ethical aspect of Gernot Böhme's aesthetics of atmospheres derives primarily from the possibility of a critique of aspects of "staging" in the political economy of late capitalism and will inform our discussion of the commercial use of ambient odors.[17] Saito has consistently probed the moral consequences of everyday aesthetic preferences, showing how our aesthetic evaluation of such ordinary things as lawns, landscapes, and laundry can have ethical consequences.[18] Surely, the billions spent in creating and maintaining lawns with their huge demand for water and fertilizers makes no sense environmentally, yet people have a strong aesthetic attachment to lawns, including their smell ("new-mown grass" frequently shows up on short lists of people's favorite smells). Compared to the waste of lawn maintenance, the use of electric driers year-round is a far less serious environmental problem, although electric dryers are often the second most demanding electrical appliances after refrigerators in many homes.[19] More serious environmental consequences follow from aesthetic preferences for the "majestic nature" of mountains over bogs and marshes, or for animals or birds that are considered conventionally beautiful or cute. The overall health of the environment depends on a biodiversity that includes "ugly" and "smelly" landscapes, animals, birds, and insects, but they are a hard sell.[20]

By explicitly including the ethical consequences of everyday aesthetic choices within aesthetic theory we take a final step in broadening the idea of aesthetic experience and judgment to serve as a guide for our discussion of various aspects of scenting. Hence, the three chapters of Part IV will explore not only the aesthetic meanings but also the ethical implications of wearing perfumes (Chapter 13), of injecting ambient odors into public spaces (Chapter 14), and of adding aromas to food and drink, especially "fast food" (Chapter 15). The postlude, "Wilderness, Gardens, and Paradise," will briefly consider the role of smell in both the aesthetics and ethics of our relation to nature, ending with the place of smell in the cultural imagination of Paradise.

As a transition to the next chapter on the morality and meaning of scenting the body, I have inserted a prelude on Huysmans's *Against Nature* and Patrick Süskind's *Perfume: The Story of a Murderer*, two powerful cautionary tales written a hundred years apart about the moral danger of a preoccupation with smell and perfumes.

Notes

1. Glenn Parsons and Allen Carlson, *Functional Beauty* (Oxford: Oxford University Press, 2008), 94–97. One part of their argument draws on Kendall Walton's seminal "categories of art" essay to describe how an object's proper function can play a role in our perception of its aesthetic characteristics depending on whether the object's features are standard, contra-standard, or variable.
2. Parsons and Carlson, *Functional Beauty*, 177–78.
3. Thomas Leddy, *The Extraordinary in the Ordinary: The Aesthetics of Everyday Life* (Peterborough: Broadview Press, 2012), 151; Yuriko Saito, *Aesthetics of the Familiar: Everyday Life and World-Making* (Oxford: Oxford University Press, 2017).
4. Leddy, *Extraordinary in the Ordinary*, 127–50. Sherri Irvin has described similar experiences in which our normal immersion in a routine, such as drinking our morning coffee, can break into a modest epiphany. "Scratching an Itch," *Journal of Aesthetics and Art Criticism* 66, no. 1 (2008): 31.
5. Saito, *Aesthetics of the Familiar*, 31.
6. Saito, *Aesthetics of the Familiar*, Chapter 5. For a more general discussion of the role of smell in everyday aesthetics see Emily Brady, "Sniffing and Savoring," in *The Aesthetics of Everyday Life*, ed. Andrew Light and Jonathan M. Smith (New York: Columbia University Press, 2005), 177–93.
7. Saito, *Aesthetics of the Familiar*, 131.
8. Parsons and Carlson, *Functional Beauty*, 189. Carlson reaffirmed this criticism more recently in "The Dilemma of Everyday Aesthetics," in *Aesthetics of Everyday Life East and West*, ed. Liu Yuedi and Curtis L. Carter (Newcastle upon Tyne: Cambridge Scholar Publishing, 2014), 62–63. Glenn Parsons has also continued the subjectivism critique in *The Philosophy of Design* (Cambridge: Polity Press, 2016), 107–8. Jane Forsey has devoted a long segment of her book *Aesthetics of Design* to similar criticisms (193–243).
9. Böhme, *The Aesthetics of Atmospheres*, 20.

10. Böhme, *The Aesthetics of Atmospheres*, 168.
11. Gernot Böhme, *Atmospheric Architectures: The Aesthetics of Felt Spaces*, trans. A.-Chr. Engels-Schwarzpaul (London: Bloomsbury, 2017), 103, 139, 163.
12. By adopting something like "aesthetic concept pluralism" one would be engaging in a cross-cultural *characterization* of various aesthetic practices rather than attempting a "pancultural definition." For the difficulty of even such capacious disjunctive definitions of art as those of Stephen Davies and Jerrold Levinson, see Daniel Wilson's case against applying attempted "pancultural" definitions to practices such as *chado* in "The Japanese Tea Ceremony," 33–42.
13. Lopes, *Being for Beauty* 109. In the case of aesthetic *experience*, a promising direction for developing a pluralistic model has been suggested by Sherri Irvin's proposed continuum of aesthetic experiences based on the role of cognitive grasp and self-awareness in aesthetic encounters. Sherri Irvin, "Is Aesthetic Experience Possible?," in Currie et al., *Aesthetics and the Sciences of Mind*, 44–49.
14. Lopes, *Being for Beauty*, 107.
15. Lopes, *Being for Beauty*, 222.
16. Parsons, *Philosophy of Design*, 132.
17. Gernot Böhme, "Contribution to the Critique of the Aesthetic Economy," *Thesis Eleven* 73, no. 1 (2003): 71–82.
18. See *Everyday Aesthetics*, Part V, and *Aesthetics of the Familiar*, Chapters 6–7. Thomas Leddy has discussed the ethical aspects of *Everyday Aesthetics* in "Everyday Aesthetics and Happiness," in Yuedi and Carter, *Aesthetics of Everyday Life*, 31–4
19. Saito, *Aesthetics of the Familiar*, 127.
20. These environmental implications are only the more obvious ethical outcomes of ordinary aesthetic choices. Saito also discusses how our everyday aesthetic experiences, both positive and negative, can lead us to "world making" at the level of everyday interactions with others. Yuriko Saito, "Everyday Aesthetics in the Japanese Tradition," in Yuedi and Carter, *East and West*, 164. In that same volume several other essays discuss the ways in which the Confucian tradition in China also reflects an attitude to daily life that is at once aesthetic and moral. Kelvin E. Y. Low has also addressed interpersonal issues related to smell in his *Scents and Scent-sibilities: Smell and Everyday Life Experiences*, (Newcastle upon Tyne, UK: Cambridge Scholars Publishing, 2009).

Prelude
Two Cautionary Tales

Hans Rindisbacher's *The Smell of Books* examines a broad swath of continental European literature from the mid-nineteenth century to the 1980s, arguing that the treatment of smell in French, German, and Russian novels during this period parallels the deodorization of Western societies and the reduction of perfume to a purely aesthetic accessory—a history we discussed in Chapter 5. As Rindisbacher sees it, by the late nineteenth century smell "emerges in literature as the instrument of exquisite individuality," a phenomenon he traces from J.-K. Huysmans's *Against Nature* (1884) to Patrick Süskind's *Perfume: The Story of a Murderer* (1988).[1] What is most significant about these two novels in relation to the aesthetics and ethics of olfactory art is the central role played in both by the intense preoccupation with smell and perfumes and the psychological and moral degradation into which this leads their respective protagonists.

In *Against Nature*, the novel's protagonist, Duc Jean des Esseintes is a world-weary aesthete who retreats from Parisian society to an isolated house on the city's outskirts to lead a solitary life devoted solely to literature, art, and sensory pleasures.[2] His dining room, for example, is made to resemble a ship's cabin, complete with compass, sextant, maps, steamship schedules, and a device that emits the odor of tar as one enters.

But soon enough des Esseintes's solitary life of aesthetic artifice begins to turn against him, as social reality intrudes in the form of smell-laden memories, leading to nightmares along with sweating, tingling, and pains. One evening, he reaches for one of his favorite scented bonbons that in the past had brought on diverting reveries. But this time the aromatic bonbon leads instead to hallucinations in the form of several former mistresses whose smells are as vivid to him as their appearance (127). At the head of this parade is Miss Urania, a muscular American circus acrobat he had liked for her "wholesome animal smell," followed by a little brunette who exuded daring and exciting perfumes (130–31). Des Esseintes emerges from these smell-infused visions "crushed, broken, almost lifeless . . . his arteries throbbing" (134).

Some days later, he awakes one morning to a true hallucination: his whole house seems pervaded by an odor of frangipani that his servants cannot detect (135). He decides to cure himself by creating a counter-perfume to drown out

Art Scents. Larry Shiner, Oxford University Press (2020) © Oxford University Press.
DOI: 10.1093/oso/9780190089818.003.0028

the frangipani. We are told that he had already mastered the art of perfumery in his Paris days and had brought with him everything he needed. Given his commitment to artifice over nature, he does not simply mix natural essences, but also uses synthetic materials that he is convinced will add the unique aspect that will turn his perfume into a work of art (136).

Des Esseintes's first attempt fails, only giving him "illusions of *Venuses* by Boucher" (140). Furious, he grabs some potent essence of spikenard and inhales deeply. He is stunned by the shock, but both the eighteenth-century vision *and* the odor of frangipani disappear. This triumph spurs him on to create several unusual perfumes, but he soon tires of the sport and decides to end his experiments with one great perfume, throwing together all sorts of essences, alcohols, and spirits until there "burst into the room" a scent at once "demented and sublime" (14). Suddenly, a sharp pain shoots through his body and he opens his eyes to find himself still sitting at his dressing table. When he goes into his study and opens a window, what blows in is not fresh air, but a wave of bergamot mixed with other odors that finally dissolve into an overpowering smell of . . . frangipani. At the return of his original hallucination, des Esseintes "faints, as if dying" (146). Although the servants find him, his physical and mental health continue to decline until a doctor is called in who orders him to give up his solitary aesthetic life.

How are we to understand *Against Nature*? Whatever else it is, it is surely a caution against the overcultivation of the senses, especially the sense of smell and of perfume in particular. The perfume chapter is a crucial turning point in revealing the futility and soul-sickness of the aesthete's sensual existence. Shifting from an active life in society, where odors play a minor, mostly unconscious, role, to an exquisite, odorously rich aesthetic environment may be dangerous to one's mental health. The psychologist Peter Köster might agree. Huysmans himself, who had once been fascinated with perfumes, later saw *Against Nature* as a turning point in his conversion to Roman Catholicism, and in his subsequent writings it is no longer perfume, but incense in the service of religion, that interests him.

Patrick Süskind's novel *Perfume: The Story of a Murderer* offers a more grizzly warning against a fascination with smell and the pursuit of a seductive perfume, a warning that is so shocking that, were it not for its postmodern, intertextual complexity, drawing on both Gothic horror and magic realism, it might be considered a parody of a cautionary tale.[3] The novel focuses on a figure at the opposite end of the spectrum from a cultured aesthete like des Esseintes. Jean-Baptiste Grenouille is an abandoned child of poverty, as a baby thrown onto a stinking garbage heap to die, then abused and mistreated as a child. The one talent that saves him is that, although he himself lacks any body odor, he has a superhuman ability to detect and identify odors of all kinds, including even the subtlest body odors of others. Sold to a perfumer who exploits his abilities and pays him nothing, Grenouille, despite his ugly appearance and lack of education, is eventually able, through many difficulties plus some luck and cunning, to open his own perfumery. But one thing obsesses him: his own lack of smell and his desire to use his olfactory talents to achieve fame and adulation. After various experiments, he concludes that the missing ingredient he must have is the odor of innocent young girls, and so the murders begin—he will eventually kill twenty-five, as he tries to accumulate enough of the essence derived from their skin to make the supreme perfume, the irresistible scent that will make him loved. Yet one of the striking things about Grenouille as an antihero obsessed with smell and perfume is that he is not only amoral, but also asexual. He is drawn to his victims not by libidinal desire but by a purely olfactory compulsion; he is only interested in gathering their beautiful scent.

Finally, just as he has completed his perfume masterpiece, he is captured by the provincial police and condemned to the gallows. But when he arrives at the scaffold outside the village where the last of his murders occurred, he is able to put on his perfume. Soon the powerful, beatific scent overwhelms everyone, and the thousands who have come to see him die, including the executioner, the bishop, and even the father of the last girl he murdered, are not only convinced this exquisitely scented man could not possibly be a murderer, but end up prostrating themselves before him. Moreover, the perfume has worked like a miraculous pheromone: people are so overcome with carnal desire that what was supposed to be an execution turns into a huge orgy. Grenouille sneers as he watches the crowd, thinking on how he, although born on a garbage heap and "surviving solely on impudence and loathing, small, hunchbacked, lame, ugly, shunned, an abomination within and without—he had managed to make the world admire him, no . . . Love him! Desire him! Idolize him!" (239). In this moment, he feels godlike, like a "more splendid God than the God that stank of incense" (240).

But Grenouille ccannot enjoy his triumph. For as soon as he realizes how irresistible his perfume is, he is filled with hatred for the orgiastic mob he has engendered (240). Suddenly the father of the last girl he murdered rushes toward him, not to attack, but to embrace him. Grenouille faints and is taken home by this fragrance-besotted man who wants to adopt him. Waiting until everyone is asleep, Grenouille slips away, walking toward Paris. He arrives on a sweltering, stinking day and after midnight joins the thieves, cutthroats, and whores around a campfire in the Cemetery of the Innocents. All anyone could remember later was that a little man had opened a small bottle, sprinkled the contents over himself, and suddenly became "bathed in beauty" (254). Filled with amazement and desire, the rabble want to touch him, no, consume him; they tear at his clothes, then his hair, then his skin. They dismember him, each devouring a piece. But rather than horror at their cannibalism, when they are finished they sit around the fire in silence, suffused with a "glow of happiness," feeling "they had done something out of love" (255).

In a novel filled with literary allusions, the final scene is a macabre parody of the Last Supper; the body and blood of Grenouille, the creator of the most powerful perfume ever achieved, is literally eaten and drunk and fills the communicants with innocent bliss and love. Instead of the "stink" of incense accompanying the Eucharist, the thieves and cutthroats receive the sublime scent of the ultimate perfume able to transform the very dregs of society. Yet we cannot forget that this beatific ending has been achieved through the murder of twenty-five young women or that the narcissistic creator of this consummate perfume could not overcome his hatred for humankind. In Süskind's dark fairy tale, the effects of fulfilling the dream of an irresistible scent suggest that the artistic pursuit of perfume (or the ultimate pheromone) may be a deeply questionable enterprise. Would we not be better off accepting our natural odor, good, bad, or indifferent, rather than play upon the senses of our fellow humans with artificial concoctions? Were not the Roman moralists and the church fathers right to condemn gaining influence over others through the sensuous means of perfume?

Notes

1. Rindisbacher, *The Smell of Books*, 186.
2. Huysmans, *Against Nature*, trans. Brendan King (Sawtry, UK: Daedalus, 2008). Subsequent citations in the text refer to this translation.
3. Patrick Süskind, *Perfume: The Story of a Murderer*, trans. John E. Woods (New York: Random House, 1986). Subsequent citations are to this edition.

13
The Meanings and Morality of Scenting the Body

If you have ever owned a dog, you may have had this experience. You have just given him a bath, brushed down his coat, and decided to take him for a walk. No sooner are you on the sidewalk than he bolts, yanking the leash out of your hand, races into your neighbor's yard, and throws himself on a pile of another dog's excrement, wallowing joyfully. After a minute or two, he leaps up, head in the air and trots proudly back as if he were wearing the finest of perfumes. Alexandra Horowitz of the Dog Cognition lab at Barnard College has canvased some of the explanations offered for this all-too-common behavior, such as a dog's instinctive drive to mask its odor from possible prey or predators, but she notes that there is no agreement among experts.[1] Although many other animals and insects use odors for masking, and some monkeys have been known to chew aromatic plants and rub them into their fur, it's the appearance of pleasure dogs derive from rolling around in smelly feces that gets our attention.[2] Even so, it is we humans who seem to be the only animal species that has deliberately and consistently through millennia adorned itself with odors.[3]

In Chapter 5 we showed that perfume/incense played important spiritual and therapeutic roles in the West for thousands of years, although from the eighteenth century on the uses and meanings radical narrowed to almost purely aesthetic ones related to adornment. Unfortunately, mainstream perfume advertising has reinforced the long-standing popular impression that perfume's primary aesthetic functions are either sexual attraction or signaling social status. Yet the motives, meanings, and moral estimates of perfume wearing have always been multiple. The first half of this chapter will explore some of the meanings of perfume wearing beginning with the ancient Hebrews, Greeks, and Romans along with the classic moral objections to perfume use from philosophers and theologians. The second half of the chapter will discuss the survivals and transformations of these traditional meanings and moral arguments in the contemporary world.

Greek Philosophers, Roman Moralists, and Church Fathers

The biblical Song of Songs opens with the woman's plea, "O that you would kiss me with the kisses of your mouth! / For your love is better than wine, your anointing oils are fragrant" (1:2–3), and a few verses later she adds, "My love is to me a sachet of myrrh lodged between my breasts" (1:14). The man then likens her to a fragrant garden of "nard and saffron, cane and cinnamon, . . . with all trees of frankincense, myrrh and aloes" (4:13–15). Rabbinic and Christian commentators have usually interpreted the lovers' declarations as an expression of the mutual love of God and Israel or of Christ and the Church. But by including the Song of Songs in the biblical canon, the rabbis and church fathers, whether they intended to or not, consecrated the mingling of the sexual and spiritual in some of the most sensual poetry ever written. Thus, despite the insistence of some priestly traditions on a radical separation of holy and profane perfume/incense to parallel the separation of spiritual and sexual love (Exodus 30:38), an equally ancient tradition has suggested they are continuous.

Among the ancient Greeks, as we noted earlier, both men and women used perfumed oils, but their use by women was socially and morally fraught since perfumes were especially associated with courtesans. The myth of the "perfumed panther" is telling in this respect. According to Greek folklore—although the philosopher Theophrastus also accepted it—the panther or leopard has a breath so sweet that it need only exhale and prey will be irresistibly drawn.[4] Given the ancient Greek prejudice that kept married women out of public life, the panther-like courtesan might attract men not only by her perfume and beauty, but also by her educated conversation. As one ancient writer famously put it: "We have courtesans for pleasure . . . and wives in order to have a legitimate posterity and a faithful guardian of the hearth."[5] The classical scholars Marcel Detienne and Jean Pierre Vernant have argued that this view of marriage involved a symbolic olfactory economy in which the use of perfume/incense in the relation between gods and men stood in opposition to the relative absence of perfume in the relation of husbands and wives.[6]

Yet Aristophanes's antiwar play *Lysistrata* suggests a more complex view. Near the climax of the play, after several days of the sex strike, the tumescent Kinesias arrives at the Acropolis to beg his wife Myrrhine for sex. She insists that he first promise to vote to end the war, but each time his assent sounds equivocal and she finds another excuse to run back inside for something—a bed, a mattress, a pillow—and finally she says she is going back for a perfume. By now Kinesias is beside himself with lust and the comic sparring that ensues over whether or not they need perfume to enhance lovemaking suggests that the audience would have understood and accepted a positive role for perfume in marriage.[7]

Whether or not Aristophanes's play reflects a positive attitude toward perfume, the most important Greek philosophers clearly held negative views. In Xenophon's *Symposium*, when the host Kallias, who has provided a cithara player and dancing girl for diversion, suggests bringing in perfumes, Socrates objects. Perfumes are not appropriate for men, he says, or even for married women.[8] And Plato's Socrates is even more dismissive than Xenophon's. In *The Republic*, Socrates contrasts the simple needs of a "true community" that exemplifies morality "writ large," to the needs of an "inflamed community" that exemplifies immorality writ large, that is, a community whose citizens want "savories, perfumes, incense, prostitutes, and pastries" (II, 373a).[9] It would be hard to denigrate and ridicule perfumes and incense more completely than by linking them with "prostitutes and pastries."

As in so much else, Aristotle follows Plato but draws a more moderate conclusion. As we noted earlier, Aristotle placed the sense of smell well beneath sight and hearing, yet above the senses of taste and touch, noting that, unlike animals, humans seem to take pleasure in certain smells just for the enjoyment of it. But he was wary of perfume wearing. In the *Nicomachean Ethics*, he claims that temperance is primarily concerned with bodily appetites or pleasures and in this respect the pleasures of vision or hearing, no matter how inordinate, cannot be said to be *either* temperate or intemperate. Thus, no one is called intemperate or undisciplined, he says, for an excessive enjoyment of paintings, music, or dramas. In the case of the sense of smell, something similar is true with things like "the smell of apples or roses or incense." But someone would be called intemperate, Aristotle believes, "for enjoying the smell of perfumes or cooked delicacies." Why? "An intemperate person enjoys them because they remind him of the objects of his appetites." A dog, Aristotle continues, does not find pleasure in the smell of a rabbit as a smell, but only because the dog anticipates eating the rabbit. Its smell simply triggers the chase, capture, and physical satisfaction. Similarly, the man or woman who responds to a perfume is intemperate because he or she is not responding to the scent itself so much as to a remembered or anticipated satisfaction of physical desire (1118a1–25).[10]

Of course, we may want to ask why Aristotle's argument does not leave open the possibility of sometimes enjoying a perfume for its pleasant aroma in the way we enjoy the pleasant smell of apples or roses or incense. The fact that Aristotle, unlike Plato, includes incense in his list of appropriate objects of smell suggests that a moderate use of perfume should be possible, especially since incense and perfumes had many overlapping ingredients in the ancient world (myrrh, for example).[11] In fact, we will later see a surprising instance of such a moderate view of perfume wearing among Christian theologians. First, however, we need to consider the Romans, both in their excesses with regard to perfume use and in the sharp criticisms of perfume wearing by Roman moralists.

In the *Natural Histories*, Pliny the Elder reserved his greatest indignation for emperors like Caligula and Nero who indulged in lavish and hugely expensive perfume displays. At one of Nero's banquets attendees were doused with perfumes sprayed from pipes in the ceiling, and it was said that for his wife's funeral he ordered as much perfume and incense as all of Arabia could produce in a year (IV.12.82–88).[12] Thus, Pliny's deepest objection to perfumes is that they are "the most superfluous of all forms of luxury" (IV.13.1). Fine clothes can be worn again and jewelry can be passed on to one's heirs, but "perfumes lose their scent at once, and die in the very hour they are used." Pliny had other complaints about perfume wearing. In addition to perfumes' ephemerality and cost, perfume's "highest recommendation," he says, seems to be "that when a woman passes by her scent may attract the attention even of persons occupied in something else." Worse yet, Pliny says, perfumes are instruments of deception, invented by the Persians to conceal their stinking bodies (IV.13.3). And now, he growls, not only do soldiers wear perfumes but "people even add them to their drinks!" (IV.13.25). These passages gather up many of the moral objections to perfumes that have continued to circulate to this day: luxury, ephemerality, waste of resources, deception, and seduction.

But we need to consider one other Roman example that provided the classic expression of the negative attitude toward women's use of perfumes. In one of Plautus's comic dramas, the courtesan Philematium is about to put on some perfume when her wise old servant woman says, "A woman smells right when she does not smell at all" (*mulier recte olet, ubi nihil olet*). If perfume is inappropriate for a courtesan, how much less so for an upright woman! Mark Bradley suggests that the phrase was already a kind of proverb, but, in any case it became one, famously repeated by Montaigne fifteen hundred years later.[13] Of course, there were also more moderate attitudes in the Greek and Roman worlds. Many people simply found perfumes pleasant and revitalizing; as Lucretius put it: wearing perfume "from time to time makes us feel fresh and new."[14]

When we turn to the Christian attitude to perfume wearing, most theologians, as one might expect, condemned it out of hand, often recycling the pagan moralists' arguments. Clement of Alexandria wrote: "Just as cattle are led by rings through their noses . . . the self-indulgent are led by odors and perfumes." Ambrose of Milan warned that "a whiff of fragrance hinders thought," and John Chrysostom denounced perfume use on grounds of both luxury and vanity, also claiming it was a deceptive masking of a body rotten within. Of course, the theologians also associated wearing perfumes with prostitutes and immoral sex, contrasting it to the "fragrance of virtue" available in holy oil and incense.[15]

Yet there was at least one moderate voice among all the condemnations. It came from the same Clement of Alexandria who warned Christian men against being led by the nose like cattle.

> Let us not develop a fear of perfume.... Let the women make use of a little of these perfumes, but not so much as to nauseate their husbands.... There are perfumes that are neither soporific nor erotic, suggestive neither of sexual relations nor of immodest harlotry, but are wholesome and chaste and refreshing[16]

Here is a theologian who was capable of seeing a kind of Aristotelian mean on the issue of perfume wearing that Aristotle himself overlooked. We find a similar moderation among the rabbis of the post-Temple period. As Deborah Green concludes from her review of Talmudic debates, "enough women wore perfume on the Sabbath" for the rabbis to feel some need to regulate it, but they were more concerned about whether putting on a perfume violated the prohibition of working on the Sabbath than in questioning perfume wearing's intrinsic morality.[17]

Despite the minority voices of Clement and the rabbis, the net effect on perfume wearing of Christianity's triumph in the late Roman Empire and its subsequent dominance during the Middle Ages was negative to say the least. But the issue became almost moot in the West with the "fall" of Rome, and the moral debate would not revive until the Crusaders returned from the Near East carrying perfumes in their baggage. As perfume wearing slowly revived among the elites from the late Middle Ages on, the various philosophical and theological arguments against it also resurfaced. For example, Plautus's line turns up in the Renaissance in Montaigne, who joins his embrace of Plautus' line "She smells best who smells not at all" with the old argument against deceptive masking: "Perfumes are rightly considered suspicious in those who use them, and are thought to cover up some natural defect."[18] In the Reformation period Calvin criticized those who "bear about them the scent of the perfumer's shop" as part of his critique of luxury among the Genevan upper classes.[19]

Interestingly, the same themes we find in Montaigne and Calvin—artifice, masking, and luxury—also turn up in eighteenth-century discussions of perfume wearing. Yet, as Corbin points out, these criticisms were applied not to all perfumes, but primarily to the heavier, animal-derived scents of musk, civet, and ambergris, whereas lighter, floral scents were accepted and flourished. Moreover, some writers could, like Lucretius, still celebrate perfume wearing as an aesthetic pleasure for the wearer and those around them. The perfumer Antoine Dejean wrote in 1764 that perfumes can "make us pleasing to ourselves," and by making "us lively in gatherings ... we please others."[20] Corbin also reminds us of Alexander Dumas's later quip about the eighteenth-century French upper classes: "Apart from philosophers ... everyone smelled nice."[21]

Perfume Pleasures in Asia and the Middle East

Rather than pursue other examples of the fortunes of the various classical arguments for and against perfume wearing, I want to turn our attention to contemporary thinking about these issues. But first, a brief comment is in order on a major difference between classical Western critiques of perfume wearing and Asian and Middle Eastern practices such as those we earlier encountered in the rich perfume and incense cultures of classical India, imperial China, and medieval Japan, something that can be found in the Arab-Islamic tradition as well. Unlike the West, where perfume has so often been dismissed as a trivial luxury or an instrument of immoral solicitation, incense/perfume in early India, China, and Japan was part of a highly sophisticated culture of pleasure. Indeed, whereas both Western philosophy and religion have at best been ambivalent about the value of sensory pleasure, in Hinduism, for example, pleasure (*kama*) is one of the three legitimate aims of life, and, as James McHugh emphasizes, is "valued as an *end in itself*." Consequently, "Perfumes were indispensable to the goal of pleasure and the informed consumption of them was a vital part of what it meant to be a cultivated person."[22]

Similarly, in Arab-Islamic traditions, sensory pleasure is not considered suspect in itself. Paradise, for example, is conceived of as a place of sensory pleasures, including those of smell. Moreover, Muhammad personally liked perfumes and recommended wearing them. For example, he advised men to "bathe, oil their hair and perfume themselves for the Friday sermon at the mosque, and both men and women could use fragrance for sexual encounters with their marriage partners."[23] Earlier we mentioned the use of fragrant oils to anoint the newborn in Arab countries and the customary scenting of both bride and groom at Arab weddings. Obviously, in Islam as in Hinduism and most other cultural traditions, excess of any kind is condemned, as is the use of perfumes in connection with the violation of moral norms, but the atmosphere of general suspicion that has dogged perfume wearing in the West is largely absent elsewhere.

There have, of course, been many exceptions in Western thought to the suspicion of perfume wearing from Lucretius on. It is worth recalling at this point Spinoza's line from the *Ethics* that we quoted earlier: "The wise man renews and refreshes himself with moderate food and drink, and also with scents, the beauty of plants in bloom, dress, music, sports, theater."[24] Spinoza's attitude toward pleasure in general and the pleasures of scents in particular grows naturally out of his positive attitude toward the body. As Chantal Jaquet writes, Spinoza lays the basis for a kind of "olfactory ethic" that "embraces a culture of the nose and the love of perfumed scents paralleling the charms of music and theater."[25] In that sense Spinoza suggests a moral basis for finding pleasure and joy in the use of perfumes not unlike the attitudes and practices of the Middle East and Asia.

Contemporary Meanings

Today, most serious commentators and social scientists agree that the meanings and motivations of perfume wearing are multiple. One can usefully group the major contemporary themes under two headings: those meanings and motives that are externally directed at affecting others, and those that are internally oriented toward satisfying oneself. The olfactory psychologist Rachel Herz makes a similar distinction, remarking that, viewed "extrinsically," the human drive toward perfume wearing aims to "manipulate the mood or behavior of others," but viewed "intrinsically" it aims at a person's own sensual pleasure.[26] Although I agree with Herz that externally directed perfume wearing sometimes aims at manipulation, the term "manipulation" unfortunately suggests morally questionable subterfuge and deceit. Certainly, not all perfume wearing that is intended to make oneself attractive to others is manipulative in that sense. Moreover, I would suggest that externally and internally directed aims are not totally immiscible and that the heuristic strength of the external/internal rubric is to show how differently the same issue looks when the behavior is understood as primarily aimed at influencing others or as primarily directed at self-satisfaction. I will begin with three themes that are typically conceived as purely externally directed: seduction, masking, and artifice.

In its simplest external version, seduction has traditionally been understood to consist in the use of perfume as a lure. It lies behind the ancient association of perfume with courtesans and the "designing woman." In such crude versions of seduction, perfume wearing is simply a utilitarian device, a manipulative trick; it's the view implicit in many of the internet ads claiming to have discovered "the human pheromone," the sure-fire scent to get someone into bed.

But a person might also put on a perfume primarily for the pleasure of how it makes *the wearer* feel, and, as Dejean already argued back in the eighteenth century, the person may become lively and attractive to others as a result. Borrowing from Gernot Böhme's aesthetic theory, we could say that such a person projects an attractive "atmosphere," yet without specifically intending sexual seduction in the manipulative sense. Some contemporary social scientists agree. The anthropologist David Le Breton remarks that one can wear a perfume "just to feel good and fit the part," letting the perfume "boost the intensity of one's aesthetic relation to the world."[27] And Rachel Herz concluded in one of her studies that "feelings of self-confidence inspired by wearing fragrance can alter the wearer's behavior in a manner that increases their attractiveness to others, independent of whether those who judge them as attractive can also smell them."[28] Indeed, in many cultures this way of experiencing perfume has often played a role in traditional "courtship," with all its complex rituals involving varying kinds of adornment on both sides.

There is a familiar philosophical contrast that might help us better understand the difference between the externalist interpretation of "seduction" as manipulation and the internalist understanding of "seduction" as attractiveness: the difference between the Platonic and Aristotelian views on rhetoric. For Plato, rhetoric is a morally questionable attempt to manipulate others into thinking the worse is the better case; for Aristotle rhetoric is the art of putting the best arguments for what one is convinced is the truth into their most attractive and persuasive form. The issue is similar to the problem of the role of cognition in emotion that we discussed in Chapter 4. There we argued that treating the effect of odors (or perfumes) as emotional in the sense of "irrational" disregards the important cognitive dimension of emotion itself. In the case of perfume and seduction, the externalist claim that perfumes are primarily used to manipulate others falsely assumes that the emotional effects of wearing a perfume, like the rhetorical enhancement of an argument, will simply overwhelm the other person's cognitive capacities and subvert rational responses.

A similar dual reading can be made of the objection that perfume wearing involves deceitful *masking*. If we understand masking solely as externally aimed, it consists in hiding some defect such as strong body odor or the odor of a bad habit or an illness—hence the often-repeated claim by a Chrysostom or a Montaigne that we should suspect anyone who wears perfume of having something to hide. No doubt there are cases where this may be true, but there are other ways of understanding masking. First, one might mask a strongly offensive odor out of consideration for the comfort of others or perhaps, less commendably, simply out of social conformity, as the convention of wearing deodorants suggests. Second, in wearing a perfume one may not so much aim at *hiding* one's natural odor as at *complementing* it. Once a perfume is on a person's skin, it blends with natural body odor to form a new odor, altering that person's olfactory signature. People with experience in wearing perfumes seek scents that enhance their natural odor profile just as they seek clothing designs and colors that complement their body's natural shape or skin tint. There is some evidence from behavioral studies to support this interpretation.[29] To insist that masking one's body odor is always a case of morally questionable deceit is to embrace a version of what the philosopher Paul Ricoeur called the "hermeneutics of suspicion," an interpretative posture that assumes there is always an unsavory hidden motive behind everything people do.

Yet modern biology has discovered something that suggests a different kind of reason for men, at least, to avoid even partially masking their natural odor during courtship. As Rachel Herz reminds us, there are a number of studies that show that the physical characteristics men find most attractive in women are visual, whereas for women a man's smell (and his ability to provide resources for a family) are often more important than looks. But given the crucial role that maximizing HLA/MHC differences plays in producing healthy offspring who will have strong immune systems, men who wear fragrances may end up masking this unconsciously perceived information. As Herz notes, by the time the relationship has gone far enough for the man's real scent to be revealed, too strong an emotional attachment may have been established.[30]

After seductive masking, the next objection to consider is the "artifice" objection. Part of the philosophical complaint implicit in Plautus's topos, "She smells best who smells not at all," is a rejection of the "artificial" in favor of the "natural." Interpreted in externalist terms, the "artifice" objection is similar to the masking objection: it is wrong in principle to cover one's natural odor with something that is an artificial human construction. One reply to this assertion is that the same objection could be made to wearing adornments of any kind, including cosmetics, jewelry, and even clothing (or in the case of many traditional cultures any body decoration from tattooing to scarification). To reject the propriety of wearing perfume on grounds of "artifice," while allowing other adornments, is illogical and may simply reflect a discomfort with odors in general.

A nineteenth-century romantic version of the artifice objection is reflected in Paul Gauguin's musings on the natural perfume of Tahitian women. He contrasts the authentic Tahitian woman who mingles her healthy animal smell with the natural scent of coconut oil and gardenias to those Tahitian women corrupted by European shopkeepers who have sold them "a frightful perfumery made of musk and patchouli. When they are gathered together in church, all these perfumes become insupportable."[31]

Yet even if one does not adopt one of these traditional versions of the artifice objection, there are more moderate and reasonable contemporary perspectives. Thus, one might refrain from wearing strong perfumes containing artificial molecules, or refrain from wearing any scent at all in certain tightly enclosed situations, out of respect for the comfort of others who may have various kinds of sensitivities or allergies. A similar, but even broader ethical case for not wearing perfume based on artifice could be made on environmental grounds, especially wearing perfumes that contain synthetic chemicals.

Many environmentalists have expressed concern at what they see as the excessive and dangerous use of artificial chemicals in contemporary society. Artificially constructed molecules are not only prominent in many perfumes and personal hygiene products, but also in most household cleaning and decorating products, as well as in nearly all clothing, furniture, automobiles, and so on. Given the possible dangers inherent in the long-term interactions among the many chemicals ingredient in products of daily use, caution might suggest erring on the side of avoidance.[32] Thus, people with a strong personal commitment to environmentalism might refrain from perfume wearing on both external and internal grounds, or at lest confine themselves to perfumes created from organic essential oils. This version of the artifice objection would also generate an updated version of the traditional luxury objection. From an environmental point of view, expenditure on perfume could be seen as a waste of resources on something ephemeral and superfluous sincemany essential oils are quite expensive. (The current political agitation for perfume bans by people who claim to suffer from multiple chemical sensitivity has different motivations and will be discussed at the end of the next chapter.)

So far we have focused on the modern forms of the classic critiques of perfume wearing based on *externalist* versions of such arguments as seduction, masking, and artifice. Now we need to discuss three themes that exemplify positive self-directed or *internal* meanings such as identity, pleasure, and spirituality.

Numerous writers on perfume have drawn attention to the way people may choose to wear a particular kind of perfume as part of establishing or expressing a personal *identity*. The philosopher Richard Schusterman has noted that "the choice of a fragrance is not simply . . . to attract others by satisfying their tastes," but, like the choice of clothes, "an assertion of one's own taste and an appeal to be appreciated not just sensually but also cognitively for . . . style."[33] The literary scholar Richard Stamelman's *Perfume: Joy, Obsession, Scandal*, on the other hand, interprets the identity motive as part of a contemporary cultural "image-system" that is based on a "network of personal and collective fantasies."[34] Thus, just as the externally oriented versions of seduction and masking can be conceived from a less morally objectionable internalist perspective, so the internally directed notion of identity may also have its more questionable externally oriented versions, for example, "I'm rich enough/stylish enough to wear this!" As the existentialist philosophers argued, too often we do not take full responsibility for our identity choices, and one could argue that the recent vogue for "celebrity perfumes" reflects the kind of "inauthenticity" the existentialists condemned. In any case, the choice of whether or not to wear a fragrance, and what kind to wear, and when and where to wear it are all opportunities, as Stamelman puts it, to change our body's "presence in the world . . . and redefine its identity."[35]

In my discussion of a possible internalist interpretation of seduction as attraction I already called attention to the fact that many people wear perfume or cologne simply for the pleasure of "how it feels." Rachel Herz has concluded from her studies of perfume wearing that "perfume is created and used for pure pleasure more than for any other function."[36] And the well-known perfumer Sophia Grojsman once remarked, "The proudest moment for me is to know that I am making some woman happy."[37] Ann Gottlieb, a consultant to perfume companies, when asked why women wear perfume, spontaneously linked identity and pleasure: "It is a projection of who you are. It makes you feel good."[38]

This connection between pleasure and identity is even recognized by some perfume advertising, which otherwise so often focuses on sexual seduction.

Some television ads have tried to project the pleasures of perfume wearing through thirty-second fantasies of liberation and transformation. In one such ad, a beautifully dressed woman wearing a choker of pearls grasps a silken scarf hanging from the oculus of what looks like a Renaissance palace, and climbs up the scarf through the opening; once on the roof, we see that the palace is in the midst of a glittering contemporary city. Then the lone name, *J'adore*, appears. What's significant about this typical piece of associative advertising (Gernot Böhme would call it atmospheric "staging") is that although the woman is certainly attractive, the focus is not primarily on sex but on a moment of aesthetic experience that includes daring, adventure, ascension, one could even say transcendence, as the woman surveys the city like a goddess.[39]

Thus, the sensory and intellectual pleasures of perfume wearing not only have an affective but an imaginative aspect. The affective aesthetic aspect is reflected in descriptions of wearing perfume as making one "feel good" (Gottlieb) or generating a feeling of "happiness" (Grojsman) or, as Lucretius put it, of feeling "fresh and new." The more imaginative aesthetic aspect is reflected in wearers' delight in the artistry of a particular fragrance and the way it blends with their own body odor to give them a psychological lift. Translated in terms of the two poles of everyday aesthetics theories (Leddy to Saito), it seems that for many men and women who wear perfume or cologne almost daily, the pleasure is part of what Saito called 'the aesthetic texture of ordinary life.' But the fantasy of the woman in the *J'adore* commercial ascending through the oculus to overlook the city suggests that wearing a perfume may from time to time produce an aesthetic experience similar to Leddy's ordinary experiences becoming extraordinary or taking on an "aura."

The anthropologist David Le Breton remarks that perfume "changes nothing in the world," but by "radically altering its atmosphere" a perfume can become "an elementary instrument of transcendence."[40] Alfred Gell has gone even farther and argued, as we saw earlier, that in small-scale, traditional societies like the Umeda in Papua, perfumes are often believed not only to alter the atmosphere of the world, but to change reality since they connect the ordinary and the spirit world. This observation led Gell to suggest that there may be an aspect of magic even in modern perfume wearing. Gell speculates that the deeper meaning of perfume wearing may not be in its intended effect on others but simply in the "*act* of putting it on," an act that symbolically gestures toward "*the transcendence of the sweet life.*" Because "it is perfume (spirit, halfway between thing and idea), it partakes of ... transcendence ... while still remaining part of the world." The modern Westerner's putting on a perfume, therefore, can sometimes be "a *magical* act," because in putting it on, the perfume's volatility seems to give access to "a charmed universe."[41]

Whether or not one agrees with the exact terms of Gell's analysis, he calls attention to a spiritual dimension of wearing perfume that raises perfume's sensual and aesthetic aspects into the world of dream and imagination and reminds one of some of the uses of incense in religious contexts and in the more sophisticated versions of kodo. No doubt our commercialized capitalist societies have managed to so trivialize perfume wearing that its spiritual dimension remains but a faint echo of the deeper possibilities that Gell and Le Breton describe or that one finds in ancient Sanskrit texts or in Spinoza. The uses of perfume are indeed multiple and obviously include manipulative seduction, masking, and status display, but also at times aesthetic pleasure in how an artfully designed scent makes one feel, and at rare moments, a fleeting sense of transcendence.

Notes

1. Horowitz, *Being a Dog*, 19–20.
2. Stoddart, *Adam's Nose*, 163–64
3. Stephen Davies has been exploring the general practice and principles of human bodily adornment for some time. For a sample see his essays "The Aesthetics of Adornments," in Yuedi and Carter, *East and West*, 124–31, and "Analyzing Human Adornment," *American Society for Aesthetics Newsletter* 38, no. 2 (Summer 2018): 1–4.
4. Annick Le Guérer, *Scent: The Mysterious and Essential Powers of Smell*, trans. Richard Miller (New York: Random House, 1992), 15–23.
5. Cited by J.-P. Vernant, "Introduction" in Marcel Detienne, *The Gardens of Adonis: Spices in Greek Mythology*, trans. Janet Lloyd (London: Harvester Press, 1977), vi.
6. Vernant, "Introduction," in Detienne, *Gardens*, xxi–xxv.
7. Aristophanes, *Lysistrata*, Douglass Parker (Ann Arbor: University of Michigan Press, 1969).
8. Xenophon, *Symposium*, A. J. Bowen (Warminster, UK: Aris & Phillips, 1998), 31.
9. Plato, *Republic*, trans. Robin Waterfield (Oxford: Oxford University Press, 1993), 64.
10. Aristotle, *Nicomachean Ethics*, trans. David Ross, Revised by J. L Ackrill and J. O. Urmson (Oxford: Oxford University Press, 1998), 72–73.
11. There is another passage in which Aristotle speaks of "every sense" being active in relation to its object, a point that seems to be more open to accepting a moderate use of perfumes. Book X, 1174b–1175a18. I am grateful to Thomas Leddy for pointing this out.
12. Pliny, *Natural History*, Vol. 4, H. Rackham (Cambridge, MA: Harvard University Press, 1945), 60–61, 100–113. Subsequent references in the text are to this version.
13. Bradley, "Foul Bodies in Ancient Rome," in Bradley, *Smell and the Ancient Senses*, 142–43.
14. Shane Butler, "Making Scents of Poetry," in Bradley, *Smell and the Ancient Senses*, 78.
15. Harvey, *Scenting Salvation*, 158–69. Clement adds that men should be "redolent, not of perfume but of perfection and women should be fragrant with the odor of Christ,

not that of powders and perfumes." Clement of Alexandria, *Christ the Educator*, trans. Simon P. Wood (New York: Fathers of the Church, 1954), 150.
16. Clement, *Christ the Educator*, 150.
17. Deborah Green, *The Aroma of Righteousness: Scent and Seduction in Rabbinic Life and Literature* (University Park: Pennsylvania State University Press, 2011).
18. Montaigne, *Essays*, 228.
19. The perfume comment is from Calvin's short essay titled "De Luxu," or "Concerning Luxury," which has been translated and commented on by Ford Lewis Battles in *Interpreting John Calvin* (Grand Rapids, MI: Baker Books, 1996), 331.
20. Corbin, *The Foul and the Fragrant*, 72.
21. Corbin, *The Foul and the Fragrant*, 76.
22. McHugh, *Sandalwood and Carrion*, 106.
23. Mary Thurlkill, *Sacred Scents in Early Christianity and Islam* (Lanham, MD: Lexington Books, 2017), 116.
24. Spinoza, *Ethics*, 260
25. Jaquet, *Philosophie de l'odorat*, 77–8.
26. Rachel S. Herz, "Perfume," in *Neurobiology of Sensation and Reward*, ed. Jay A. Gottfried (Boca Raton, FL: Taylor & Francis, 2011), 381.
27. David Le Breton, "Brief Anthropology of Perfume in the Western World," in Grasse, *Perfume*, 219.
28. Herz, "Perfume," 378.
29. Stoddart, *Adam's Nose*, 190–91.
30. Herz, "Perfume," 377–78.
31. Quoted in Drobnick, "Towards an Olfactory Art History," 199.
32. I am grateful to my colleague Peter Wenz, who has written extensively on environmental ethics, for pointing this out.
33. Richard Schusterman, "Somatic Style," *Journal of Aesthetics and Art Criticism* 69, no. 2 (2011): 153.
34. Richard Stamelman, *Perfume: Joy, Obsession, Scandal, Sin. A Cultural History of Fragrance from 1750 to the Present* (New York: Rizzoli International Publications, 2006), 21.
35. Stamelman, *Perfume*, 18.
36. Herz, "Perfume," 383.
37. Cathy Newman, *Perfume: The Art and Science of Scent* (Washington, DC: National Geographic Society, 1998), 52.
38. Newman, *Perfume,* 53.
39. A reviewer of the manuscript of this book suggested that the whole scene is erotically charged. Indeed, the ad associates the perfume with a woman who is very attractive, but her fascination for the viewer derives as much from her self-confidence and strength (she climbs the scarf/rope hand over hand) as from her physical beauty.
40. Le Breton, "Brief Anthropology of Perfume," 219.
41. Gell, "Magic, Perfume, Dream," 405–6.

14
Ambient Scenting, Architecture, and the City

In 2013, the philosopher Marta Tafalla organized a colloquium in Barcelona titled "Scent, Science & Aesthetics: Understanding Smell and Anosmia." One of the colloquium speakers, the urban design theorist Victoria Henshaw, subsequently led us on a "Smellwalk" through old Barcelona.[1] The walk began at the big market near the south end of Las Ramblas, the broad avenue that bisects the center of the old city. As we strolled through the market, the dominant scent notes shifted with almost every step: intense fish odors, pungent hams hanging overhead, and the cold smell of poultry gave way to the myriad scents of exotic fruits and vegetables and, finally, the aroma of warm chocolate being swirled on a slab in front of us. When we left the market and turned toward the port, a gentle breeze from the Mediterranean cleared our nostrils before we turned into a narrow, medieval street whose rough, musty stonewalls exuded a mild limey odor, punctuated from time to time by a whiff of moss—and urine. Rounding a corner, we caught the aroma of a bakeshop and a little farther on the smell of halal meats grilling, accompanied by the sounds of Middle Eastern music. For a smellwalk, of course, is always a multisensory experience.[2]

On the Barcelona smellwalk, I found that focusing on smells made me more alert to all the other senses, especially to sounds and textures, but also to the senses that involve balance (vestibular) and body orientation (proprioceptive). One listens through the traffic noise to the sound of feet on the pavement and catches half-heard voices from courtyards, one feels the smoothness of asphalt give way to the precariousness of cobblestones, one gingerly touches the rough stone walls and the moss between the stones and feels the air caressing face and hands, bearing with it ever-changing scents. A smellwalk makes you aware of how deeply intertwined the senses are and that, although vision dominates our waking hours, we can learn to attend to each of our other senses and their interactions, letting our bodies experience something far richer than mere seeing. Gernot Böhme's notion of "atmospheres" can be useful here in that an atmosphere is by nature multisensory and it is experienced holistically in terms of a certain feeling or "mood." Böhme points out that "odors are an essential element of the atmosphere of a city,"

Art Scents. Larry Shiner, Oxford University Press (2020) © Oxford University Press.
DOI: 10.1093/oso/9780190089818.003.0030

perhaps even the most essential, for odors are, like almost no other sensible phenomenon, atmospheric . . . they envelop, cannot be avoided; they are that quality of a surroundings which most intensely allows us to sense through our disposition (*Befindlichkeit*) *where* we are. Odors enable us to identify places and to identify ourselves with places.[3]

Smellwalking and Scent Arts in the City

Yet despite the importance of odors in establishing the atmosphere of a city and its neighborhoods, most of us are so oriented to visual impressions that the smells around us, as Köster points out, remain mostly an unconscious background. Thus, smellwalks are especially useful for attuning us to the role of smell in our everyday aesthetic experience of the city. Accordingly, it is no surprise that an artist such as Jenny Marketou would create a participatory artwork that involved self-conducted smellwalks around parts of Philadelphia. Nor is it a surprise that other artists and designers such as both Sissel Tolaas and Kate McLean have undertaken smellwalks with volunteers as part of their interest in cataloguing the distinctive smellscapes of cities around the world and have generated artworks in the process.

Tolaas, who has been studying and collecting city smell profiles longer than almost anyone (she has done around thirty-five of them, from Kansas City to Singapore), reminds us that each major city's distinct odor is a mix that can vary enormously from neighborhood to neighborhood and that changes over time. From 2002 to 2004 she carried out a socially focused study of the smells of Berlin, resulting in an exhibition that featured four perfume-like vials with scents she formulated and named for the characteristic mix of smells of each of the four quadrants of the city, NE, NW, SE, SW, each quadrant inhabited by a different economic and ethnic mix. In the exhibition that grew out of her study called *Without Borders NOSOEAWE*, Tolaas's vials of scent and her wall texts drew attention to the city's economic and ethnic divisions and revealed some of the prejudices based on smell. As Jim Drobnick has commented, "Tolaas challenged a static and inevitable understanding of olfactory habitus . . . exposed its ethical dimension and opened it to critical evaluation."[4]

In a 2016 interview, Tolaas lamented that the smells of Berlin seemed less diverse to her in 2016 than they had been only a dozen years earlier. One exception, she suggested, might be the Jannowitzbrücke underground station in the former East German sector, and her comment is especially interesting in light of Helgard Haug's recreation of the Alexanderplatz scents. "If you remove a couple of tiles from the wall, the smell of the German Democratic Republic would come off . . . lignite and a detergent, which I suspect was used in all the public buildings, probably supplied by the same state-owned company."[5] Tolaas's guess about the detergent is confirmed by the reminiscence of the British journalist Neal Ascherson, who worked as a correspondent in East Berlin: "East Germany vanished from the atlas in 1990 . . . yet it did have its own authentic scent, a spicy reek brewed out of People's Cleaning Fluid, two-stroke petrol, brown-coal briquettes and cheap police tobacco."[6]

Although Kate McLean's smellwalks are also meant to arouse awareness of the importance of smell, they are even more explicitly conceived as a prelude to design artworks. As McLean remarks, "My artistic practice maps smell from a human-centered perspective" that aims not only to enhance understanding, but also to provoke opinion and discussion.[7] The smellwalks and studies conducted by Henshaw, Tolaas, McLean, and others not only help people cultivate their everyday aesthetic appreciation of city smells, but also demonstrate that the "deodorization" of cities we discussed back in Part II has hardly reached completion. Most major population centers still have neighborhoods with distinctive smells, if only from ethnic cuisines and a few open-air markets. And many smaller cities are still known for particular industrial smells, such as Hershey, Pennsylvania, of chocolate fame, or Decatur, Illinois, where the odor of soybean processing often hangs heavy in the air, or the infamous timber-processing "aroma of Tacoma" that even showed up in a rock song.[8]

Although the tendency to deodorize cities has not been completely successful, the dialectic of deodorization continues, partly because many public officials think of odors only as something to be controlled or eliminated and give more attention to complaints than to calls for encouraging a richer olfactory environment. Yet there are also commentators who perceive the general deodorizing tendency as leading to cultural loss. In *Glasgow Smells: A Nostalgic Tour of the City* (2008), Michael Meighan laments the disappearance of most of the distinctive smells of the neighborhood where he grew up.[9] Similarly, two Indian authors have written of the disappearance of many of the distinctive smells (*gandh*) of the city of Jaipur as a sign of a threatened cultural heritage. "Once we forget our 'gandh' we forget our roots."[10]

Smellwalks and smell mapping are not the only way artists have called aesthetic attention to a city's smell profile past and present. For her project called *Scent of Sidney*, as part of the Sidney Festival in January 2017, the artist Cat Jones interviewed ten prominent cultural leaders on their memories and perceptions of Sidney's smells. In Jones's installation work, visitors could listen to recordings of each interview as they sniffed a scent she created from essential oils to reflect each narrative. The scent that Jones called "Icons of a Lost Economy" invoked the mix of oil refinery and brewery smells described by a sociologist, Michael Dary, who grew up in Sidney's industrial suburbs. Jones named another scent "Dharawal" after the olfactory memories of the aboriginal elder Aunty Fran

Bodkin, who grew up in the Dharawal homelands. The elements of that scent included the smell of pink boronia flowers and the fragrance of frangipani, a tree that was traditionally planted by the Dharawal people each time a female child was born.[11] Jones's *Scent of Sidney* installation shows how a conceptually oriented, participatory artwork can transcend the borders separating art, design, and social science.

A more pointedly political example of using a work of scent art to deal directly with the threat to olfactory diversity is Michael Rakowitz's *Rise* (2001). Back in 2001, the Chinatown area of New York was being steadily gentrified and many tenants evicted as developers turned old buildings into upscale apartments and condos. But sometimes the developers would open space in one of their buildings for temporary art exhibitions, hoping to attract attention and add cachet. Rakowitz figured out a way to make visitors to one such exhibition more aware of what was happening to the ethnic diversity of the neighborhood. For *Rise*, he ran a vent pipe from the Chinese bakery next door up nine floors and into the rooms where the exhibition was being held. Along with the aromas of the bakery that filled the gallery, he also made available samples of the baked goods, and as a result, some visitors dropped in on the Fei Dar bakery as they left, met the owners, and discussed the fate of the neighborhood.[12] The works of Jones and Rakowitz illustrate the way works of scent art can both celebrate the olfactory diversity that remains in cities and awaken awareness of the need for action when it is further threatened.

Smell and Unban Design

As Henshaw points out few urban planners and designers treat the olfactory aspect of city life as something that could make a positive contribution to health and aesthetic satisfaction.[13] The neglect of the positive aspects of the urban smells is no doubt partly a result of the traditional disparagement of the sense of smell, but is also an extension of the modern sanitary campaigns that led to the dialectic of deodorization. Unfortunately, some people confuse any odor they find personally distasteful with pollution. But one needs to distinguish pollutants—chemicals in the air that can cause actual harm, but may or may not be detectable by their odor, such as carbon monoxide—from odorants—molecules that are by definition detectable by their smell and may cause no harm at all. Certainly, there are odors that are disliked and found discomforting by many individuals, and some of these smells end up being classified by officials as "nuisance" odors, such as the smell of frying grease emitted by fast-food chains.

Among the unfortunate effects of urban planners' and legislators' focus on deodorization and control is an increasing olfactory blandness within certain upscale areas of cities and the diminishing sense of place we mentioned earlier. A. A. Gill has remarked on the economic aspect of this. "The world can be split into all sorts of haves... and have-nots. But here is a new source of division: smell. And we in the rich half are the ones who are the have-nots."[14] Of course, because such losses are often gradual, people may not notice them unless, like Meighan, they leave and return to their old neighborhood after deodorization has taken its toll. On the other hand, some longtime residents of neighborhoods may feel their own sense of place threatened by the arrival of immigrants who bring new smells through exotic cuisines or street festivals such as the Hindu Ganesh parade, which is often accompanied by camphor torches.

But suppose urban planners and designers did begin to take a proactive rather than purely reactive approach to smells. What would change? In the most general terms, it would mean that planners would have to become alert to the potentially positive contribution of distinctive smells to a sense of well-being and aesthetic satisfaction, and this might lead to them to regard many existing sources of odors as potential assets rather than potential nuisances. Henshaw speaks of the health and quality-of-life benefits of olfactory planning as part of the "restorative" use of odors, a concept familiar from environmental psychology studies showing the benefits of exposure to nature, whether large parks, the countryside, the woods, or wilderness.[15] Such studies obviously suggest preserving and enhancing the number of trees, green spaces, parks, waterways, ponds, fountains, and so on. In the past theses features have been treated primarily as visual objects and even then often resisted because of their maintenance expense. But by taking a multi-sensory approach that includes smell along with sound and touch, the planning and design of such restorative features would begin to consider not just the look of trees and flowers or their ease of maintenance, but also their smell as part of their larger health and aesthetic value.

More controversial is the possibility of extending such a positive approach to actually scenting public spaces. In light of the individual variability of odor preferences and sensitivities, as well as the ethical controversy over private businesses using ambient odors to improve sales, this may not be a good idea. The one public scenting possibility Henshaw thinks could be acceptable would parallel what several cities in Italy and Sweden have tried as a way of countering intrusive traffic noise in parks: these parks have speakers scattered in them that play nature sounds (water, birds) that partially mask traffic noise. Yet given studies that show that almost any pleasant odor raised to sufficient intensity can become noxious, to spread a fragrance of sufficient intensity to mask traffic smells might make the odor as unpleasant to many people as the exhaust fumes. Whatever interventions planners and designers make, they will obviously have to consider what is consistent with the locale and will be acceptable to a majority of residents. Virginia Postrel made some wise comments on the need to avoid "design tyranny" when she wrote that the aim of planners and designers should be "to discover rules that preserve aesthetic discovery and diversity, accommodating plural identities and tastes while still allowing the pleasure of consistency and coherence."[16] That is a tall order, but given our concern with understanding the aesthetic potential of the sense of smell and the use of odors in various arts, her emphasis on "aesthetic discovery and diversity" is exactly what is called for in the olfactory aspect of city planning—and also in architects' design of buildings and public spaces.

Aromatic Architecture

The heavy emphasis on visual appearance in modern architectural theories and philosophies is as woefully inadequate to the bodily and sensual experience of architecture as it is to the experience of the urban environment as a whole. A vision-centric aesthetic has often led architectural critics not only to write reviews concerned solely with a building's visual aspects, but also to accompany their reviews with photographs of the building empty of people and furnishings. In such criticism, architecture is discussed as if buildings were merely sculptures to be looked at rather than places to be inhabited. This sort of vision-centric bias is one reason for the sensory poverty of some recent and contemporary architecture, especially major corporate developments. As Juhani Pallasmaa remarks, instead of offering multisensory experience, too many of our urban buildings today have turned into "image products."[17]

Barbara Erwine suggests that the vision-centric bias has also been reinforced by the modern schism between what she calls "Architecture with a capital 'A'" and engineering. Architects are typically trained to focus on the visual aspects of design and leave structural issues as well as sensory matters such as air quality and odors to engineers. And engineers, in turn, are trained to think primarily in terms of *controlling* sound, temperature, humidity, air quality, and odors according to minimal standards. But this control- and minimum standards-based approach has led engineers to focus almost exclusively on "uniformity and not experiential delight."[18] Accordingly, most engineers pay little attention to the positive aesthetic possibilities of odors, and most architects simply ignore smell. The leading architect Elizabeth Diller, for example, admits that, "as an architect, smell is not something that I consciously design."[19]

Fortunately, the vision-centric bias of so many architects and theorists has been challenged in recent decades. A number of architects and architecture theorists such as Pallasmaa and Steven Holl have embraced the idea of "embodied knowledge," developed by the twentieth-century phenomenologist Maurice Merleau-Ponty, whose work as a whole attempted to make the "lifeworld," or lived space and time, the starting point of philosophical reflection rather than an already abstracted space and time presumed to be its basis.[20] In doing so Merleau-Ponty anticipated aspects of Gernot Böhme's notion of "atmospheres," which Böhme makes a key to understanding architecture: "Architecture produces atmospheres in everything it creates."[21] The philosopher Mădălina Diaconu, in turn, has used the notion of atmospheres to argue for the importance of explicitly including smell in theorizing about architecture. Diaconu wants us to replace a vision-centric approach, which treats space as an abstract object, with an understanding of lived space that is sensitive to both smell and touch. An architectural theory based on embodied thinking, she writes, would replace the "'perspectives' of visual space" with the "directionality of olfactory space (trails, tracks and traces)," and replace the abstract idea of the "order" of visual space with the "quality we call atmosphere."[22]

Jenifer Robinson, a philosopher working in the analytic rather than phenomenological tradition, has also written that good architecture "invites or compels multisensory experiences and ways of moving and acting that can be felt in a bodily way."[23] But whatever philosophical framework we choose, if we do approach the aesthetics of architecture from an embodied and multisensory perspective, we will no longer be satisfied with appreciating buildings by just *looking* at them as if their spaces were empty volumes without sound, texture, or smell. Instead, we will have to learn to walk through them with all our senses alert, taking in sounds, textures, smells, temperature, humidity, and the way the spaces affect our balance, orientation, and sense of proximity. All these aspects working together, as the architect Peter Zumthor says, create the overall "feel" or "atmosphere" of a building.[24]

But what can architects actually *do* with smell? Pierre von Meiss has succinctly sketched the design implications inherent in a multisensory and purpose-rooted architectural aesthetic in his book *The Elements of Architecture*:

> The aesthetic experience of our environment is an all-embracing affair, and there are certain situations where hearing, smell, and touch are engaged even more intensely than sight. Let us, therefore, try to imagine the echo in spaces we design, the smells given off by the materials used, the activities likely to take place in them, and the tactile experiences they will be producing.

As von Meiss suggests, a major way architects can attend to smell in their designs is in the choice of materials. Although seldom acknowledged in architectural histories, part of the delight of traditional stone, brick, or wooden buildings has always been their smell along with their auditory and tactile qualities. Many ancient palaces used aromatic woods like cedar or cypress for their fragrance as well as their insect-repelling qualities.[25] Other kinds of building materials, such as mortar or mud blocks, were in some cases made fragrant by infusing them with aromas. Islamic builders, for example, sometimes mixed rose water with the mortar used for mosques so that under the full intensity of the sun, the walls would emit a light fragrance. Naturally, the aromas given off by buildings of stone, masonry, or wood gradually loose much of their original smell over time, although they may develop other more complex scents, some of them from their subsequent uses. Of course, in much contemporary urban architecture, especially high-rises, most buildings are made of glass, steel, aluminum, and newer alloys that give off little or no odor at normal temperatures. Hence, the aromas of such buildings are usually incidental rather than intentional. Even so, the use of aromatic materials like woods for their fragrance has continued in smaller structures such as houses or pavilions.[26]

Another way architects of the past attended to the olfactory dimension of their designs was in the provision of air. Before the advent of the modern sealed building, the capture of air was sometimes influenced by the desire to bring in fresh air bearing the ambient odors of the larger geographical setting, whether the sea, mountains, forest, countryside, or garden. Traditional Japanese houses, for example, often had sliding panels that could open rooms onto a fragrant garden. But in dry climates such as the Middle East and parts of South Asia, traditional ventilation and cooling devices such as wind towers on public buildings (Persia) or directional wind catchers (Hyderabad) were intended as much to dilute interior odors as bring in new ones, thus functioning more like modern mechanical ventilation.

Many contemporary buildings go even farther in their effort to eliminate odors and are completely sealed to ensure that external air is mechanically processed to remove particles and noticeable smells as well as to control fluctuations in temperature and humidity. When sealed buildings are first completed, they are often left empty for at least two weeks, with their ventilation and exhaust systems going full tilt to rid them of the noxious VOCs (volatile organic compounds) that come from of caulk, adhesives, paints, carpet, vinyl tiles, and so on. Sometimes these initial measures are inadequate, or the ventilation system subsequently falters, allowing the volatiles to generate what is called "sick building syndrome," requiring further deodorizing efforts.[27] In sum, a good deal of the most notable contemporary urban high-rise architecture, whether deliberately or unintentionally, provides an odorless and, apart from its visual aspects, a largely sterile sensory environment.

Yet the contemporary sealed building also offers another possibility for an aromatic architecture, although it usually occurs as an occupant adaptation rather that as part of the original design. This is the injection of ambient odors using room diffusers or even parts of the ventilation system to offer mood enhancement for the occupants, whether residents, workers, or customers. The paternalism of these uses obviously raises important ethical issues.

The Ethics of Ambient Scenting

In his 2016 manifesto *Liberté, Egalité, Fragrancité*, the perfumer/artist Christophe Laudamiel proclaimed that henceforth all buildings should get "several rotating scents over the course of a day, so that you can feel you are inside a breathing building full of soul."[28] Although the commercial use of ambient scents in hotels and stores is intended to create a pleasant atmosphere for lingering, whether it makes patrons feel they are in a "breathing building full of soul" is an open question. Certainly, the motives of ambient scenting have been more economic than aesthetic from the beginning. Over three decades ago the Japanese scent firm Shiseido began touting the positive effects of injecting scents into factories or offices, claiming they would improve worker productivity by stimulating alertness (citrus, peppermint) or reducing anxiety (lavender). Some extended care facilities for the elderly have experimented with giving each residential corridor a signature odor to aid residents in finding their way. Although the number of residential complexes and workplaces that use ambient scenting remains limited, ambient scenting in the marketplace has become a global urban phenomenon, with hotel chains such as Marriot and Westin adopting "signature scents" and innumerable retail stores from Abercrombie & Fitch to IKEA using proprietary fragrances. One could see ambient scenting as a reflection of the larger trend that Böhme discusses under the rubric of "staging." He argues that late capitalism in developed countries has created an aesthetically oriented economy (some economists call it "the experience economy"), in which "staging-value," that is, the staging of desires through the creation of atmospheres, has now surpassed both "use" value and "exchange value as they were envisaged in classical Marxian economics.[29] Ambient scenting in the workplace and marketplace has met with mixed responses. For many people with no particular aversion to perfumes or other scented products, the deliberate use of ambient odors to create a pleasant atmosphere smacks of unethical "manipulation." Most people accept the idea that bookstores naturally smell like books, leather shops like leather, gift boutiques like scented candles, pet stores like straw and animals. But an artificial odor introduced into an office, waiting room, or hotel lobby or a "signature" scent floating in a clothing store, although it seems on the one hand like nothing more than olfactory muzak, when it is called to people's attention, sometimes leads to indignation at what is perceived as an attempt to subliminally influence them.

There are two interrelated issues involved in the debate over ambient scenting that need to be distinguished: (1) the factual question of whether and how odors modify behavior and just how effective they are, and (2) the ethical issue of whether employers or merchants should attempt to influence behavior by modifying an environment. The ethical issue in turn takes somewhat different forms for the workplace and the marketplace. The management professors Samantha Warren and Kathleen Riach, writing about the use of ambient scents in the workplace, have questioned the "ethics of manipulating people" with scents, including the right to "provoke or evoke feelings."[30] In the case of scent marketing, Kevin Bradford and Debra Desrochers argue in the *Journal of Business Ethics* that there exists an ethical norm governing the use of information in market exchange that requires the consumer be made aware of the source of attempts at persuasion and of how those sources operate.[31] The factual and ethical issues in the case of both the workplace and marketplace are interdependent. Thus, if ambient odors are in fact able to influence behavior without people being able to defend themselves, there would seem to be a strong case that they constitute an instance of unethical manipulation and should not be used. Unfortunately, neither the factual nor the ethical issues are straightforward.

Let's consider first the factual question of the power of odors to influence behavior. As we saw in Part I, this is a complex issue, with some research suggesting that smells operate as immediate emotional triggers, bypassing our cognitive capacities, but other research and arguments suggesting there is a cognitive element to most of our smell responses. Not surprisingly, both the scent-marketing companies who sell ambient scenting systems and the critics who condemn ambient scenting tend to emphasize the immediate emotional effects. But, as I argued in Part I, our sense of smell is not purely emotional and unconscious in all situations, and that means that our total response is not always an immediate and unthinking one. And even if some odors may trigger a visceral avoidance response important for survival, most of the pleasant odors of the kind used in ambient scenting do not trigger an irresistible attraction response powerful enough to overwhelm judgment. Based on the neuroscience and psychology studies that have been done so far, the worst fears of subliminal persuasion seem highly exaggerated (as are the optimistic promises of the firms that sell ambient scenting systems).[32] As part of atmospheres, ambient odors do have a general effect on mood and also a tendency to arouse emotion-laden memories, and to that extent may make a store's or hotel's atmosphere seem pleasanter (or not, if they are incongruent with other sensory information, or with an individual's past associations). But ambient scenting is no more likely to cause people to lose self-control and take foolish actions than using pleasing colors, textures, and sounds is likely to do. Because we respond multimodally to most environments, simply adding a scent to the atmospheric mix will not automatically force a particular behavior.[33] Once we put the use of ambient scents in the larger multisensory framework of atmospherics, it becomes even clearer that scents by themselves do not subvert rationality, but are normally only one part of a total sensory complex. Indeed many more companies have "signature colors" and "signature sounds" intended to influence our behavior than have "signature scents." Thus, once again, the facile claim that the sense of smell alone of all the senses triggers purely emotional reactions simply won't hold up.

Turning now from the factual to the ethical aspect of ambient scenting, even if ambient odors do not possess a unique power to subvert rationality and cannot easily be used to "manipulate" us, there remains the more general moral objection to ambient scenting that people should not be subjected to an altered and unexpected environment without their consent, especially in the workplace. This is the particular concern raised by Warren and Riach on the grounds that a business organization is a power hierarchy in which those who exercise control ought to consult their employees on the work environment. It could be argued that the workplace differs in this respect from the marketplace in that there is an implied contract involved and that changing the environment of the workplace by adding ambient scenting, even if it is meant to be beneficial to all concerned (as supposedly are changes in lighting, colors, tactile surfaces, the soundscape, etc.), the decision should involve all stakeholders. Moreover, given the problems that can be posed for people suffering from particular aversions or allergies, such consultation would seems to be both morally advisable and wise, whether or not it becomes legally obligatory. But I see no reason to reject the use of ambient scenting in the workplace so long as there is appropriate consultation.

Ambient scenting in the marketplace raises similar issues involving consent, especially given Bradford and Desroches's point about the norm of providing relevant information for market exchange. They suggest there is a parallel to an FCC notice dating back to 1974 concerned with subliminal visual persuasion. It declared that any attempt to transmit visual messages below the threshold of normal awareness is a deceptive act contrary to the public interest.[34] Obviously, the use of truly subliminal (i.e., below the threshold of conscious detection) ambient scenting would ipso facto be a deceptive practice and morally condemnable. There is clear evidence that some scents that are below the threshold of detection (like some sounds or visual images) can indeed influence preferential responses, for example, make respondents in an experiment prefer certain faces to others at a higher than chance rate.

But how does one apply the ethical principle of providing information for marketplace exchange in the more typical situation where the ambient scenting is done at a detectable level? Bradford and Desroches suggest that even if the scents are clearly detectable, given how new the practice of ambient scenting is, most people are likely to be unaware that the intention of such scents is a form of persuasion, and that makes ambient scenting tantamount to deception.[35] But the term "persuasion" here is being used ambiguously. Providing a detectable scent as part of the creation of a pleasant multisensory atmosphere that also includes pleasant sounds, lighting, colors, and textures, is a highly indirect form of "persuasion." No doubt customers will spend more time in atmospheres they find pleasant, and the longer they stay, the more opportunity there will be for them to spend money. But it would be a gross exaggeration to say that the scent aspect of such multisensory atmospheres deceives people into parting with their money. The "persuasion" here falls into the category of the kinds of accompaniments that go into any instance of public speaking, such as using an appropriate tone of voice, type of gesture and facial expression, and choice of clothing. If the same speaker were also to wear pleasant-smelling cologne, would that be a bit of "deceptive persuasion?" The ethical disagreement over the creation of such atmospherics in the case of ambient scents or a speaker's self-presentation is again similar to the difference between Plato and Aristotle over rhetoric, with Aristotle recognizing both the inevitability and necessity of what Plato feared as inessential and devious atmospherics.

But here again, as in the case of the ambient scenting of the workplace, what is ethically permissible in a general way is not always morally commendable or wise in certain situations. Given the variability in people's responses to odors, which depends on gender, age, and personal history (something also true, by the way, of responses to colors and music), and also given the fact that a small minority of the population suffers from allergic reactions to some scents, it would be appropriate to offer notification when feasible. Of course, defenders of ambient scenting might argue that simply by entering the premises of a given store or entertainment venue one is tacitly accepting the atmosphere offered since one is always free to leave. This may be true of retail stores, where walking out the door is relatively easy, but hotels would do well to provide some notification that ambient scenting is being used, and also offer rooms that are not scented. Both the Westin and Marriott hotel chains make their signature scents along with diffusers available for purchase online, so that giving customers a notice of their scenting practice at the time of registration would alert the hypersensitive without unduly burdening the hotel or unduly alarming other patrons. My general conclusion on the ethics of ambient scenting is that we should neither condemn it outright nor fully embrace it. In the case of the workplace, appropriate consultation is the right way to proceed; in the case of the marketplace, appropriate notification is the right thing to do in situations like hotels where customers will be spending a longer period of time than happens in most retail situations. Moreover, if both the trend toward ambient scenting and the adoption of "signature scents" continues in the retail and hospitality sectors, more and more people will become aware of the practice and there will be less need for providing information for market exchange. Even so, the comfort and health of the minority who have serious problems with some smells still need to be addressed through a program of notification.

So far we have discussed the aesthetics and ethics of ambient scenting as if it were a purely contemporary commercial phenomenon, but it obviously had important historical precedents prior to the "deodorization" process, in the deliberate scenting of homes and shops with fragrant boughs, incense, or spices, as well as the use of various scents as plague preventives and in a variety of medical cures. Today's interest in aromatherapy harks back to some of these older medical practices, although the term itself only appeared in 1937 when a French chemist published a book using "aromatherapy" in the title and claimed curative powers for essential oils.[36] If some aromatherapy advocates see profound health benefits in the use of odors and in that sense want to reverse some aspects of the "deodorization" process, there is another contemporary movement that seeks to extend "deodorization" even farther, namely, those who believe that what they call multiple chemical sensitivity (a term first proposed in 1950) is an actual disease that requires bans on the use of perfumes and other scented products in certain public places. Both the exaggerated claims made for aromatherapy and the often-unreasonable demands for fragrance bans made on behalf of MCS sufferers raise ethical issues we should address.

Aromatherapy versus Fragrance Bans

"Aromatherapy" today covers a confusing span of activities. About the only thing common to all of its forms is the use of natural essential oils derived from plant sources (in contrast to the synthetic substances also used in contemporary perfumery). Unfortunately, one widespread way of practicing aromatherapy makes it part of the kind of spa beauty and relaxation treatments that combine elements of cosmetics, massage, pop psychology, and wellness claims, all of which is touted in New Age jargon. Aromatherapy's image suffers even more from the marketing hype of companies that push "home cures" via essential oils. The company called "21 Drops™ Essential Oil Therapy," for example, sells a set of twenty-one individual oils with names like Detox, Invigorate, Headache, Pain Relief, Decongest, Sleep, each one containing three or four different oils. The vial of Calm, we are told, contains sweet orange to alleviate fear, jasmine to relieve depression, and vetiver to restore nerves. And the set as a whole is claimed to be not only "scientific" and "natural" but also to offer a "customized solution for everything from headache to heartache."[37] It was similar wild claims in the past that led the Warwick University psychologist J. B. King to dismiss the entire aromatherapy movement as "a shadowy world of romantic illusion, its magic easily dispelled by the harsh light of science."[38]

But before we dismiss everything going under the name "aromatherapy," we need to acknowledge the existence of "clinical aromatherapy," a serious enterprise that is recognized by some health professionals as one of several kinds of "complementary medicine." A few physicians are sympathetic to clinical aromatherapy, such as the surgeon Mehmet Oz, who has written an enthusiastic forward to Jane Buckle's 2015 nursing textbook, *Clinical Aromatherapy*. He asserts, for example, that "when nausea is relieved through inhalation of peppermint and insomnia is alleviated through the inhalation of lavender . . . we are witnessing clinical results, not just the 'feel good' factor."[39] Unfortunately, such specific effects of odors as Oz mentions have been disputed by other health professionals and psychologists because the mechanisms by which the effects are achieved are not fully understood, and because patient expectation or caregiver suggestion may be at work rather than any direct chemical action of the odors.[40] Of course, even if the use of aromatherapy to achieve specific therapeutic effects remains controversial, aromas might still be used in healthcare settings as one part of multisensory atmospheres for healing. A supportive general atmosphere is often a key component in the recovery process and is an essential part of hospice treatment. This is where aromatherapy might have a clinical use along with attention to other sensory components of general atmospheres, although such interventions would need to be tailored to people's individual sensitivities and preferences.

Something similar is true of the atmospheric use of scents in everyday contexts such as the home. These more down-to-earth uses of aromas focus on the way scents can be deployed to create a pleasant, stimulating, or even spiritual atmosphere, such as diffusing essential oils or burning incense. Such practices can offer the kind of sensory enrichment that Saito and other philosophers of everyday aesthetics have called to our attention. The moderate use of scents in the home is an excellent example of Böhme's idea of staging aesthetic atmospheres. Indeed, scenting a dwelling using flowers, herbs, or tree boughs is an age-old practice in many cultures. Such everyday aesthetic experiences are small pleasures, although not without their depth and dignity, as Spinoza observed, and surely can make a general contribution to the sense of physical and spiritual well-being for those who take the time to learn how to use them. Of course, here too there are ethical implications since atmospheric scenting using natural sources is one thing, whereas environmentalists would caution against the repeated use of commercial room sprays and other products containing synthetic scents.

Unfortunately, one person's everyday aesthetic delight in a fragrance could turn out to be another person's poison. For a few people even a slight whiff of a scent that is indiscernible and harmless to most people can trigger physical symptoms including any combination of headache, fatigue, confusion, dizziness, nausea, shortness of breath, pounding heart, memory problems, depression, or anxiety. The condition has long been called "multiple chemical sensitivity," although more recently some scientists have preferred the term "idiopathic environmental intolerance" (idiopathic signifying "of unknown cause"). Whichever term is used, MCS/IEI has yet to be recognized as a specific disease by the larger scientific and medical community and remains highly controversial.[41]

Lacking both a consistent symptom profile and a proven relationship to some specific physiological cause(s), MCS is often regarded as psychogenic.[42] Various psychological mechanisms have been explored, but no consensus reached.[43] Of course, even if the vast majority of cases of MCS have psychological rather than physiological causes, the subjective experience of sufferers is real enough. Thus, it is no surprise that MCS advocacy groups want to banish perfumes and other scented products from workplaces, schools, and eventually from all public spaces, as happened in Halifax, Nova Scotia.

But I believe most workplace fragrance bans, with the possible exception of portions of hospitals and clinics, are not justified. First of all, there is the problem of how many people actually suffer from MCS.[44] Moreover, the call for bans is often based on fallacious arguments, analogies, and misleading rhetoric, such as the catchphrase "Perfume is the new secondhand smoke."[45] This phrase overlooks the fact that the chemicals in secondhand smoke are a long proven and direct physical threat to almost everyone, whereas most perfumes do not elicit painful symptoms in the vast majority of people. Another fallacious argument sets up a rhetorical choice between "aesthetics" and "health" that transforms the issue into a simple choice between physical survival and trivial preferences, a version of the age-old argument over needs versus wants. But as Aristotle noted long ago, in addition to our basic needs without which it is not possible to live, such as "breathing and nourishment," there are those things necessary "for good to exist or come to be," that is, things that serve other legitimate ends.[46] As we saw in the previous chapter, for many people their use of scent is not a frivolous adornment, but may be tied to identity or to a desire to be sociable or may even have spiritual meanings.[47] Certainly, the concerns of MCS sufferers in the workplace should be taken seriously and their needs accommodated appropriately within reason, but this can be done without instituting impractical total bans.

Although a nationwide ban on wearing fragrances may seem a distant and unlikely possibility, the handful of legal victories in favor of scent-free workplaces, and the few bans on perfumes in all public places like that of Halifax, show that unless those who enjoy the scented life are not careful to find ways of accommodating the legitimate concerns of the MCS minority, the history of deodorization could have one more chapter.[48] In her book on smell in urban design, Victoria Henshaw asked if such contemporary hypersensitivities to odors are not "an inevitable response to the increasing olfactory sterility of modern cities and buildings."[49] This suggests one more reason to preserve and cultivate aesthetically rich smellscapes for our cities.

Notes

1. In addition to Tafalla, the philosophers included Emily Brady, Barry Smith, and Cain Todd; the other participants besides Henshaw were the olfactory art critic Jim Drobnick and the biologist Laura López-Mascaraque, who put us through a scent identification experiment.
2. At one point Henshaw stopped to suggest some rules for doing smellwalks on our own: (1) Don't go when you have a cold or hangover and keep well hydrated. (2) Seek as much variety as possible, including alleys and trash cans. (3) Use your other senses: vision to identify promising businesses or find sound of exhaust fans. (4) Minimize conversation to keep focused on smelling, and when strong odors that start to fade from adaptation, give your smell receptors a rest by sniffing the crook of your arm. (5) Don't be afraid to look silly or suspicious.
3. Böhme, *Aesthetics of Atmospheres*, 125.
4. Jim Drobnick, "The City, Distilled," in *Senses and the City: An Interdisciplinary Approach to Urban Sensescapes*, ed. Mădălina Diaconu et al. (Vienna: LIT Verlag, 2011), 272.
5. Quoted in Jordan Todorow, "This Artist Used Over 6500 Scents to Recreate the Smell of 35 World Cities," *Atlas Obscura*, November 18, 2006, https://www.atlasobscura.com.
6. Quoted in Victoria Henshaw, *Urban Smellscapes: Understanding and Designing City Smell Environments* (New York: Routledge, 2014), 222.
7. Kate McLean, "Communicating and Mediating Smellscapes: The Design and Exposition of Olfactory Mappings," in Henshaw, *Designing with Smell*, 69.
8. As might be expected, both Hershey's and Tacoma's aromas have modified with time and internet postings about Hershey in late 2017 confirmed the individuality of odor sensitivities, since some who have recently visited claim that Hershey longer smells of chocolate, but others claim its does.
9. Michael Meighan, *Glasgow Smells: A Nostalgic Tour of the City* (Stroud, UK: Tempus Publishers, 2008).
10. Quoted in Reinarz, *Past Scent*, 208.
11. Tania Leimbach, "Scent, Sensibility, and the Smell of a City," *The Conversation*, January 16, 2017, http://www.theconversation.com.
12. Jim Drobnick, "Eating Nothing: Cooking Aromas in Art and Culture," in Drobnick, *Smell Culture Reader*, 349–50.
13. Henshaw, *Urban Smellscapes*, 15.
14. Quoted in Henshaw, *Urban Smellscapes*, 100.
15. Henshaw, *Urban Smellscapes*, 174.
16. Virginia Postrel, *The Substance of Style: How the Rise of Aesthetic Value Is Remaking Commerce, Culture and Consciousness* (New York: Harper Collins, 2003), 123.
17. Juhani Pallasmaa, *The Eyes of the Skin: Architecture and the Sense* (West Sussex: John Wiley & Sons, 2005), 30.
18. Barbara Erwine, *Creating Sensory Spaces: The Architecture of the Invisible* (New York: Routledge, 2017), 13.
19. Anna Barbara and Anthony Perliss, *Invisible Architecture: Experiencing Places through the Sense of Smell* (Milan: SKIRA Press, 2006).

20. Steven Holl, Juhani Pallasmaa, and Alberto Pérez Gómez, *Questions of Perception: Phenomenology of Architecture* (San Francisco: William K. Stout Publishers, 2006).
21. Quoted in Christian Borch, "The Politics of Atmospheres: Architecture, Power and the Senses," in *Architectural Atmospheres: On the Experience and Politics of Architecture*, ed. Christian Borch (Basel: Birkhäuser, 2014), 82.
22. Diaconu, "Mapping Urban Smellscapes," in Diaconu, *Senses and the City*, 234.
23. Jenifer Robinson, "On Being Moved by Architecture," *Journal of Aesthetics and Art Criticism* 70, no. 4 (2012): 342.
24. Peter Zumthor, *Atmospheres: Architectural Environments, Surrounding Objects*, trans. Iain Galbraith (Basel: Birkhäusaer, 2006).
25. Mira Lochner, *Traditional Japanese Architecture: An Exploration of Elements and Forms* (Tokyo: Tuttle Publishing, 2010), 70–73.
26. A good example of an aromatic wood construction is Zumthor's Swiss Pavilion, made of unseasoned larch and pine for the Hanover Exposition of 2000.
27. A sign of "sick building" syndrome is that people's experiences of headache, nose or throat irritation, coughing, itching, etc. are related to the amount of time spent in the building and clear up when they leave.
28. The manifesto is available at the website of Laudamiel's company Dream Air, http://www.dreamair.mobi/christophe-laudamiel.
29. Böhme, *Aesthetics of Atmospheres*, 33, 76–86.
30. Samantha Warren and Kathleen Riach, "Olfactory Control, Aroma Power and Organizational Smellscapes," in Henshaw et al., *Designing with Smell*, 151.
31. Kevin Bradford and Debra Desrochers, "The Use of Scents to Influence Customers: The Sense of Using Scents to Make Cents," *Journal of Business Ethics* 90, no. 2 (2009): 141–53
32. Gilbert, *What the Nose*, 170–83.
33. Charles Spence et al. "Store Atmospherics: A Multisensory Perspective," *Psychology and Marketing* 31, no. 7 (2014): 472.
34. Bradford and Descroches, "Use of Scents to Influence," 146–47.
35. Bradford and Descroches, "Use of Scents to Influence," 148–49.
36. René-Maurice Gattefosse, *Gattefosse's Aromatherapy*, ed. Robert B. Tisserand (London: Random House, 1996).
37. The quotations are drawn from an insert that came with the purchase of the oils.
38. J. R. King, "The Scientific Status of Aromatherapy," *Perspectives in Biology and Medicine* 37, no. 3 (1994): 413.
39. In Jane Buckle, *Clinical Aromatherapy: Essential Oils for Heath Care* (St. Louis, MO: Elsevier, 2015), v–vi.
40. Herz, *Scent of Desire*, 92–9.
41. The term "idiopathic environmental intolerance" came into use after a meeting of the World Health Organization in 1996. Omer Van den Bergh et al., "Idiopathic Environmental Intolerance: A Comprehensive Model," *Clinical Psychological Science* 5, no. 3 (2017): 551, https://doi.10.1177/2167702617693327.

42. Christopher H. Hawkes and Richard L. Doty, *Smell and Taste Disorders* (Cambridge: Cambridge University Press, 2018), 220–21.
43. Herz, *Scent of Desire*, 104–14; Gilbert, *What the Nose Knows*, 112–13, 118–19.
44. A thorough literature search done by the Australian Department of Health and Ageing found that actual medical diagnoses of MCS range from 0.2% to 4% globally. See "A Scientific Review of Multiple Chemical Sensitivity: Identifying Key Research Needs," November 2010, 55, https://www.health.gov.au/.
45. Antifragrance activists have spread this catchphrase across the internet and often equate all fragrances and perfumes with toxic chemicals.
46. For a discussion of the needs vs. wants issue from a design perspective see Parsons, *Philosophy of Design*, 133–39.
47. The drive to impose perfume bans also overlooks the problems related to the perception of reasonableness and enforcement, e.g., who will police the workplace, what punishments would be appropriate, what should be done about clients and visitors?
48. A successful lawsuit in Detroit, for example, led to an award of $100,000 and a total fragrance ban in one Detroit city building. See *McBride vs. City of Detroit*, February. 12, 2010. Such bans usually include fragrances from essential oils.
49. Henshaw, *Urban Smellscapes*, 40.

15
Enhancing Flavors with Scents in Contemporary Cuisine

> What a top taster really needs to worry about is their nose.
> —Charles Spence

Perhaps nothing reveals the aesthetic importance of smell in our daily lives so much as its role in producing the experience of flavor. When the Oxford psychologist Charles Spence wrote that professional tasters need to worry more about their nose receptors than their taste buds, he was referring to the fact that the tasting expert Eleanor Freeman, of the British online health company Graze, had insured her taste buds for as much as 3 million pounds sterling.[1] Although losing the sensitivity of one's taste buds would throw off the perception of flavor, by themselves taste buds are of little use for either detecting or creating complex flavors. Today, scientists generally restrict the term "taste" to the five qualities that are registered by our tongues (sweet, sour, bitter, salty, umami), whereas "flavor" is used to describe what results from combining the information from the taste buds with the more extensive information coming from retronasal smell. You may recall from Chapter 2 that, although we initially enjoy the aroma of foods orthonasally via the molecules that reach our nasal receptors as we breathe in, once the food is in our mouth and we begin to chew and swallow, its scents also reach our nasal receptors retronasally as we breathe out, via the opening to the nose at the back of the mouth. As Spence puts it, "Taste constitutes only a very small part of our multisensory flavor experiences . . . all those fruity, floral, meaty, herbal notes that we enjoy while eating and drinking really come from the nose."[2]

The fact that our brains think flavors are coming from our mouth is called "oral referral," or "mouth-capture," a phenomenon analogous to the fact that we experience the sound of movie actors' voices as coming from their lips on the screen rather than from speakers strategically located elsewhere in the theater. Similarly, the illusion that flavor is in our mouths results from the rapidity with which our brains integrate the signals coming from the taste buds with the signals coming from our nasal receptors. Hence, Spence's suggestion that professional tasters should be more worried about the smell receptors in their nose than the taste buds on their tongue.

Art Scents. Larry Shiner, Oxford University Press (2020) © Oxford University Press.
DOI: 10.1093/oso/9780190089818.003.0031

People who have lost their sense of smell often lament that foods have lost their "taste." Due to oral referral, that is exactly how it feels. But in most cases their taste buds may be functioning perfectly well; what they have lost by losing their sense of smell is the ability to perceive and enjoy flavors. When wine connoisseurs counsel us to see, swirl, sniff, sip, and swallow, it is not only the sniff (the orthonasal experience) that informs us of the wine's structure, but also the breathing out that accompanies the swallowing (the retronasal experience). People who have experienced an onset of anosmia in midlife from an illness or accident will often lose their appetite and become depressed since food has lost its savor along with so much else, the smell of flowers, of their homes, of friends and loved ones. Bonnie Blodgett's memoir *Remembering Smell* offers some poignant descriptions of what it is actually like to lose one's sense of smell.[3]

Yet the phenomenon of "oral referral" doesn't mean that in everyday conversation we have to quit using the term "taste" for the flavors of food or drink. For one thing, "taste" has too long a history of meaning "flavor" and is so embedded in our everyday usages it would be hard to root out, and using it for flavor works well enough in most situations. Moreover, as Carolyn Korsmeyer points out, the use of "taste" as a metaphor for aesthetic preferences is also firmly rooted in both everyday and scholarly usages and offers rich opportunities for philosophical analysis. Even so, the philosopher Louise Richardson has argued against intruding scientific accounts of retronasal smell even into our philosophical reflections on perception, defending the commonsense view that treats smell as purely orthonasal.[4] But I think she goes too far since the role of retronasal smell in taste/flavor perception was actually noticed long before the scientific explorations in the 1980s began. The famous gastronomist Brillat-Savarin wrote in 1825: "I am . . . convinced that there is no full act of tasting without the participation of the sense of smell."[5]

Brillat-Savarin not only presciently observed the crucial role of what we now call "retronasal" smell, but also developed a sophisticated aesthetics of food that answered several of the traditional philosophical objections to treating taste and smell as the basis of serious aesthetic judgments. As we have seen, most of the Western philosophical tradition has treated taste and smell as affording only immediate hedonic reactions of liking or disliking and incapable of contributing to intellectual or imaginative insight. But, as Kevin Sweeney points out in *The Aesthetics of Food*, Brillat-Savarin argued that the appreciation of fine cuisine could be cognitively complex, temporally extended, and both reflective and imaginative.[6] According to Brillat-Savarin flavor perception has three stages, an initial one involving our first perception of aroma in the nostrils and taste on the tip of the tongue, a second stage when we chew and savor, and a final stage that includes swallowing and smelling that he called "reflective." Here is his illustration of these stages as he describes the eating of a peach:

> He who eats a peach . . . is first of all agreeably struck by the perfume which it exhales; he puts a piece of it into his mouth, and enjoys a sensation of tart freshness which invites him to continue; but it is not until the instant of swallowing, when the mouthful passes under his nasal channel, that the full aroma is revealed to him, and this completes the sensation which the peach can cause. Finally, it is not until it has been swallowed that the man, considering what he has just experienced, will say to himself, "Now there is something really delicious!"[7]

Clearly, Brillat-Savarin was acutely aware of the role of both orthonasal *and* retronasal smell in the flavor experience.

But equally important for developing an olfactory aesthetics is that Brillat-Savarin's account of the role of taste and smell in flavor perception combines the cognitive and imaginative dimension in a way not unlike Kant's idea of the metaphorical "taste of reflection." The kind of reflective experience Brillat-Savarin has in mind is similar to Kant's, Sweeney argues, because it describes cuisine as affording us a complex, developing encounter "worthy of imaginative involvement and reflective enjoyment."[8] That description of the dining experience recalls the similar description of the stages in the aesthetic appreciation of perfumes we encountered in Chapter 11. There, you will remember, I argued that the appreciative experience of the best perfumes is cognitively complex, temporally extended, and reflective. Thus both these arts that involve smell, perfume, which is primarily orthonasal, and cuisine, which combines orthonasal and retronasal smell with the sense of taste, are capable of eliciting the kind of intellectual, imaginative, and emotional responses that have traditionally been called aesthetic. Of course, the aesthetic appreciation of dining, as Charles Spence reminds us, involves not only taste and the two kinds of smell in constituting flavor, but the textures and sounds of our foods, as well as its visual appearance along with the setting or atmosphere of the place where we eat.[9]

As leading chefs have begun to exploit the multisensory aspects of the dining experience, smell has increasingly been acknowledged as playing an important role. This may begin with the use of ambient odors in the dining area, followed by the aroma of the beverage that precedes the meal. When the food arrives, diners inhale the aromas coming off the plate, and these scents, Spence remarks, "end up anchoring, and hence disproportionately influencing, the tasting experience that follows."[10] In fact, some experimental chefs add scents to food and drink at the moment they are served. The London chef Jozef Youssef, for example, enhanced the flavor of a lobster poached in butter sauce by spraying it with the scent of saffron.[11] Once diners begin to eat and drink, part of their aesthetic pleasure is in finding their flavor and texture expectations either confirmed or surprised, each outcome affording a differently slanted aesthetic experience, although surprise usually engages a greater level of interest in formal and expressive qualities. Thus, diners might be presented with something that looks and smells (orthonasally) like a strawberry, "until you swallowed a mouthful, whereupon it suddenly, magically transformed into the retronasal flavor of pineapple."[12] Of course, traditionalists may balk at such aromatic enhancements, insisting that the food or drink should "speak for itself" (remember Pliny's indignant exclamation at those Romans "who even add perfume to their drinks!"). Although this kind of experimental cuisine, with its enhanced aromas and flavors, surely gives an artistic flair to culinary creation, do such odor-enhanced meals qualify as works of (fine) art?

Avant-Garde Cuisine: Fine Art or Design Art?

The question of whether experimental cuisine is fine art might seem at first glance tangential to the aesthetics of smell. But, as we will see, a consideration of some philosophical attempts to answer the question of the art status of fine cuisine will allow us to test and extend our earlier answer to the question "Is perfume fine art?" as well as our proposal for a pluralistic concept of the arts as a nonhierarchical continuum.

Certainly, the most adventurous high-end restaurants and celebrity chefs today are treating the dining experience as a sort of participatory art experience, and the art world has sometimes endorsed the claim. In 2007, Farran Adrià, one of the pioneers of the new cuisine, was invited to participate in the famous cutting-edge art biennial Documenta. As Spence points out, many of these chefs are taking advantage of the new scientific knowledge of the multimodal nature of eating and drinking to offer up an aesthetic dining experience as a kind of *Gesamtkunstwerk*. Consider this account of an opening course at Paul Pairet's Ultraviolet restaurant in Shanghai: an apple wasabi sorbet was served in frozen slices as images of a Gothic abbey flashed on the walls, incense wafted over the table, and AC/DC's "Hell's Bells" rang in the diners' ears.[13] And this was only the first course!

Kevin Sweeney devotes two chapters of his recent book *The Aesthetics of Food* to the controversy over the art status of food and of wine, and in the course of his analysis also stresses the importance of both orthonasal and retronasal smell in the contemporary appreciation of cuisine. Sweeney views the experimental approaches of such chefs as Adrià, Grant Achaz, Jozef Youssef, or Hans

Blumenthal as "postmodernist" in their willingness to disguise, deconstruct, or transform ingredients in ways that challenge diners' expectations.[14] This desire to challenge expectations in high-end cuisine strikes me as similar to the aim of some niche perfumers for whom wearability is less important than creating aesthetic interest through unusual smell combinations.

But does the conceptual and "dematerializing" tendency of contemporary avant-garde cuisine actually turn the food it into fine art? As we saw in Chapters 11 and 12 on perfume, if one adopts an aesthetic definition of fine art or a combination aesthetic/institutional (disjunctive) definition *and* accepts the argument that taste and smell can afford cognitively informed and imaginative aesthetic experiences, there is no reason to exclude complex perfumes or, in this case, smell-enhanced haute cuisine, from the fine arts. Yet to the extent that in addition to all its pyrotechnics and surprises, experimental cuisine still tries to offer *nourishment* and takes place in restaurant settings rather than in art galleries or museums, one could, as in the perfume case, claim such cuisine for the category of the design or responsive arts rather than the fine or free arts. Yet, as in the perfume case, we might also imagine the emergence of an art cuisine along the lines of our imagined art perfume. Of course, just as there have been artists who create perfume-like works (Clara Ursitti's *Self-Portraits in Scent*) or commission perfumes as part of installations (Lisa Kirk's *Revolution Pipe Bomb*), so artists have not only made artworks out of food (Kara Walker's *Marvelous Sugar Baby* [2014]), but have cooked and served food in art museums (Rirkrit Tiravanija) or even opened actual restaurants that integrate art and food in some way (Carsten Höller's *Double Club*).[15] Whether or not something like an autonomous art cuisine would ever emerge and what its distinguishing characteristics might be is an open question, but what is clear is that in the realm of cuisine, as in the case of perfume creation, there are many types of cuisine and cuisine-like works that overlap with (fine) art practices so that what seems like a fixed boundary is a relatively open borderland. Sweeney himself thinks the crucial issue is not whether some of the most interesting contemporary experimental cuisine merits the appellation "fine art," but whether "the creations of talented chefs . . . can involve diners in having a rich aesthetic experience . . . that would have an imaginative and emotional component."[16]

In *Making Sense of Taste*, Carolyn Korsmeyer similarly rejects the traditional hierarchical concept of art and minimizes the importance of the fine art issue, although the focus of her concern is ordinary cuisine rather than avant-garde experiments. She argues that the actual aesthetic experience of food is far more important that the question of the fine art label because food and eating in general play a key symbolic role in nearly all cultures, not only in festivals or commemorations such as Diwali, Ramadan, Passover, or Christmas, but also in the meanings that permeate sharing an ordinary meal. Thus, a good part of the aesthetic experience of food and dining depends on its significance in various narrative contexts, whether personal, familial, or communal, and these narratives in turn modulate the purely sensory satisfactions of the food's culinary qualities. Hence, to separate a particular food or kind of cuisine from its larger social context and treat it as an artwork would, in Korsmeyer's view, actually "impoverish its aesthetic import," a point similar to the one I made about treating standard perfumes as fine art.[17] Moreover, she suggests that a preoccupation with categorical ranking often diverts us from seeking a deeper philosophical understanding of the complex roles that the arts of taste and smell in general actually do play. One function of reflecting on the meaning of everyday arts such as cooking is to "direct attention to the supposedly 'lower' aspects of being human—the fact that we are animal and mortal."[18] As we saw earlier, the association of smell with our animal and bodily nature in contrast to mind and spirit has been a leitmotif of the Western intellectual tradition, a motif attacked by Nietzsche in his encomium on the sense of smell with which I opened this book.

I agree with Korsmeyer and Sweeney that the question "Is fine cuisine fine art?" like the question "Is fine perfume fine art?" perpetuates a set of outmoded hierarchical polarities. That is why I have proposed instead that we adopt a pluralistic approach to the concepts of both art and aesthetic. The philosophers Raymond Boisvert and Lisa Heldke have taken a similar pluralistic approach to the question of whether fine cuisine should be considered fine art, and it applies to the question of the art status of perfume as well. They argue that traditional aesthetics' way of thinking about arts like cuisine was based on a negative polarity whose key concept "was the *not*":

> *not* in everyday experience, *not* related to the proximal senses, *not* useful, *not* ethical. Our alternative will emphasize *and*: delightful, *and* useful, *and* beautiful, *and* good. A well-written, well-performed play can pull this off. So can a well-designed building, a poem and a painting. So, too, can a good meal. Any creative endeavor might achieve these levels. Any of these creative forms can also *fail* to achieve the highest levels of excellence; no single kind of endeavor is guaranteed to achieve those levels. The rigid boundary separating art from not art has just been rendered a permeable membrane.[19]

The Everyday Aesthetics of Aromas

From the point of view of everyday aesthetics, one of the many virtues of Korsmeyer's book on gustatory taste is that it focuses on such everyday matters as sharing a family meal on a major holiday or being offered a bowl of chicken soup when we are sick (whose aroma is often as comforting as its flavor). One of the most rewarding everyday aesthetic experiences is to come home after a long day, open the door, and be greeted by the rich aromas of cooking. Even if both partners work and have little time for food preparation, and even if home cooking or baking takes place mostly on weekends or holidays, the aromas that fill an apartment or a house can be among the signal aesthetic experiences of everyday life. In temperate climates, food aromas can be especially comforting in the fall when the air is crisp and our noses are caressed by the warm, moist smells of a simmering sauce or roasting vegetables. In a conversation about favorite childhood memories of coming home to cooking aromas, a friend mentioned his grandmother's pies baking in the oven, but immediately added that the question also reminded him of the detested smell of liver and onions. Such aroma memories nicely illustrate both the commonalities and the individual differences in people's odor sensitivities and associations. Probably most of us appreciate the aromas of baking bread or pies, but when I was in middle school and came home from football practice, I especially loved those days when the smell of liver and onions greeted me as I opened the back door, just as my friend found a similar smell repulsive.

Of course, whether we respond with delighted anticipation or "Oh no, liver and onions again," the phenomenon of habituation means that the intensity of the aroma fades quickly once the door is shut, and after a while we may hardly notice it. Yet if we go back outside for a few minutes, we discover it anew when we return. Cooks will also have had the pleasure of using their sense of smell to check ingredients for freshness, enjoy the volatiles released during peeling and chopping (except maybe the onions), then monitoring the aromas released during mixing, blending, and cooking. One of the virtues of the "everyday aesthetics" movement has been to make us aware that we don't have to seek out works of "postmodernist" cuisine to have deeply satisfying aesthetic experiences with aromas and flavors.

The Ethics of Aromas and Flavors in Fast Food

For many years, the Cinnabun Company had an outlet at the juncture of three departure corridors at the St. Louis, Missouri, airport. The powerful odor of baking cinnamon rolls drifted down each corridor and called many of us with a force no amount of signage or catchy jingles could have done. Many food franchises in airports and malls use this kind of strategy, making sure that their exhaust fans are set at the lowest permissible power. My impression is that for the most part the purveyors of healthier foods haven't caught on to the importance of sensory marketing and tend to rely too much on verbal claims. Moreover, many people are suspicious of anything labeled "healthy" or "low calorie," assuming it will have a mediocre flavor and lack the tactile satisfactions of richer foods, so that foods heavy in calories, fat, and sugar or salt continue to win out.

To get an idea of how temptingly varied the sensory satisfactions of fast food can be, let's consider the multisensory aesthetic experience of eating some French fries with a hamburger and soda at a McDonald's.[20] As we enter and stand in line to order, our nostrils usually get a strong whiff of the somewhat meaty smell of the vegetable oil bubbling in the frying baskets that are typically located up front. The smell is "meaty" for a reason. Back in 1990, under public pressure to reduce the amount of cholesterol in its foods, McDonald's, which had been cooking its fries in an oil of 93% beef tallow, switched to a vegetable oil mix, but had a major flavor house develop an additive to give the vegetable oil a strong meat flavor.[21] I well remember my teenage age daughter coming home at night after her part-time job at McDonald's, reeking of French fry grease.

When our fries arrive on the tray with the hamburger and soda, their aroma rises to greet us and the trigeminal nerves in our nose and face feel the slight wave of heat as we anticipate a familiar and satisfying taste/flavor about to be enjoyed. Of course, our expectations are also stoked by the golden color that is a signal of fries cooked just right and by the tactile feeling as we pick up the first one, our fingers feeling the salt clinging to them and sensing already whether they are hot, light, and crisp enough. As we raise the first one toward our lips, their aroma gets stronger, and the moment the golden fry touches our tongue we are gratified with the confirmed pleasure of the salty taste. As we bite down, we experience the slight sweetness of the potato flavor from the soft center at the same time our tactile sensors register the firm edges. But the crucial pleasure comes as we chew and breathe out, when the salty/sweet messages coming from our tongues are seamlessly integrated with the retronasal potato and meaty vegetable oil smells reaching our nasal receptors. At the same time all this is happening, our tongues are busy continuing to test the texture of the fries by feeling for the firm edges and the appropriate softness of the center, and our hearing is registering the sound of these movements as we work the first mouthful of fries into a paste we can swallow. As we swallow, our breathing automatically pauses so the food will not go down our windpipe and we get one last flavor shot as our nose receptors catch the retronasal smell message from the thin layer of food still coating the back of our mouth and throat.

All these movements and perceptions happen so quickly and unconsciously that we are hardly aware of the enormous complexity of the mixture of sensory signals that are being integrated in our brains as we enjoy bite after bite of French fries. Of course, the above description has left out the way the aromas and flavors, and the tactile, motor, auditory, and visual experiences get even more complicated as we dip the fries in ketchup with its spicy/sweet smoothness, begin to bite into our hamburger (which affords the additional aromas, flavors, and textures of cheese, lettuce, onion, pickle, and tomato), and then, as we take sips of our soft drink, enjoying the cold/sweet contrast with the hot/salty fries. It is no wonder that the hamburger, French fries, and soda meal is so popular in the United States and has even been successfully exported to countries around the world. Many of us have been eating this meal with relish since we were kids; in fact a least three generations of Americans have been raised on hamburgers, fries, and sodas since the end of World War II. But if we add up the calories, the fats, and the sugars and salt we can see why this kind of food tradition could contribute to overeating and even to obesity, diabetes, and high blood pressure. Its not just that even a "medium"-sized burger, fries, and soda adds up to over 1,100 calories (almost half the recommended daily allotment), but that, as David Kessler of the Yale Medical School points out, the combination of salt, fat, and sugar makes it a meal hard to resist.[22]

But current food and nutrition science is showing ways not only to give "health" foods a more appealing look, texture, sound, and taste, but also to use aromas as an encouragement. Of course, many of the currently most familiar and potent odors are precisely those attached to foods heavy in fats and sugars, and our preference for them is learned early in life. Moreover, those preferences are reinforced by powerful corporations that bombard us daily in print, on television, and even on our phones, with "food porn," those alluringly colorful close-ups of foods with their seductive sound effects and music. As Spence points out, studies have shown that watching such ads increases hunger, correlates with an increase in body mass index, wastes mental resources, and too often promotes unhealthy foods.[23] Gordon Shepherd puts the blame partly on the manipulation of flavors and identifies "retronasal smell and its associated multisensory brain mechanism . . . as underappreciated factors."[24] Its not going to be easy to change deeply engrained cultural habits and—despite the importance of the smell factor—we certainly can't change habits by aromas and flavors alone.

Although public criticism and pressure have led some fast-food corporations to make caloric and fat content information available and to add more salads to their menus, more needs to be done to nudge companies and consumers in the right direction. Some readers may be familiar with "nudge theory" in political philosophy and the social sciences, which argues that rather than using some form of coercion, government and public service agencies should encourage setting up conditions that make healthier choices easier, and unhealthy choices more difficult. Some food pantries for the poor, for example, put fruits and vegetables up near the cash registers where they can be seen and smelled (unlike the typical grocery store that locates an array of candy at the checkout).

In 2016, a general review of social science studies on the efficacy of nudge strategies in the United States and several other developed countries concluded that nudges, in the studies reviewed, had resulted in an average 15.3% increase in healthier dietary choices.[25] Although some libertarians object that nudge strategies amount to a kind of manipulation, surely corporate food advertising, store display and shelf placement choices, as well as fast-food restaurant menu designs also attempt to "manipulate" us. Why should agencies charged with promoting public health remain inactive and allow the most powerful for-profit interests to dominate the field? No doubt there are further issues to be addressed in adjudicating the political and ethical concerns related to nudge strategies, but to do nothing in the face of what is widely agreed to be a serious health problem seems a worse choice.

Apart from the possibilities of using the knowledge we are gaining of the interactions among the senses, including both orthonasal and retronasal smell, to encourage healthier food choices, there are some lighthearted approaches to using aromas as a way of cutting down on calories and other unwonted aspects of certain foods and drinks. A Harvard professor of biomedical engineering, David Edwards, has helped create a device called Le Whiff that allows you to inhale the aromas of coffee, chocolate, and other flavors. The coffee inhaler gives you the sensation of a caffeine boost without the stomach acid, and the chocolate version avoids a large dose of sugar and fat. More recently Edwards has come up with LeWaff, a handsome glass carafe that when tilted on its side uses ultrasound to turn liquids into a cloud of tiny droplets that are then poured into a glass. The mist is meant to be consumed with a thick glass straw, thus providing the mouth with a full flavor experience through both taste and retronasal smell, with the caloric and/or alcohol intake minimized.[26]

As it turns out, the idea of not only enjoying the flavors of food without the heavy calories, but even temporarily living off aromas, goes all the way back to the fifth century B.C.E. According to legend, the philosopher Democritus survived the last four days of his life on the aroma of hot bread alone. Marsilio Ficino adds that by some accounts the bread had honey and wine poured on it to enrich the aroma and that, had Democritus wished, he could have lived even longer on the vapors. As we noted earlier, Ficino himself believed that scent and spirit have a commonality that allows odors to nourish our spirits, which in turn nourish the body. For that reason, he claims that "people who wish to lengthen their life in the body, should especially cultivate the spirit . . . foment it always with choice air; feed it daily with sweet odors; and delight it with sound and song."[27] Good advice for any age.

Notes

1. Charles Spence, *Gastrophysics: The New Science of Eating* (New York: Penguin Random House, 2017), 20. I am grateful to Carolyn Korsmeyer for drawing my attention to Spence's work.

2. Spence, *Gastrophysics*, 34.
3. Bonnie Blodgett, *Remembering Smell* (Boston: Houghton Mifflin, 2010).
4. Louise Richardson, "Flavour, Taste and Smell," *Mind and Language* 28, no. 3 (2013): 322–41.
5. Jean Antheleme Brillat-Savarin, *The Physiology of Taste, or Meditations on Transcendental Gastronomy*, trans. M. F. K. Fisher (Washington, DC: Counterpoint, 1998), 41.
6. Sweeney, *Aesthetics of Food*, 127–28.
7. Brillat-Savarin, *Physiology of Taste*, 43.
8. Sweeney, *Aesthetics of Food*, 132.
9. Spence's *Gastrophysics* devotes a chapter each to the role of smell, sight, sound, and touch in the perception of flavor and the other sensory aspects of food and drink.
10. Spence, *Gastrophysics* 29, 25.
11. Spence, *Gastrophysics*, 30.
12. Charles Spence and Betina Piqueras-Fiszman, *The Perfect Meal: The Multisensory Science of Food and Drink* (Hoboken, NJ: Wiley Blackwell, 2014), 231.
13. Spence, *Gastrophysics*, 207.
14. Sweeney, *Aesthetics of Food*, 159–63.
15. There have been many types of artist's restaurants over the decades, most short-lived experiments. For Höller's *Double Club* see Mark Clintberg, "Gut Feeling: Artists' Restaurants and Gustatory Aesthetics," *Senses and Society* 7, no. 2 (2012): 209–10.
16. Sweeney, *Aesthetics of Food*, 214.
17. Korsmeyer, *Making Sense of Taste*, 141.
18. Korsmeyer, *Making Sense of Taste*, 145.
19. Raymond D. Boisvert, and Lisa Heldke, *Philosophers at Table: On Food and Being Human* (London: Reaktion Books, 2016), 87.
20. My reflections here are inspired by the work of Gordon Shepherd and Charles Spence as well as many visits to McDonald's and other burger franchises over the years.
21. Shepherd, *Neurogastronomy*, 186.
22. Shepherd, *Neurogastronomy*, 184–91.
23. Spence, *Gastrophysics*, 58–60.
24. Shepherd, *Neurogastronomy*, 191.
25. Anneliese Arno and Steve Thomas, "The Efficacy of Nudge Theory Strategies in Influencing Adult Dietary Behavior: A Systematic Review and Meta-analysis," *BioMed Central Public Health* 16, no. 676 (2016), https://doi.a0.1186/s12889-016-3272-x.
26. Sarah Sweeney, "A 'Whiff' of a Breakthrough," *Harvard Gazette*, October 21, 2010, https://news.harvard.edu/gazette/story/2010/10/a-whif-of-a-breakthrough/; Sarah Wells, "Mad Science with a Mission," *Boston University News Service*, April 13, 2018, https://bunewsservice.com/mad-science-with-a-mission-how-david-edwards-is-changi. Charles Spence mentions another entertaining experiment with an orthonasal way to enjoy alcohol. Spence, *Gastrophysics*, 35.
27. Ficino, *Three Books on Life*, 225.

Postlude

Wilderness, Gardens, and Paradise

There are certain places on California's Central Coast where the scent from stands of eucalyptus can penetrate your car even with the windows closed, although the smell is so inviting you are tempted to open them a bit.[1] You can have equally interesting scent experiences driving east through the California and Nevada deserts after a rain when you can inhale the pungent smell of sage and creosote bush. Or consider the fact that sometimes you can smell rain before it comes, first from the ozone in the air produced by electrical discharges, and then, especially if you are in arid regions, from the smell of geosmin released from the earth. As Cynthia Barnett points out, you can inhale an especially intense version of earth odors in some rural areas of India, West Africa, or Australia that experience the climatic extremes of months of no rain followed by stretches of monsoon. Back in 1964 two Australian scientists discovered that a major source of this odor were geosmin, a soil-dwelling bacteria, and terpenes secreted by plants. These kinds of molecules are absorbed by rock and clay during hot dry periods, building up great quantities that are then released by the sudden rise in humidity. The scientists nicknamed the smell "petrichor," from *petra* the Greek for rock and *ichor*, the blood of the gods. In India where perfumery goes back to the period of the Vedas, adepts in the ancient city of Kannauj have even discovered how to capture these natural smells by distilling them from chunks of earth and turning them into a perfume, called *mitti attar*, or "earth scent."[2]

Nature and Environmental Aesthetics

This book has focused on smell in the human arts, yet in this final interlude I want to call attention to the place of smell in the aesthetics of nature. In some contemporary writing on the aesthetics of the environment, despite the frequent criticisms of "scenic" attitudes, there has been a default vision-centric perspective in which sounds, textures, and smells play only a minor role. Even so, one cannot criticize the pioneers of the environmental aesthetics movement for not giving a more prominent place to proximal senses such as smell, since even the visual aspects of nature were almost completely ignored in favor of the fine arts by philosophical aesthetics until late in the twentieth century. Yet as the biologist E. O. Wilson remarks, when we pay close attention to the natural world, we find that "it is held together by odors."[3] There are some physiological as well as historical reasons for the neglect of smells in nature such as the phenomenon of habituation and Westerners' default state of being "unconscious" of odors much of the time. In the woods or forests as in the city or our own houses, we are likely not to notice the ambient odor mix around us as we focus on other activities, unless something happens to get our attention by standing out from the background. As Wilson points out, "Even a trained naturalist walking through forest . . . has no idea of the thunderous round-the-clock chorus of olfactory signals upon whose perception the forest dwellers' lives depend."[4]

Theoretical approaches to the aesthetics of nature are often seen as falling roughly into two groups: on one side, those that emphasize the cognitive role of scientific knowledge in aesthetic response, pioneered by Allen Carlson, and on the other side a variety of multifaceted or "integrative" views like those of Arnold Berleant or Emily Brady that emphasize some combination of general knowledge, sensory perception, emotion, and imagination. Despite these general theoretical differences, both Carlson and Brady take a multisensory approach to the aesthetic appreciations of nature that includes smell. Carlson describes the way that "common-sense/scientific knowledge" leads us to use all our senses to draw different aspects of the natural background into focus, depending on the setting. In the case of a prairie, for example, we not only engage in visual surveying, but in "feeling the wind blowing across the open space, and smelling the mix of prairie grasses and flowers," whereas in a dense forest, we not only scrutinize more closely, but engage in "listening carefully for the sounds of birds, and smelling carefully for the scent of spruce and pine."[5] Emily Brady evocatively writes that when "standing on a rock on the edge of a turbulent sea," we will not only have the spectacle and sound of crashing waves, but will feel "the mist thrown up by the waves, the wet, fresh smell of the sea, . . . the taste of salt."[6] Like E. O. Wilson, she also points out that smells, like sounds in nature, tend to "constitute a sensuous backdrop" and are often not explicitly noticed. Thus we may become habituated to the fragrance of a pine forest against which more specific scents like "the peaty, moist smell of a rotting log, or the putrid odor of an animal carcass" will stand out.[7]

Wilderness

A forest or isolated beach, of course, may be more or less "wild." What we normally consider "wilderness," that is, relatively unaltered natural environments, has sometimes been thought of as a kind of secular equivalent to the original

Paradise. Yet in the great classical religions of the West, Paradise has been likened not to what is wild and untamed, but to a fragrant garden (the Persian root of the term "paradise" meant a garden or park). Our modern idealized wilderness is partly a child of the Romantic imagination, whose American versions owe much to Transcendentalists like Thoreau. But no matter how wilderness is defined, the drive to preserve it has most often imagined wilderness in visual terms, with an occasional nod to sounds and textures, but has seldom discussed the importance of natural features experienced through smell. Perhaps it is time that our laudable efforts to preserve physical and species diversity in nature should also take into account places, flora, and fauna worthy of preservation for their smell. The Japanese Ministry of Environment has actually created a list of "One Hundred Sites of Good Fragrance," which includes both natural and cultural sources of scents that should be cherished and protected, such as the dogtooth violets of Mount Kenashigasen.

Of course, we should not overlook ethical issues related to preservation whether we are preserving mountain vistas, mountain lions, or mountain smells. First, although not directly related to smell, there is the knotty problem that in less developed countries, such preservation often comes at the expense of hunter-gatherers and poor subsistence farmers. Then there is the more general issue of whether and to what extent aesthetic appreciation should lead to a moral obligation to preserve nature.[8] A more specific aspect of the relation of aesthetics to the ethics of preservation that does relate to smell is the fact we noted earlier: people's aesthetic preferences often lead them to disregard the preservation of animals, plants, and landscapes that are deemed either "ugly" or "smelly."

Emily Brady, Isis Brooks, and Jonathan Prior have recently argued that given environmental aesthetics' traditional focus on wild nature, aestheticians need to develop sharper tools for theorizing the aesthetics of "modified" natural environments. They remind us that most of our experiences of nature are not of wilderness, but of nature altered by human presence, whether the cultivated countryside, preserves, parks, or gardens.[9] And they urge that these human modifications not be treated as "poor cousins of wild places but aesthetic places that inspire a love of nature in their own right."[10] Certainly, smell historically played an important role in the human experience of cultivated nature, although the role of scent in gardens has diminished since the eighteenth century, as we will see in the next section.

Gardens

Gardens are an ancient part of civilization, ranging from forest clearings used to grow food to the many types of pleasure gardens in highly urban cultures, some functioning as symbols of status and power, others reserved as places of socializing and play or for retreat and reflection. The Greek philosopher Epicurus conducted his philosophical school in a garden, a place emblematic of his ideal of inner harmony and equanimity and also conducive to its achievement. Down into the eighteenth century at least, the inclusion of fragrant plantings was an important part of the atmosphere of most gardens, with a few exceptions such as Zen rock gardens. In ancient Egypt and Persia most gardens were walled, a fact that kept the scents of plants from dispersing. In the West during the Roman imperial era, the wealthy created gardens within their villas in the form of peristyle courtyards surrounded with murals, the central area open to the sky and filed with statuary, fountains, and fragrant flowers. With the triumph of Islam in the eastern Mediterranean and North Africa, palaces and mosques began to include walled courtyard gardens, but unlike the Roman gardens, these "eschewed vibrant images and statues and instead relied upon trellises, perfumes, fountains and textiles for aesthetic grandeur."[11] In medieval Europe, fragrant walled gardens were often a part of cloisters, houses, or palaces. Albertus Magnus wrote of "pleasure gardens . . . mainly designed for the delight of two senses, sight and smell."[12] Throughout the ancient, medieval, and Renaissance periods, then, gardens were not only something to be looked at and walked through or reposed in, but enjoyed for their smells.

Things began to change in the seventeenth and eighteenth centuries as more open gardens became prevalent, whether the formal Italian and French styles or the more informal English landscape type. Some philosophers such as Kant included gardens on their lists of the fine arts, although hardly for their fragrance. In her book *The Meaning of Gardens*, the philosopher Stephanie Ross takes as her starting point Horace Walpole's declaration, "Poetry, Painting, and Gardening . . . will forever by men of Taste be deemed Three Sisters."[13] Ross's book, like Mara Miller's study *The Garden as Art*, develops a case for understanding how the more complex gardens deserved their place among the fine arts. What is interesting about the case for the garden as fine art given our concern with olfaction is that from the mid-eighteenth century on, smell was increasingly ignored in both garden design and in aesthetic theories related to gardens.[14] There seem to be two reasons smell began to play a diminished role by the eighteenth century. First, the older walled gardens concentrated scents and kept them from dissipating, whereas the newer, more open and expansive gardens of the nobility on the continent and the landscape gardens of English aristocracy, despite the occasional inclusion of a few "fragrant islands," allowed most scents to be carried away by air currents.[15] A second and more important reason for the greater importance of smell in gardening up to the eighteenth century was that the smell of flowers and plants was valued equally with their visual beauty. As we saw earlier in the case of the rose, which was prized as much for its odor as for its appearance in Shakespeare's day, by the nineteenth century, gardeners cultivated and organized new varieties of roses and other flowers primarily for their colors, shapes, and sizes, increasingly ignoring smell. Moreover, from the late nineteenth century on, the arrival of inexpensive photographic reproduction intensified the tendency to think of gardens largely in visual terms. As Miller emphasizes, a two-dimensional image, whether an engraving, painting, or a photograph, only tells us what a garden *looks* like and leaves out the sounds and fragrances as well as the tactile and temporal aspects of experiencing a garden by walking through it.[16] Today, most of the gardens that are designed with a special consideration for the smell of flowers, bushes, and trees seem to be gardens intended for the sight impaired.[17]

Although David Cooper's book *A Philosophy of Gardens* is even less concerned than Ross and Miller with fragrance, and Cooper even rejects their focus on aesthetics in favor of the garden's role in the classic project of the "good life," his reflections intersect with our next topic, the place of smell in the religious imagination of the garden. The true appreciation of gardens, Cooper writes, is less as "an aesthetic spectacle" than in the way they exemplify the dependence of human creative activity on the natural world and at the same time the way they reveal nature's dependence on human creativity to manifest its deepest meaning.[18] Gardens, Cooper believes, ground what he calls "epiphanies of co-dependence" and ultimately "of man's relationship to mystery."[19] Cooper's emphasis on the spiritual aspect of the garden experience makes a good introduction to considering the place of the fragrant garden in the religious imagination.

Paradise

In three of the major ancient Mediterranean religions—Judaism, Christianity, and Islam—the Paradise from which humans came and, in the case of Christianity and Islam, the Paradise to which believers yearn to return, was often conceived of as a fragrant garden. In an earlier chapter we showed that Christianity by the fourth century had developed a rich olfactory piety whose sensibility was displayed, as Susan Harvey says, "in the extensive ritual uses of incense and holy oils . . . in the sensed fragrance of divine presence, in the 'fragrance' of virtue, the 'odor of sanctity,' the sweet scent of relics . . . the perfumed delights of heaven."[20] There is no finer expression of the latter than Ephraim Syrus's set of *Hymns on*

Paradise, which refer to the ultimate destination of the saved as "that treasure of perfumes / the storehouse of scents," a place where

> A vast censer
> Exhaling fragrance
> Impregnates the air
> With its odiferous smoke,
> Imparting to all who are near it
> A whiff from which to benefit

But as Harvey points out, the scents Ephraim describes are not merely for the pleasure of smelling: they are the very stuff of eternal life, since, as Ephraim puts it: "Instead of bread, it is the very fragrance of Paradise that gives nourishment. Instead of liquid, this life-giving breeze does service." Moreover, what Ephraim calls the "Fragrance of Life" was believed by many to be the same fragrance that nourished Adam and Eve before the Fall.[21] Christian theology, of course, developed a notion of the resurrected body as a "spiritual" body, a point of view from which such references to fragrance and nourishing smells would be largely metaphorical. Nevertheless, what I want to underline here is that the bodily metaphors and images Ephraim chooses for Paradise are not visual or auditory, but olfactory.

When we turn to Islam, the images of the bodily pleasures of Paradise are more literal. The Koran describes the Garden of Paradise as containing rivers of milk and honey where the righteous may eat and drink their fill. Unfortunately, non-Muslims too often tend to overemphasize these material descriptions along with the mention of sexual rewards that will be offered male martyrs. But for our purposes what is most interesting about traditional descriptions of the Islamic Garden of Paradise is that its air is filled with sweet scents, including camphor, musk, ginger, and saffron. And as in the Christian tradition, Muslim writers claim that Adam originally enjoyed these fragrances before the Fall; in fact one of the *Tales of the Prophets* says that among the things Adam most regretted leaving behind when he and Eve were expelled from Paradise was the sweet "smell of the Garden and its perfume." Indeed, the *Tales* credit Adam with bringing perfume to earth. In one version of these stories perfume (musk) springs from the tears he wept for the loss of Paradise.[22]

As Mary Thurlkill argues, despite the differences between the Christian and Islamic images of the garden of Paradise,

> Both traditions rely upon scent to depict paradisiacal favors reserved for pious men and women. Paradise's landscape offers abundant sweet smells ultimately

associating this archetypal Garden with its earthly echoes. . . . Fragrance, in its unique way, thus extends beyond paradisiacal boundaries into earthly time and space, reminding humanity not only of perfection lost but also perfection promised.

Whether one thinks of nature in terms of religious imagery or in terms of the garden as an epiphany of our relation to nature, the richest and most complete experience of both wild and cultivated nature will always include the dimension of scent.

Notes

1. One can have an experience of fragrant stands of eucalyptus along US Highway 1 south of Gilroy or the rows of eucalyptus along either side of a stretch of State Route 291 between Santa Barbara and Carpinteria.
2. Cynthia Barnett, *Rain: A Natural and Cultural History* (New York: Crown Publishers, 2015), 212–15.
3. Edward O. Wilson, *The Origins of Creativity* (New York: Norton, 2017), 62. For a philosophical attempt to redress the slighting of smell in environmental aesthetics see Daniel Press and Steven C. Minta, "The Smell of Nature: Olfaction, Knowledge and the Environment," *Philosophy and Geography* 3 no. 2 (2002): 173–186, https://doi.org/10.1080/13668790008573711.
4. Wilson, *Origins of Creativity*, 62.
5. Allen Carlson, *Aesthetic and the Environment: The Appreciation of Nature, Art and Architecture* (London: Routledge, 2000), 51–52.
6. Emily Brady, *Aesthetics of the Natural Environment* (Tuscaloosa: University of Alabama Press, 2003), 65.
7. Brady, *Aesthetics of the Natural Environment*, 126.
8. Allen Carlson surveys this issue in "Environmental Aesthetics, Ethics, and Ecoaesthetics," *Journal of Aesthetics and Art Criticism* 76, no. 4 (2018): 399–410.
9. Emily Brady, Isis Brook, and Jonathan Prior, *Between Nature and Culture: The Aesthetics of Modified Environments* (London: Rowman and Littlefield, 2018). In their endeavor to develop an account of the positive aesthetic aspects of appreciating nature in its many "modified" forms, the authors revisit the much-maligned eighteenth-century theory of the "picturesque" and attempt to rehabilitate it.
10. Brady et al., *Between Nature and Culture*, 27.
11. Thurlkill, *Sacred Scents*, 138.
12. Mara Miller, *The Garden as Art* (Albany: State University of New York Press, 1993), 32.
13. Stephanie Ross, *What Gardens Mean* (Chicago: University of Chicago Press, 1998), xiii. Of course, Walpole is playing off the more familiar list of poetry, painting, and music as the three sister arts.

14. Both Miller and Ross acknowledge that the experience of gardens is always multisensory, but they also accept the fact that most of us today treat them as primarily visual experiences. Ross, *What Gardens Mean*, 157; Miller, *Garden as Art*, 128.
15. Barbara Lange, "A Shift of Senses: A Brief Foray into the History of Scents and Odors in Art," in Wetzel et al., *Belle Haleine*, 97–104.
16. Miller, *The Garden as Art*, 47–50.
17. Of course, for anosmics and the smell impaired, the garden that focuses solely on visual aspects is just fine, although a thoughtful anosmic like Marta Tafalla understands the importance of smell in the garden and what the anosmic misses. See her "Smell and Anosmia in the Aesthetic Appreciation of Gardens," *Contemporary Aesthetics* 12 (2014), https://www.contempaesthetics.org/.
18. Cooper argues that gardens are best understood neither as art nor as nature, but in terms of a "total perception" similar to Merleau-Ponty's version of "atmosphere." David Cooper, *A Philosophy of Gardens* (Oxford: Oxford University Press, 2006), 47–51.
19. Cooper, *Philosophy of Gardens*, 48, 158.
20. Harvey, *Scenting Salvation*, 223.
21. Harvey, *Scenting Salvation*, 235–36.
22. Thurlkill, *Sacred Scents*, 138.

16
An Invitation to Discovery

> Epigraphs in an undecipherable language . . . this is what you will be,
> *O perfumeries*, for the noseless man of the future.
> —Italo Calvino, "The Name, the Nose"

We began this book with examples of the remarkable variety of contemporary olfactory arts that call for the attention of aesthetic theorists. But we immediately had to confront intellectual traditions going back to Plato that said only the objects of sight and hearing can be beautiful, a view reinforced in modern aesthetics from Kant to the present by the belief that smell lacks the cognitive capacity to be a vehicle for serious art making or reflective aesthetic experience. We also noted that this philosophical tradition has been supported by a wider intellectual disparagement of smell, exemplified by Darwin's belief, shared by Freud and by others closer to our time, that smell is an evolutionary vestige of little use. Italo Calvino's speculation about the "noseless man of the future" who will no longer know even the most basic terms for expressing smell seems to reflect a similar pessimism about smell, but it was penned in 1975, a decade before the sensory turn in the sciences, social sciences, and humanities rediscovered the importance of smell. Drawing on evidence from a variety of contemporary biocultural sources ranging from evolutionary theory, neuroscience, and psychology to history, anthropology, linguistics, and literature, the remainder of Parts I and II countered these negative claims to show that the sense of smell, despite its limitations, does have sufficient cognitive capacity to support an olfactory aesthetics.

With this more accurate understanding of smell's strengths and limitations established, Part III began to sketch the outline of an olfactory aesthetics. Given the immense variety of olfactory arts, I characterized the category "olfactory arts," for purposes of this book, as embracing any artwork that uses *actual* odors *intentionally* in a *distinctive-making* manner. I noted that most olfactory arts under this description, with the exception of perfume and incense, are hybrids, such as the use of odors with theater, film, or music, or the use of odors with various visual art media such as sculpture or installations, or the use of odors to enrich architecture, urban environments, and cuisine. After considering the representation of smell in literature and painting, including some rare hybrids of odors with poems or paintings, I assessed the potential of odors to enhance theater (generally positive) and film (generally negative), leaving open the possibility of successful experimental adventures. With respect to music, I gave particular attention to *Green Aria: A Scent Opera*, perhaps the most important and promising experimental work of olfactory art so far in this century. I also considered the ingenious scent organ *Smeller 2.0*, before turning to what I called "scent art" or "olfactory art" in the singular, works typically created by professional artists that are hybrids of odors with various visual art forms such as sculpture, installation, performance, or participatory art, and intended for art galleries and museums. Despite the hybrid nature of such works, I argued that scent or olfactory art merited being considered an independent category analogous to sound art.

Part III then turned to question of the art status of incense and perfume, both of which have an ancient social pedigree. We began with kodo, the Japanese art of incense that is now undergoing a revival and which some aesthetic theorists believe should be considered a unique fine art form, although accepting kodo as an art form would require a radical revision of most Western concepts of art. The question of whether perfume should be considered one of the fine arts led us into current philosophical debates on defining art, and I showed that although aesthetic definitions of art can lead to a positive answer, contextual or historical definitions lead to the conclusion that perfume is a design art, not a fine art. I resolved the impasse between the results of the aesthetic and contextual/historical approaches by developing arguments analogous to those once used for admitting photography and quilts into the fine art club. Yet this solution had the effect of dividing perfume into a handful of possible "art perfumes" intended for art world contexts and the bulk of standard or "design perfumes" intended for wear. I accepted this solution but denied that placing perfumes in the category of design gave them less aesthetic value than they would have if labeled "fine art," and I proposed an "art concept pluralism" to replace traditional invidious polarities such as "major" versus "minor" that set a small group of fine art forms above all the applied, decorative, and design arts.

Part IV, "The Aesthetics and Ethics of Scenting," began by arguing for an "aesthetic concept pluralism" that would treat aesthetic experience and judgment in a nonhierarchical way similar to what I had proposed for art concept pluralism. I illustrated the idea of aesthetic concept pluralism by drawing on theoretical approaches appropriate respectively to the design arts ("functional beauty"), to ordinary objects and activities ("everyday aesthetics"), and to atmospheres ("aesthetics of atmospheres"). We then considered the aesthetic and ethical aspects of perfume wearing, of ambient scenting in architecture and urban design, and of both orthonasal and retronasal smell in the aesthetics and ethics of haute cuisine, everyday cooking, and fast food. In a brief interlude, we considered some aspects of the role of the sense of smell in the appreciation of nature, focusing on the olfactory aspect of gardens as an art form and their counterpart in the religious imagination of Paradise

Although this book has attempted to survey most of the major olfactory arts and the aesthetic issues they raise (with the exception of digital and virtual reality experiments), there is an important philosophical question that remains.[1] When I presented a sketch of some of the ideas developed here at a conference a couple of years ago, I was asked, "But do you think olfactory art can be profound?" I was not quick witted enough to ask back, "What do mean by "profound?" But the question has stayed with me. I believe Sibley was getting at something like the "profundity" issue when he spoke of the arts of taste and smell as belonging to the realm of aesthetics, but at the lower end of the aesthetic spectrum and "trivial" compared to arts based on vision and hearing. If "profound" means something like able to deal in an illuminating way with the most serious human questions, as Sibley suggested—truth, justice, love, death, hope—it would be ridiculous to ask if a single perfume or even a single olfactory art installation could illuminate these in a way comparable to the great novels, operas, or historical paintings that have stood the test of time. But I think a hybrid work like *Green Aria* suggests a first step in a direction that might lead to an experience both moving and thought provoking. Yet for any such work to achieve something approaching "profundity" would depend not only on an artist or artists of genius, but also on an audience able to follow it and critics able to interpret it. My aim in this book has not been to propose that the scent arts or perfumery as they currently exist are on the brink of creating works that will equal Shakespeare's *Lear*, Mozart's *Don Giovanni*, Rembrandt's *Night Watch*, or Welles's *Citizen Kane*. It may be that given such factors as odors' volatility and the sense of smell's tendency to rapid habituation, we will have to be satisfied in the olfactory arts with more modest works. Even so, I believe today's olfactory arts have already far exceeded anything Beardsley, Sibley, Scruton, or Dutton thought possible. In any case, I find the game of ranking whole art forms or art practices largely uninformative.[2] Moreover, we should not forget that there is a depth to be found in rightly experiencing ordinary things, including their smell. This is something that writers on everyday aesthetics have reminded us, Zen masters have taught us, mystics have described, and Proust discovered when he found that a smell or a taste, fragile and impalpable as it might be, could at certain moments liberate the essence of things.

Such smell experiences will be genuinely rewarding aesthetic experiences when their emotional, imaginative, intellectual, and ethical dimensions are intertwined in ways that can foster reflection and conversation. And reflection and conversation about our sensory engagement with the world is a large part of what aesthetics is about and what this book has attempted to describe and interrogate with respect to smell and the olfactory arts. As part of that inquiry I have tried to overcome some of the intellectual barriers to our thinking it is worth the effort to develop knowledge of olfaction and the olfactory arts and to cultivate our sense of smell. At the end of Part II, I mentioned the retired civil servant, Barney Shaw, who sought to learn about smell and cultivate his sense of smell with interesting results. Another such account from a different perspective can be found in Alexandra Horowitz's *Being a Dog: Following the Dog into a World of Smell*.

Horowitz interweaves a fascinating story of how she began cultivating her own sense of smell at the same time she was investigating dogs' phenomenal smelling abilities. Among other things, she went on a "smellwalk" through Brooklyn with the redoubtable artist/designer Kate Mclean, volunteered as a subject in a months-long research project on human olfactory ability at Rockefeller University, and worked her way through a set of fifty-four wine notes intended to sharpen one's appreciation of wine bouquet. But most of all, she simply began to seek out odors and keep a record of her smell experiences in order to develop a vocabulary and imagery for expressing and remembering different scents. "Ultimately what I have learned to do is simply to bother to attend to smells . . . by making associations—with words and with images—to fix my mind on a smell and then curl it into a slip of memory."[3] Horowitz's adventures, like those of Barney Shaw, jibe with my own experience. Since the time I began this book project several years ago, I have gone from largely ignoring the smells around me to gradually learning to pay attention to odors and even to seek out opportunities for smelling. I now sometimes draw close to things or draw them close to me and actively sniff. Yet I have no illusion that I have any particular talent for smelling or that I could identify many smells accurately in an experiment, but my olfactory world has enormously enlarged and offered new sources of intellectual interest and physical pleasure. This book has been meant, then, not only as a theoretical exploration of the aesthetics of smell and the olfactory arts, but also as an invitation to discovery.

Notes

1. The field of olfactory technology is so vast and rapidly changing that I decided not to include it in this book. Debora Parr has an excellent essay on technologies for creating and controlling scent, in which she briefly discusses various digital devices for smell communication. Debora Riley Parr, "Indeterminate Ecologies of Scent," in Henshaw, et al., *Designing with Smell*, 259–269.
2. Dominic Lopes makes a similar point about ideal critics' ordinal ranking across practices, calling it "the view from aesthetic nowhere." *Being for Beauty*, 203–5.
3. Horowitz, *Being a Dog*, 272.

Bibliography

Aftel, Mandy. "Perfumed Obsession." In Drobnick, *Smell Culture Reader*, 212–15.
Ahlers, Lisa Anette and Annja Müller-Alsbach, eds. *Belle Haleine: The Scent of Art, Interdisciplinary Symposium*. Basel: Museum Tinguely, 2015.
Allégret, Marina Jung. *Aux Sources du Parfum: Un Essai pour tout comprendre sur cet art merveilleux*. Paris: Vérone editions, 2016.
Antunes, Luis Rocha. *The Multisensory Film Experience: A Cognitive Model of Experiential Film Aesthetics*. Bristol, UK: Intellect, 2016.
Aristophanes. *Lysistrata*. Translated by Douglass Parker. Ann Arbor: University of Michigan Press, 1969.
Aristotle. *The Basic Works of Aristotle*. Edited by Richard McKeon. New York: Random House, 1968.
Aristotle. *Nichomachean Ethics*. Translated by David Ross, Revised by J. L Ackrill and J. O. Urmson. Oxford: Oxford University Press, 1998.
Aristotle. *Parva Naturalia*. Commentary by David Ross. Oxford: Clarendon Press, 1955.
Arno, Anneliese, and Steve Thomas. "The Efficacy of Nudge Theory Strategies in Influencing Adult Dietary Behavior: A Systematic Review and Meta-analysis." *BioMed Central Public Health* 16, no. 676 (2016). https://doi.a0.1186/s12889-016-3272-x.
Arshamian, Artin, and Maria Larsson. "Same Same but Different: The Case of Olfactory Imagery." *Frontiers in Psychology* 5, no. 34 (2014): 1–8. https://doi.10.3389fpsyg.2014.00034.
Aubaile-Sallenave, Françoise. "Bodies, Odors, and Perfumes in Arab-Muslim Societies." In Drobnick, *Smell Culture Reader*, 391–99.
Baars, Bernard J. "Multiple Sources of Conscious Odor Integration and Propagation in Olfactory Cortex." *Frontiers in Psychology* 4, no. 930 (2013): 1–4. https://doi. 103389/fpsyg.2013.00930.
Bacci, Francesca. "Scent-ific Art in Context: Developing a Methodology for a Multisensory Museum." In Ahlers et al., *Belle Haleine*, 126–36.
Bacci, Francesca, and David Melcher, eds. *Art and the Senses*. Oxford: Oxford University Press, 2011.
Banes, Sally, and André Lepecki, eds. *The Senses in Performance*. New York: Routledge. 2007.
Barbara, Anna, and Anthony Perliss. *Invisible Architecture: Experiencing Places through the Sense of Smell*. Milan: SKIRA Press, 2006.
Barkat-Defrades, Melissa and Elisabeth Motte-Florac. *Words for Odours: Language Skills and Cultural Insights*. Newcastle upon Tyne, UK: Cambridge Scholars Publishers, 2016.
Barnes, David S. "Confronting Sensory Crisis in the Great Stinks of London and Paris." In *Filth: Dirt, Disgust, and Modern Life*, edited by William A. Cohen and Ryan Johnson, 103–29. Minneapolis: University of Minnesota Press, 2005.
Barnett, Cynthia. *Rain: A Natural and Cultural History*. New York: Crown Publishers, 2015.
Barthes, Roland. *Sade, Fourier, Loyola*. Translated by Richard Miller. Paris: Éditions du Seuil, 1971.

Barwich, Ann-Sophie. "Up the Nose of the Beholder? Aesthetic Perception in Olfaction as a Decision-Making Process." *New Ideas in Psychology* 47 (2017): 157–65. https://philpapers.org/rec/BARUTN-2.
Battles, Ford Lewis. *Interpreting John Calvin*. Edited by Robert Benedetto. Grand Rapids, MI: Baker Books, 1996.
Batty, Clare. "A Representational Account of Olfactory Experience." *Canadian Journal of Philosophy* 40, no. 4 (2010): 511–38.
Batty, Clare. "Smelling Lessons." *Philosophical Studies* 153 (2011): 161–74.
Baudelaire, Charles. *Les Fleurs du Mal*. Paris: Poulet-Malassis et de Broise, 1857.
Baudelaire, Charles. *Les Fleurs du Mal*. Paris: Calmann-Lévy, 1866.
Bayne, Tim, and Charles Spence. "Multisensory Perception." In Matten, *Oxford Handbook of Philosophy of Perception*, 603–20.
Beardsley, Monroe. *Aesthetics: Problems in the Philosophy of Criticism*. New York: Harcourt, Brace & World, 1958.
Beer, Betina. "Boholano Olfaction: Odor Terms, Categories, and Discourses." *Senses and Society* 9, no. 2 (2014): 151–73.
Bendeth, Marian. "Interview with 'Les Christophs': Christophe Laudamiel and Christoph Hornetz: Perfumers of Le Coffret." *Basenotes*, January, 2007. https://www.basenotes.net.
Berleant, Arnold. *Art and Engagement*. Philadelphia: Temple University Press, 1991.
Berntsen, Dorthe. "Spontaneous Recollections: Involuntary Autobiographical Memories Are a Basic Mode of Remembering." In *Understanding Autobiographical Memory*, edited by Dorthe Berntsen and David C. Rubin, 290–310. Cambridge: Cambridge University Press, 2012.
Bishop, Claire. *Installation Art: A Critical History*. London: Tate Publishing, 2005.
Bishop, Claire, ed. *Participation*. London: Whitechapel Gallery, 2006.
Blodgett, Bonnie. *Remembering Smell: A Memoir of Losing—and Discovering—the Primal Sense*. Boston: Houghton Mifflin Harcourt, 2010.
Böhme, Gernot. *The Aesthetics of Atmospheres*. Edited by Jean-Paul Thibaud. New York: Routledge, 2017.
Böhme, Gernot. *Atmospheric Architectures: The Aesthetics of Felt Spaces*. Edited and Translated by A.-Chr. Engels-Schwarzpaul. London: Bloomsbury Publishing, 2017.
Böhme, Gernot. "Contribution to the Critique of the Aesthetic Economy." *Thesis Eleven* 73, no. 1 (2003): 71–82.
Boisson, Claude. "La dénomination des odeurs: Variations et régularités linguistiques." *Intellectica* 1, no. 24 (1997): 29–49.
Boisvert, Raymond D., and Lisa Heldke. *Philosophers at Table: On Food and Being Human*. London: Reaktion Books, 2016.
Borch, Christian, ed. *Architectural Atmospheres: On the Experience and Politics of Architecture*. Basel: Birkhäuser, 2014.
Bouasse, Sarah. "Is the Perfumer an Artist?" *NEZ: The Olfactory Magazine* 4 (Autumn–Winter 2017): 87–92.
Boudonnat, Louise de, and Harumi Kushizaki. *La Voie de l'encens*. Paris: Esteban, 2000.
Boyd, Robert, and Peter J. Richeson. *Not by Genes Alone*. Chicago: University of Chicago Press, 2004.
Bradford, Kevin, and Debra Desrochers. "The Use of Scents to Influence Customers: The Sense of Using Scents to Make Cents." *Journal of Business Ethics* 90, no. 2 (2009): 141–53.
Bradley, Mark, ed. *Smell and the Ancient Senses*. London: Routledge, 2015.

Brady, Emily. *Aesthetics of the Natural Environment*. Tuscaloosa: University of Alabama Press, 2003.

Brady, Emily. "The Negative Aesthetics of Odor." Paper presented at the workshop "Scent, Science and Aesthetic: Understanding Smell and Anosmia." Barcelona, May 22–24, 2013.

Brady, Emily. "Smells, Tastes, and Everyday Aesthetics." In *The Philosophy of Food*, edited by David M. Kaplan, 69–86. Berkeley: University of California Press, 2012.

Brady, Emily. "Sniffing and Savoring: The Aesthetics of Smells and Tastes." In *The Aesthetics of Everyday Life*, edited by Andrew Light and Jonathan M. Smith, 177–93. New York: Columbia University Press, 2005.

Brady, Emily, Isis Brooks, and Jonathan Prior. *Between Nature and Culture: The Aesthetics of Modified Environments*. London: Rowman and Littlefield, 2018.

Brady, Michael S. *Emotional Insight: The Epistemic Role of Emotional Experience*. Oxford: Oxford University Press, 2013.

Brand, Peg Zeglin, ed. *Beauty Unlimited*. Bloomington: Indiana University Press, 2012.

Brakel, Marcel van, Wander Eikelboom, and Frederik Duerinck, eds. *Sense of Smell*. Breda: Eriskay Connection, 2014.

Brewer, Warwick, David Castle, and Christos Pantelis. *Olfaction and the Brain*. Cambridge: Cambridge University press, 2006.

Brier, Bob, and Hoyt Hobbs. *Ancient Egypt: Everyday Life in the Land of the Nile*. New York: Sterling, 2009.

Bright, Susan. *Art Photography Now*. London: Thames & Hudson, 2011.

Brillat-Savarin, Jean Antheleme. *The Physiology of Taste, or Meditations on Transcendental Gastronomy*. Translated by M. F. K. Fisher. Washington, DC: Counterpoint, 1998.

Buckle, Jane. *Clinical Aromatherapy: Essential Oils for Health Care*. St. Louis, MO: Elsevier, 2015.

Bullough, Edward. "Psychical Distance as a Factor in Art and an Aesthetic Principle." In *Art and Its Significance*, edited by Stephen David Ross, 458–67. Albany: State University of New York Press, 1994.

Burke, Edmund. *A Philosophical Enquiry into the Origin of our Ideas of the Sublime and Beautiful*. Notre Dame: University of Notre Dame Press, 1968.

Burr, Chandler. *The Art of Scent, 1889–2012*. New York: Museum of Arts and Design, 2012.

Burr, Chandler. "Fragrance Market Is Establishing a Foothold in China." *New York Times*, May 10, 2008. https://www.nytimes.com/2008/05/10/business/worldbusiness /10perfume.html.

Burr, Chandler. *The Perfect Scent: A Year Inside the Perfume Industry in Paris and New York*. New York: Henry Holt, 2007.

Bushdid, C., Marcelo O. Magnasco, Leslie B. Vosshall, and Andreas Keller. "Humans Can Discriminate More Than One Trillion Olfactory Stimuli." *Science* 21, no. 343 (2014): 1370–72. http://doi: 10.1126/science.1249168.

Butler, Shane. "Making Scents of Poetry." In Bradley, *Smell and the Ancient Senses*, 74–89.

Calkin, Robert R., and J. Stephan Jellinek. *Perfumery: Practice and Principle*. New York: John Wiley & Sons, 1994.

Calvino, Italo. *Under the Jaguar Sun*. Translated by William Weaver. New York: Harcourt Brace, 1988.

Campen, Chretien van. *The Proust Effect: The Senses as Doorways to Lost Memories*. Oxford: Oxford University Press, 2014.

Carlisle, Janice. *Common Scents: Comparative Encounters in High-Victorian Fiction.* Oxford: Oxford University Press, 2004.
Carlson, Allen. *Aesthetics and the Environment: The Appreciation of Nature, Art and Architecture.* London: Routledge, 2000.
Carlson, Allen. "The Dilemma of Everyday Aesthetics." In Yuedi and Carter, *Aesthetics of Everyday Life East and West*, 48–64.
Carlson, Allen. "Environmental Aesthetics, Ethics, and Ecoaesthetics." *Journal of Aesthetics and Art Criticism* 76, no. 4 (2018): 399–410.
Carlson, Allen, and Sheila Lintott. *Nature, Aesthetics, and Environmentalism: From Beauty to Duty.* New York: Columbia University Press, 2007.
Carné, Violaine de. "Théâtre olfactif: Un itinéraire singulier." In Jaquet, *L'Art olfactif*, 255–70.
Carra, Carlo. "The Painting of Sounds, Noises, and Smells." In *Futurism: An Anthology*, edited by Lawrence Rainey, Christine Poggi, and Laura Wittman, 156–59. New Haven: Yale University Press, 2009.
Carroll, Noël. *Art in Three Dimensions.* Oxford: Oxford University Press, 2010.
Carroll, Noël. "Historical Narratives and the Philosophy of Art." *Journal of Aesthetics and Art Criticism* 51, no. 3 (1993): 313–26.
Carroll, Nöel. *Theorizing the Moving Image.* Cambridge: Cambridge University Press, 1996.
Carvallho, F. "Olfactory Objects." *Disputatio* 6, no. 38 (2014): 45–66.
Castro, J. B., and William P. Seeley. "Olfaction, Valuation, and Action: Reorienting Perception." *Frontiers in Psychology* 5, no. 299 (2014): 1–4.
Cather, Willa. *Death Comes for the Archbishop.* New York: Alfred A. Knopf, 1927.
Chu, Simon, and John Joseph Downes. "Proust Nose Best: A Scientific Investigation of a Literary Legend. Odors Are Better Cues to Autobiographical Memory." *Memory and Cognition* 30, no. 4 (2002): 511–18.
Classen Constance, David Howes, and Anthony Synnott. *Aroma: The Cultural History of Smell.* London: Routledge, 1994.
Clement of Alexandria. *Christ the Educator.* Translated by Simon P. Wood. New York: Fathers of the Church, 1954.
Clements, Ashley. "Divine Scents and Presence." In Bradley, *Smell and the Ancient Senses*, 46–59.
Coles, Alex. *DesignArt.* London: Tate Publishing, 2005.
Collingwood, R. G. *The Principles of Art.* Oxford: Oxford University Press, 1938.
Condillac, Étienne Bonnot de. *Philosophical Writings of Étienne Bonnot, Abbé de Condillac.* Vol. 1. Translated by Franklin Philips and Harlan Lane. Hillsdale, NJ: Lawrence Erlbaum, 1982.
Cooper, David. *A Philosophy of Gardens.* Oxford: Oxford University Press, 2006.
Corbin, Alain. *The Foul and the Fragrant: Odor and the French Social Imagination.* Translated by Miriam L. Kochan Cambridge, MA: Harvard University Press, 1985.
Cox, Christophe. "From Music to Sound: Being as Time in the Sonic Arts." In Kelly, *Sound*, 80–87.
Cupere, Peter de. *Scent in Context: Olfactory Art.* Antwerp: Stockmans Art Publishers, 2016.
Currie, Greg, Mathew Kieran, Aaron Meskin, and Jon Robson, eds. *Aesthetics and the Sciences of Mind.* Oxford: Oxford University Press, 2014.
Danto, Arthur C. *The Abuse of Beauty.* Chicago: Open Court, 2003.
Danto, Arthur C. *What Art Is.* New Haven: Yale University Press, 2012.

Darwin, Charles. *Charles Darwin's Notebooks, 1836–1844: Geology, Transmutation of Species, Metaphysical Inquiries.* Cambridge: Cambridge University Press, 1987.

Darwin, Charles. *The Descent of Man, and Selection in Relation to Sex.* Princeton: Princeton University Press, 1981.

David, Sophie, Melissa Barkat-Defradas, and Catherine Rouby. "A Contrastive Study of French and Arabic Olfactory Lexicons." In *Words for Odours*, edited by Barkat-Defrades. Newcastle upon Tyne, UK: Cambridge Scholars Publishers, 2016, 167–188.

Davies, David. *Art as Performance.* Oxford: Oxford University Press, 2004.

Davies, Stephen. "The Aesthetics of Adornments." In Yuedi and Carter, *East and West*, 124–31.

Davies, Stephen. "Analyzing Human Adornment." *American Society for Aesthetics Newsletter* 38, no. 2 (Summer 2018), 1–4.

Davies, Stephen. *The Artful Species.* Oxford: Oxford University Press, 2012.

Davies, Stephen. "Defining Art and Artworlds." *Journal of Aesthetics and Art Criticism* 73, no. 4 (2015): 375–84.

Deák, František. "Symbolist Staging at the Théâtre d'Art." *Drama Review* 20, no. 3 (1976): 120–22.

Deigh, John. "Concepts of Emotion in Modern Philosophy and Psychology." In *The Oxford Handbook of Philosophy of Emotion*, edited by Peter Goldie, 17–40. Oxford: Oxford University Press, 2010.

Delau, Robert, and Jean-Robert Pitte. *Géographie des Odeurs.* Paris: Éditions L'Harmattan, 1998.

Deleuze, Gilles. *Proust et les signes.* Paris: Presses Universitaires de France, 2014.

Delon-Martin, Chantal, Jane Plailly, Pierre Fonlupt, Alexandra Veyrac, and Jean-Pierre Royet. "Perfumers' Expertise Induces Structural Reorganization in Olfactory Brain Regions." *NeuroImage* 68 (2013): 55–62. https://doi.10.1016/j.neuroimage.2012.11.044.

Damasio, Antonio R. *Descartes' Error: Emotion, Reason and the Human Brain.* New York: G. P. Putnam's Sons, 1994.

Detienne, Marcel. *The Gardens of Adonis: Spices in Greek Mythology.* Translated by Janet Lloyd. Princeton: Princeton University Press, 1994.

Devlin-Glass, Frances. "Armpits and Melons: An Olfactory Reading of James Joyce." *The Conversation*, June 15, 2017. http://www.theconversation.com/armpits-and-melons-an-olfactory-reading-of-james-joyce-78832.

Dewey, John. *Art as Experience.* New York: G. P. Putnam's Sons, 1934.

Diaconu, Mădălina. "Mapping Urban Smellscapes." In Diaconu, *Senses and the City*, 223–38.

Diaconu, Mădălina, Eva Heuberger, Ruth Mateus-Berr, and Lukas Marcel Vosicky, eds. *Senses and the City: An Interdisciplinary Approach to Urban Sensescapes.* Vienna: LIT Verlag, 2011.

Di Benedetto, Stephen. *The Provocation of the Senses in Contemporary Theater.* New York: Routledge, 2010.

Dickie, George. *The Art Circle: A Theory of Art.* New York: Haven, 1984.

Dickinson Emily. *Complete Poems.* Edited by Thomas H. Johnson. Boston: Little Brown and Company, 1960.

Digonnet, Rémi. *Métaphore et olfaction: Une approche cognitive.* Paris: Honoré Champion, 2016.

Domisseck, Sophie, and Roland Salesse. "Le spectateur olfactif: La reception de la pièce *Les Parfums de l'âme* par le public." In Jaquet, *L'Art olfactif*, 287–98.

Donne, John. *The Poems of John Donne*. Vol. 1. Edited by Herbert J. C. Grierson. Oxford: Oxford University Press, 1912.
Doty, Richard L., ed. *Handbook of Olfaction and Gustation*. 3rd ed. Hoboken, NJ: Wiley Blackwell, 2015.
Doty, Richard L., and E. L. Cameron. "Sex Differences and Reproductive Hormone Influences on Human Odor Perception." *Physiology and Behavior* 97, no. 2 (2009): 213–28. https://doi.10.1016/j.physbeh.2009.02.032.
Dowling, Christopher. "The Aesthetics of Daily Life." *British Journal of Aesthetics* 50, no. 3 (2010): 226–38.
Drobnick, Jim. "The City, Distilled." In Diaconu et al., *Senses and the City*, 257–76.
Drobnick, Jim. "Clara Ursitti: Scents of a Woman." *Tessera*, June 2002, 85–97.
Drobnick, Jim. "Eating Nothing: Cooking Aromas in Art and Culture." In Drobnick, *Smell Culture Reader*, 342–56.
Drobnick, Jim. "Inhaling Passions: Art, Sex and Scent." *Sexuality and Culture* 4, no. 3 (2000): 37–56.
Drobnick, Jim. "Reveries, Assaults and Evaporating Presences: Olfactory Dimensions in Contemporary Art." *Parachute* 89 (Winter 1998): 10–19.
Drobnick, Jim, "Smell: The Hybrid Art." In Jaquet, *L'Art olfactif*, 173–89.
Drobnick, Jim, ed. *The Smell Culture Reader*. Oxford: Berg, 2006.
Drobnick, Jim. "Towards an Olfactory Art History: The Mingled, Fatal, and Rejuvenating Perfumes of Paul Gauguin." *Senses and Society* 7, no. 2 (2012): 196–208.
Dubois, Danièle, and Catherine Rouby. "Names and Categories for Odors: The Veridical Label." In Rouby et al., *Olfaction, Taste and Cognition*, 47–66.
Ducey, Carolyn, and Amy Shell. *Perspectives: Art, Craft, Design and the Studio Quilt*. Lincoln, NE: International Quilt Study Center & Museum, 2009.
Dugan, Holly. *The Ephemeral History of Perfume: Scent and Sense in Early Modern England*. Baltimore: Johns Hopkins University Press, 2011.
Dutton, Dennis. *The Art Instinct: Beauty, Pleasure and Human Evolution*. New York: Bloomsbury, 2009.
Eco, Umberto. *The Aesthetics of Thomas Aquinas*. Cambridge, MA: Harvard University Press, 1988.
Elsaesser, Thomas, and Malte Hagener. *Film Theory: An Introduction through the Senses*. New York: Routledge, 2015.
Ellena, Jean Claude. *Le parfum*. Paris: Presses Universitaires de France, 2007.
Ellena, Jean Claude. *Perfume: The Alchemy of Scent*. New York: Arcade Publishing, 2011.
Endrissat, Nada, and Claus Noppeney. "Materializing the Immaterial: Relational Movements in a Perfume's Becoming." In *How Matter Matters: Objects, Artifacts, and Materiality in Organization Studies*, edited by Paul R. Carlisle, Davide Nicolini, Ann Langley, and Haridimos Tsoukas, 58–91. Oxford: Oxford University Press, 2013.
Erwine, Barbara. *Creating Sensory Spaces: The Architecture of the Invisible*. New York: Routledge, 2017.
Fareed, Faisal. "Scent of a City." *Indian Express*, March 27, 2016. https://indianexpress.com/Lifestyle.
Faulkner, William. *The Sound and the Fury*. New York: Random House, 1984.
Feagin, Susan. "Olfaction and Space in the Theater." *British Journal of Aesthetics* 58, no. 2 (2018): 131–46.
Ficino, Marsilio. *Three Books on Life*. Translated by Carol V. Kaske and John R. Clark. Binghamton, NY: Renaissance Society of America, 1989.

Fjellestad, Danuta. "Towards an Aesthetics of Smell, or, the Foul and the Fragrant in Contemporary Literature." *Cauce* 24 (2001): 637–51.
Fleischer, Mary. "Incense and Decadents: Symbolist Theatre's Use of Scent." In Banes and Lepecki, *Senses in Performance*, 105–14.
Forsey, Jane. *The Aesthetics of Design*. Oxford: Oxford University Press, 2013.
Frasnelli, Johannes, Julie A. Boyle, Johan N. Lundström, and Jelana Dordjevic. "Neuroanatomical Correlates of Olfactory Performance." *Experimental Brain Research* 201, no. 1 (2010): 1–11. https://doi.10.1007/s00221-009-1999-7.
Frasnelli, Johannes, and Genevieve Charbonneau, Olivier Collignon, and Franco Lepore. "Odor, Localization and Sniffing." *Chemical Senses* 34, no. 2 (2008): 129–44. https://doi.org/10.1093/chemse/bjn068.
Freud, Sigmund. *Civilization and Its Discontents*. Translated by James Strachey. New York: Norton, 1961.
Freeland, Cynthia. "Gustatory Film Perception." Paper presented at the Annual Meeting of the American Society for Aesthetics, New Orleans, November 2017.
Friedman, Emily C. *Reading Smell in Eighteenth-Century Fiction*. Lewisburg, PA: Bucknell University Press, 2016.
Frost, Laura. *The Problem with Pleasure: Modernism and Its Discontents*. New York: Columbia University Press, 2013.
Gardner, Howard. *Frames of Mind: The Theory of Multiple Intelligences*. New York: Basic Books, 1993.
Gattefosse, René-Maurice. *Gattefosse's Aromatherapy*. Edited by Robert B. Tisserand. London: Random House, 1996.
Gatten, Aileen. "A Wisp of Smoke: Scent and Character in the Tale of Genji." In Drobnick, *Scent Culture Reader*, 331–41.
Gauguin, Paul. *Noa Noa: A Journal of the South Seas*. Translated by O. F. Theis. New York: Farrar, Straus and Giroux, 1952.
Gavrilyuk, Paul L., and Sarah Coakley. *The Spiritual Senses: Perceiving God in Western Christianity*. Cambridge: Cambridge University Press, 2012.
Gefter, Amanda. "*Green Aria*: An Opera for Your Nose." *New Scientist*, June 5, 2009. www.newscientist.com/article/dn17236-green-aria-an-opera-for-your-nose.html.
Gell, Alfred. "Magic Perfume, Dream . . . " In Drobnick, *Smell Culture Reader*, 401–10.
Gelstein, S., Y. Yeshurun, L. Rozenkrantz, S. Shushan, I. Frumin, Y. Roth, and N. Sobel. "Human Tears Contain a Chemosignal." *Science* 331 (2011): 226–30. https://doi.10.1126/science.1198331.
Georgsdorf, Wolfgang. *Osmodrama: Storytelling with Scents*. Website. https://www.osmodrama.com.
Georgsdorf, Wolfgang. *Osmodrama via Smeller 2.0: Storytelling with Smells*. Website. https://www.smeller.net › about.
Gilbert, Avery. *What the Nose Knows: The Science of Scent in Everyday Life*. New York: Crown Publishers, 2008.
Gilbert, Avery. "Green Aria—a Scent Opera: Bravo!" *First Nerve: Taking a Scientific Sniff at the Culture of Smell*, June 3, 2009. https://www.firstnerve.com › 2009/06.
Goeltzenleuchter, Brian. "Scenting the Antiseptic Institution." In Henshaw et al., *Designing with Smell*, 2–18.
Goldie, Peter. *The Emotions: A Philosophical Exploration*. Oxford: Oxford University Press, 2000.
Goldie, Peter, ed. *The Oxford Handbook of Philosophy of Emotion*. Oxford: Oxford University Press, 2010.

Goldie, Peter, and Elisabeth Schellekens. *Who's Afraid of Conceptual Art?* London: Routledge, 2009.

Goldman, Alan. "The Broad View of Aesthetic Experience." *Journal of Aesthetics and Art Criticism* 71, no. 4 (2013): 323–33.

Gombrich, Ernst. *Art and Illusion: A Study in the Psychology of Pictorial Representation.* Princeton: Princeton University Press, 1961.

Gottfried, Jay A., Dana M. Small, and David H. Zald. "The Chemical Senses." In *The Orbitofrontal Cortex*, edited by David H. Zald and Scott L. Rauch, 125–45. Oxford: Oxford University Press, 2006.

Gottfried, Jay A. *Neurobiology of Sensation and Reward.* Boca Raton, FL: Taylor & Francis, 2011.

Gottfried, Jay A., and Keng Nei Wu. "Perceptual and Neural Pliability of Odor Objects." *Annals of the New York Academy of Sciences* 1170, no. 1 (2009): 324–32. doi: 10.1111/j.1749-6632.2009.03917.x.

Gottfried, Jay A. "Structural and Functional Imaging of the Human Olfactory System." In *Handbook of Olfaction and Gustation*, edited by Richard L. Doty, 279–304. Hoboken, NJ: Wiley Blackwell, 2015.

Gottfried, Jay A. "What Can an Orbitofrontal Cortex Endowed Animal Do with Smells?" *Annals of the New York Academy of Sciences* 1121, no. 1 (2007): 102–20. https://doi.10.1196/annals.1401.018.

Gottfried, Jay A., and Christina Zelano. "The Value of Identity: Olfactory Notes on Orbitofrontal Cortex Function." *Annals of the New York Academy of Sciences* 1239 (2011): 138–48. https://doi: 10.1111/j.1749-6632.2011.06268.x.

Grasse, Marie-Christine, ed. *Perfume: A Global History From the Origins to Today.* Paris: Somogy Art Publishers, 2012.

Green, Coady. "Scriabin in the Himalayas." *Classical-Music.com*. September 29, 2015. http://www.classical-music.com/article/scriabin-himalayas.

Green, Deborah A. *The Aroma of Righteousness: Scent and Seduction in Rabbinic Life and Literature.* University Park: Pennsylvania State University Press, 2011.

Green, Deborah A. "Fragrance in the Rabbinic World." In Bradley, *Smell and the Ancient Senses*, 146–57.

Grinand, Michel. "Une Nuit américaine à écouter et humer." *AvantChœur* 3951 (June 4, 2015). https://www.avantchoeur.com.

Guyer, Paul. *A History of Modern Aesthetics.* Vol. 1: *The Eighteenth Century.* Cambridge: Cambridge University Press, 2014.

Harris, John. "Oral and Olfactory Art." *Journal of Aesthetic Education* 13, no. 4 (1979): 5–15.

Harris, Jonathan Gil. *Untimely Matter in the Time of Shakespeare.* Philadelphia: University of Pennsylvania Press, 2009.

Harvey, Susan. *Scenting Salvation: Ancient Christianity and the Olfactory Imagination.* Berkeley: University of California Press, 2006.

Hawkes, Christopher H., and Richard L. Doty. *The Neurology of Olfaction.* Cambridge: Cambridge University Press, 2009.

Hawkes, Christopher H., and Richard L. Doty. *Smell and Taste Disorders.* Cambridge: Cambridge University Press, 2018.

Heaney, Shamus. *Opened Ground: Selected Poems, 1966–1996.* New York: Farrar, Straus and Giroux, 1998.

Hegel, G. W. F. *Aesthetics: Lectures on Fine Art.* Vol. 1. Translated by T. M. Knox. Oxford: Clarendon Press, 1975.

Henshaw, Victoria. *Urban Smellscapes: Understanding and Designing City Smell Environments*. New York: Routledge, 2014.

Henshaw, Victoria, Kate McLean, Dominic Medway, Chris Perkins, and Gary Warnaby, eds. *Designing with Smell: Practices, Techniques and Challenges*. New York: Routledge, 2018.

Herz, Rachel S. "Odor Memory and the Special Role of Associative Learning." In Zucco et al., *Olfactory Cognition*, 95–114.

Herz, Rachel S. "Perfume." In *Neurobiology of Sensation and Reward*, edited by Jay A. Gottfried, 371–89. Boca Raton, FL: Taylor & Francis, 2011.

Herz, Rachel S. *The Scent of Desire: Discovering Our Enigmatic Sense of Smell*. New York: Harper, 2008.

Higgins, Kathleen M., Shakti Maira, and Sonia Sikka. *Artistic Visions and the Promise of Beauty: Cross-Cultural Perspectives*. Cham, Switzerland: Springer, 2017.

Hobbes, Thomas. *The Elements of Law Natural and Politic: Part I, Human Nature*. Oxford: Oxford University Press, 1994.

Holl, Steven, Juhani Pallasmaa, and Alberto Pérez Gómez. *Questions of Perception: Phenomenology of Architecture*. San Francisco: William K. Stout Publishers, 2006.

Holley, André. "Cognitive Aspects of Olfaction in Perfumer Practice." In Rouby et al. *Olfaction, Taste and Cognition*, 21.

Holley, André. "Les trois piliers de l'art du parfum." In Jaquet, *L'Art olfactif*, 55–61.

Horkheimer, Max, and Theodore W. Adorno. *Dialectic of Enlightenment*. Translated by John Cumming. New York: Continuum, 1987.

Horowitz, Alexandra. *Being a Dog: Following the Dog into a World of Smell*. New York: Scribner, 2016.

Howes, David. "Futures of Scents Past." In Smith, *Smell and History*, 203–18.

Howes, David. "Olfactory Art: Introduction." In Henshaw et al., *Designing with Smell*, 5–8.

Howes, David. "Nose-Wise: Olfactory Metaphors in Mind." In Rouby et al., *Olfaction, Taste, and Cognition*, 67–81.

Hsu, Hsuan L. "Olfactory Art, Transcorporeality, and the Museum Environment." *Resilience: A Journal of the Environmental Humanities* 1, no. 1 (2016): 1–24.

Huxley, Aldous. *Brave New World*. New York: Bantam Books, 1949.

Huysmans, J.-K. *Against Nature*. Translated by Brendan King. Sawtry, UK: Daedalus, 2008.

Illich, Ivan. *H2O and the Waters of Forgetfulness: Reflections on the Historicity of "Stuff."* Berkeley: Heyday Books, 1985.

Irvin, Sherri. "Is Aesthetic Experience Possible?" In Currie et al., *Aesthetics and the Sciences of Mind*, 44–49.

Irvin, Sherri. "The Artist's Sanction in Contemporary Art." *Journal of Aesthetics and Art Criticism* 63, no. 4 (2005): 315–26.

Irvin, Sherri. "Installation Art and Performance: A Shared Ontology." In *Art and Abstract Objects*, edited by Christy Mag Uidhir, 242–62. Oxford: Oxford University Press, 2013.

Irvin, Sherri. "Scratching an Itch." *Journal of Aesthetics and Art Criticism* 66, no. 1 (2008): 25–35.

Irwine, Barbara. *Creating Sensory Spaces: The Architecture of the Invisible*. New York: Routledge, 2017.

Iseminger, Gary. *The Aesthetic Function of Art*. Ithaca: Cornell University Press, 2004.

Iwasaki, Yoko. "Art and the Sense of Smell: The Traditional Japanese Art of Scents (ko)." *Aesthetics* (Japanese Society for Aesthetics) 11 (2004): 62–67.

Iwasaki, Yoko. "La reconnaissance de l'espace et l'art olfactif Japonais." In Jaquet, *L'Art olfactif*, 223–29.
Jacob, Tim. "The Science of Taste and Smell." In *Art and the Senses*, edited by Francesca Bacci and David Melcher, 183–204. Oxford: Oxford University Press, 2013.
Jaquet, Chantal, ed. *L'Art olfactif contemporain*. Paris: Classiques Garnier, 2015.
Jaquet, Chantal. *Philosophie de l'odorat*. Paris: Presses Universitaires de France, 2010.
Jaquet, Chantal. *Philosophie du Kodo: L'esthétique japonaise des fragrances*. Paris: Presses Universitaires de France, 2018.
Jellinek, Paul. *The Psychological Basis of Perfumery*. 4th ed. Translated by J. Stephan Jellinek. London: Blackie Academic and Professional, 1997.
Jenner, Mark S. R. "Civilization and Deodorization? Smell in Early Modern English Culture." In *Civil Histories: Essays Presented to Sir Keith Thomas*, edited by Peter Burke, Brian Harrison, and Paul Slack, 127–44. Oxford: Oxford University Press, 2000.
Jha, Radhika. *Smell*. New York: Viking, 1999.
Johansen, Thomas K. "Aristotle on the Sense of Smell." *Phronesis* 41, no. 1 (1996): 1–19.
Johnson, Anthony ed. *Writing and Religion in England, 1558–1689: Studies in Community Making and Cultural Memory*. Farnham, UK: Ashgate, 2009.
Johnstone, Sam. "Osmodrama/Review." *The Cusp*, September 21, 2016. https://www.thecuspmagazine.com/reviews/osmodrama-.
Jones, Caroline A. *Eyesight Alone: Clement Greenberg's Modernism and the Bureaucratization of the Senses*. Chicago: University of Chicago Press, 2005.
Joyce, James. *A Portrait of the Artist as a Young Man*. Harmondsworth, UK: Penguin, 1976 (1916).
Joyce, James. *Ulysses*. New York: Random House, 1986.
Jütte, Robert. "Reodorizing the Modern Age." In Smith, *Smell and History*, 170–86.
Kaeppler, Kathrin, and Friedrich Mueller. "Odor Classification: A Review of Factors Influencing Perception-Based Odor Arrangements." *Chemical Senses* 18, no. 3 (2013): 189–209. https://doi.org/10.1093/chemse/bjs141.
Kames, Henry Home, Lord. *The Elements of Criticism*. Vol. 1. Hildesheim: Georg Olms Verlag, 1970.
Kant, Immanuel. *Critique of Judgment*. Translated by Werner S. Pluhar. Indianapolis: Hackett, 1987.
Kant, Immanuel. *Lectures on Anthropology*. Edited by Allen W. Wood and Robert B. Louden. Cambridge: Cambridge University Press, 2012.
Kelly, Caleb, ed. *Sound*. Cambridge, MA: MIT Press, 2011.
Kelly, Michael. *The Encyclopedia of Aesthetics*. 2nd ed. Oxford: Oxford University Press, 2014.
Kiechle, Melanie A. *Smell Detectives: An Olfactory History of Nineteenth-Century Urban America*. Seattle: University of Washington Press, 2018.
Kincaid, Jamaica. *The Autobiography of My Mother*. New York: Farrar, Straus and Giroux. 1996.
King, J. R. "The Scientific Status of Aromatherapy." *Perspectives in Biology and Medicine*, 37, no. 3 (1994): 409–15.
Korsmeyer, Carolyn. *Making Sense of Taste: Food and Philosophy*. Ithaca: Cornell University Press, 1999.
Korsmeyer, Carolyn. *Savoring Disgust: The Foul and the Fair in Aesthetics*. Oxford: Oxford University Press, 2011.Korsmeyer, Carolyn, ed. *The Taste Culture Reader: Experiencing Food and Drink*. 2nd ed. London: Bloomsbury, 2017.

Korsmeyer, Carolyn. *Things: In Touch with the Past*. Oxford: Oxford University Press, 2019.
Kostelanetz, Richard. *A Dictionary of the Avant-Garde*. New York: Schirmer Books, 2000.
Köster, Egon Peter. "The Specific Characteristics of the Sense of Smell." In Rouby, *Olfaction, Taste, and Cognition*, 27–44.
Köster, Egon Peter, Per Møller, and Jozina Mojet. "A 'Misfit' Theory of Spontaneous Conscious Odor Perception (MITSCOP): Reflections on the Role and Function of Odor Memory in Everyday Life." *Frontiers in Psychology* 5, no. 64 (2014): 1–12. https://doi.10.3389/fpsyg.2014.00064.
Kraft, Philip. "The Odor Value Concept in the Formal Analysis of Olfactory Art." *Helvetica* 120, no. 1 (January, 2019). https://doi.org/10.1002/hlca.201800185.
LaBelle, Brandon. *Background Noise: Perspectives on Sound Art*. New York: Continuum, 2006.
Lamarque, Peter. "Wittgenstein, Literature, and the Idea of a Practice." *British Journal of Aesthetics* 50, no. 4 (2010): 375–388.
Lange, Barbara. "A Shift of Senses: A Brief Foray into the History of Scents and Odors in Art." In Wetzel et al., *Belle Haleine*, 97–104.
Larsson, Maria, Johan Willander, Kristin Karlsson, and Artin Arshamian. "Olfactory Lover: Behavioral and Neural Correlates of Autobiographical Odor Memory." *Frontiers in Psychology* 5, no. 214 (2014): 1–4. https://doi: 10.3389/fpsyg.2014.00312.
Laska, Matthias, Daria Genzel, and Alexandra Wieser. "The Number of Functional Olfactory Receptor Genes and the Relative Size of Olfactory Brain Structures Are Poor Predictors of Olfactory Discrimination Performance with Enantiomers." *Chemical Senses* 30, no. 2 (2005): 171–75. https://doi.org/10.1093/chemse/bji013.
Laskiewicz, Zachar. "From the Hideous to the Sublime: Olfactory Processes, Performance Texts and the Sensory Episteme." *Performance Research* 8, no. 3 (2003): 55–65.
Le Breton, David. "Brief Anthropology of Perfume in the Western World." In Grasse, *Perfume*, 218–21.
Le Breton, David. *La Saveur du monde: Une anthropologie des sens*. Paris: Éditions Métailié, 2006.
Le Guérer, Annick. *Scent: The Mysterious and Essential Powers of Smell*. Translated by Richard Miller. New York: Random House, 1992.
Leddy, Thomas. "Everyday Aesthetics and Happiness." In Yuedi and Carter, *Aesthetics of Everyday Life*, 31–34.
Leddy, Thomas. *The Extraordinary in the Ordinary: The Aesthetics of Everyday Life*. Toronto: Broadview Press, 2012.
Leimbach, Tania. "Scent, Sensibility, and the Smell of a City." *The Conversation*, January 16, 2017. http://www.theconversation.com.
Lengyel, Olga. *Five Chimneys: The Story of Auschwitz*. Translated by Clifford Coch and Paul P. Weiss. Chicago: Zoff-Davis, 1947.
Lenkowsky, Kate. *Contemporary Quilt Art: An Introduction and Guide*. Bloomington: Indiana University Press, 2008.
Levi, Primo. *Survival in Auschwitz and The Reawakening: Two Memoires*. Translated by Stuart Woolf. New York: Summit Books, 1985.
Levinson, Jerrold. *Aesthetic Pursuits: Essays in Philosophy of Art*. Oxford: Oxford University Press, 2016.
Levinson, S. C., and A. Majid. "Differential Ineffability and the Senses." *Mind and Language* 29, no. 4 (2014): 407–27. https://doi-org.ezproxy.uis.edu/10.1111/mila.12057.

Li, Wen, Leonardo Lopez, Jason Osher, James D. Howard, Todd B. Parrish, and Jay A. Gottfried. "Right Orbitofrontal Cortex Mediates Conscious Olfactory Perception." *Psychological Science* 21, no. 10 (2010): 1454–63.
Licht, Alan. *Sound Art: Beyond Music, between Categories*. New York: Rizzoli International, 2007.
Livermore, Andrew, and David G. Laing. "Influence of Training and Experience on the Perception of Multicomponent Odor Mixtures." *Journal of Experimental Psychology: Human Perception and Performance* 22, no. 2 (1996): 267–77. http://dx.doi.org/10.1037/0096-1523.22.2.26.
Livingston, Paisley, and Carl Plantinga, eds. *The Routledge Companion to Philosophy and Film*. London: Routledge, 2009.
Lochner, Mira. *Traditional Japanese Architecture: An Exploration of Elements and Forms*. Tokyo: Tuttle Publishing, 2010.
Lopes, Dominic McIver. *Aesthetics on the Edge: Where Philosophy Meets the Human Sciences*. Oxford: Oxford University Press, 2018.
Lopes, Dominic McIver. *Being for Beauty: Aesthetic Agency and Value*. Oxford: Oxford University Press, 2018.
Lopes, Dominic McIver. *Beyond Art*. Oxford: Oxford University Press, 2014.
Lotze, M., G. Scheler, H.-R. M. Tan, C. Braun, and N. Birbaumer. "The Musician's Brain: Functional Imaging of Amateurs and Professionals during Performance and Imagery." *NeuroImage* 20 (2003): 1817–29. https://doi.10.1016/j.neuroimage.2003.07.018.
Low, Kelvin E. Y. *Scents and Scent-sibilities: Smell and Everyday Life Experiences*. Newcastle upon Tyne, UK: Cambridge Scholars Publishing, 2009.
Lucretius. *On the Nature of the Universe*. Translated by James H. Mantinband. New York: Frederick Ungar Publishing,
Luther, Martin. *Luther's Works*. Vol. 53. Edited by Ulrich S. Leupold. Philadelphia: Fortress Press, 1955.
Lycan, William G. "The Intentionality of Smell." *Frontiers in Psychology* 5, no. 436 (2014): 1–8. https://doi.org/10.3389/fpsyg.2014.00436.
Lycan, William G. "The Slighting of Smell." In *Of Minds and Molecules: New Philosophical Perspectives on Chemistry*, edited by N. Bhushan and S. Rosenfeld, 273–89. Oxford: Oxford University Press, 2000.
Lynn, Gwenn-Aël. "On Olfactory Space." In Henshaw et al., *Designing with Smell*, 26–28.
Mace, John H. "Involuntary Memories Are Highly Dependent on Abstract Cuing: the Proustian View Is Incorrect." *Applied Cognitive Psychology* 18, no. 7 (2004): 893–99. https://doi.org/10.1002/acp.1020.
MacIntyre, Alasdair. *After Virtue*. Notre Dame: University of Notre Dame Press, 1981.
Macpherson, Fiona, ed. *The Senses: Classic and Contemporary Philosophical Perspectives*. Oxford: Oxford University Press, 2011.
Mag Uidhir, Christy, and P. D. Marcus, "Art Concept Pluralism." *Metaphilosophy* 42, nos. 1–2 (2011): 83–97.
Mailer, Norman. *An American Dream*. New York: Dial Press, 1965.
Majid, Asifa, and Niclas Burenhult. "Odors Are Expressible in Language, as Long as you Speak the Right Language." *Cognition* 130 (2014): 266–70. http://dx.doi.org/10.1016/j.cognition.2013.11.004.
Man, Kingston, Joseph Kaplan, Hannah Damasio, and Antonio Damasio. "Neural Convergence and Divergence in the Mammalian Cerebral Cortex: From Experimental Neuroanatomy to Functional Neuroimaging." *Journal of Comparative Neurology* 521, no. 18 (2013): 4097–111. https://doi: 10.1002/cne.23408.

Manniche, Lise. *Sacred Luxuries: Fragrance, Aromatherapy and Cosmetics in Ancient Egypt*. Ithaca: Cornell University Press, 1999.
Marinval, Philippe. "Fragrances from Prehistoric Times to Ancient Gaul." In Grasse, *Perfume*, 31–35.
Martin, G. Neil. 2013. *The Neuropsychology of Smell and Taste*. London: Psychology Press.
Matsui, Atsushi, Yashuhiro Go, and Yoshihito Niimura. "Degeneration of Olfactory Receptor Gene Repertories in Primates: No Direct Link to Full Trichromatic Vision." *Molecular Biology and Evolution* 27, no. 5 (2010): 1192–200. https://doi.org/10.1093/molbev/msq003.
Matten, Mohan, ed. *The Oxford Handbook of Philosophy of Perception*. Oxford: Oxford University Press, 2015.
McHugh, James. *Sandalwood and Carrion: Smell in Indian Religion and Culture*. Oxford: Oxford University Press, 2012.
McGann, John P. "Poor Human Olfaction Is a 19th Century Myth." *Science* 356, no. 597 (2017): 1–6. https://DOI: 10.1126/science.aam7263.
McGinley, Michael, and Charles McGinley. "Olfactory Design Elements in the Theater: The Practical Considerations." In Henshaw et al., *Designing with Smell*, 219–26.
McLean, Kate. "Communicating and Mediating Smellscapes: The Design and Exposition of Olfactory Mappings." In Henshaw et al., *Designing with Smell*, 69.
McQueen, Donald. "Aquinas on the Aesthetic Relevance of Tastes and Smells." *British Journal of Aesthetics* 33, no. 4 (1993): 346–356.
Meighan, Michael. *Glasgow Smells: A Nostalgic Tour of the City*. Stroud, UK: Tempus Publishers, 2008.
Meiss, Peter. *Elements of Architecture: From Form to Place + Tectonics*. 2nd Edition. Translated by Theo Hakola. Milton Park, UK: Routledge, 2013.
Merrick, Christina, Christine A. Godwin, Mark W. Geisler, and Ezequiel Morsella. "The Olfactory System as the Gateway to the Neural Correlates of Consciousness." *Frontiers in Psychology* 4, no. 1011 (2014): 1–15. https://doi.org/10.3389/fpsyg.2013.01011.
Meskin, Aaron. "From Defining Art to Defining the Individual Arts: The Role of Theory in the Philosophies of Arts." In *New Waves in Aesthetics*, edited by Kathleen Stock and Katherine Thomson-Jones, 125–49. New York: Palgrave Macmillan, 2008.
Meskin, Aaron, and Roy T. Clark, eds. *The Art of Comics: A Philosophical Approach*. Malden, MA: Wiley-Blackwell, 2012.
Métailié, George. "Fragrances in Medieval China." In Grasse, *Perfume*, 120–23.
Miller, Mara. *The Garden as Art*. Albany: State University of New York Press, 1993.
Montaigne, Michel de. *The Complete Essays of Montaigne*. Translated by Donald M. Frame. Stanford, CA: Stanford University Press, 1957.
Moore, Paul A. *The Hidden Power of Smell: How Chemicals Influence Our Lives and Behavior*. Cham, Switzerland: Springer, 2016.
Morita, Kiyoko. *The Book of Incense: Enjoying the Traditional Art of Japanese Scents*. Tokyo: Kodansha International, 1992.
Morris, Edwin T. *Scents of Time: Perfume from Ancient Egypt to the 21st Century*. New York: Metropolitan Museum of Art, 1999.
Morrison, Toni. *Sula*. New York: Alfred A. Knopf, 1973.
Morse, Margaret. "Burnt Offerings (Incense): Body Odors and the Olfactory Arts in Digital Culture." 2000. http://www.thing.net/~jmarketo/textsby/incense.shtml.
Nanay, Bence. *Aesthetics as Philosophy of Perception*. Oxford: Oxford University Press, 2016.
Negrin, Llewellyn. "Fashion and Aesthetics: A Fraught Relationship." In *Fashion and Art*, edited by Adam Geczy and Vicki Karaminas, 43–54. London: Berg, 2012.

Newman, Cathy. *Perfume: The Art and Science of Scent*. Washington, DC: National Geographic Society, 1998.
Niehaus, Max. "Sound Art?" In Kelly, *Sound*, 72–73.
Nietzsche, Friedrich. *Basic Writings of Nietzsche*. Translated by Walter Kaufman. New York: Random House, 1968.
Nietzsche, Friedrich. *The Portable Nietzsche*. Translated by Walter Kaufman. New York: Viking Penguin, 1954.
Nietzsche, Friedrich. *Werke in Drei Bänden*. Edited by Karl Schlecta. Munich: Carl Hanser Verlag, 1960.
Oberinger, Fédéric. "Introduction." In *Parfums de Chine: La Culture de l'encens au temps des empéreurs*, edited by Éric Lefebvre, 10–27. Paris: Paris Museés, 2018.
Olofsson, Jonas K., Nicholas E. Bowman, Katherine Khatibi, and Jay A. Gottfried. "A Time-Based Account of the Perception of Odor Objects and Valences." *Psychological Science* 23, no. 10 (2012): 1224–32. https://doi.10.1177/0956797612441951.
Olofsson, Jonas K., and Jay A. Gottfried. "The Muted Sense: Neurocognitive Limitations of Olfactory Language." *Trends in Cognitive Sciences* 19, no. 6 (2015): 314–21. https://doi: 10.1016/j.tics.2015.04.007.
Onfray, Michel. *L'Art de Jouir: Pour un materialisme hédoniste*. Paris: Bernard Grasset, 1991.
Orwell, George. *The Road to Wiggan Pier*. London: Victor Gollancz, 1937.
Osborne, Harold. "Odors and Appreciation." *British Journal of Aesthetics* 17, no. 4 (1977): 37–48.
Osman, Ashraf. *Stop and Smell the (Olfactory) Art* (blog). https://www.scentart.net/tag/ashraf-osman.
Pallasmaa, Juhani. *The Eyes of the Skin: Architecture and the Senses*. West Sussex, UK: John Wiley & Sons, 2005.
Palmer, Richard. "In Bad Odour: Smell and Its Significance in Medicine from Antiquity to the Seventeen Century." In *Medicine and the Five Senses*, edited by W. F. Bynum and Roy Porter, 61–68. Cambridge: Cambridge University Press, 1993.
Paquet, Dominique. "L'acteur olfactif." In Jaquet, *L'Art olfactif*, 237–54.
Pardo, Carmen. "The Emergence of Sound Art: Opening the Cages of Sound." *Journal of Aesthetics and Art Criticism* 75, no. 1 (2017): 35–48.
Parr, Debora Riley. "Indeterminate Ecologies of Scent." In Henshaw et al. *Designing with Smell*, 259–269.
Parsons, Glenn, and Allen Carlson. *Functional Beauty*. Oxford: Oxford University Press, 2008.
Parsons, Glenn. *The Philosophy of Design*. Cambridge: Polity Press, 2016.
Patel, Aniruddh D. *Music, Language, and the Brain*. Oxford: Oxford University Press, 2008.
Pelosi, Paolo. 2016. *On the Scent: A Journey through the Science of Smell*. Oxford: Oxford University Press.
Pickett, Holly Crawford. "The Idolatrous Nose: Incense on the Early Modern Stage." In *The Performance of Religion on the Renaissance Stage*, edited by Jane Hwang Degenhardt and Elizabeth Williamson, 19–37. Farnham, UK: Ashgate, 2011.
Plailly, Jane, Chantal Delon-Martin, and Jean-Pierre Royet. "Experience Induces Functional Reorganization in Brain Regions Involved in Odor Imagery in Perfumers." *Human Brain Mapping* 33, no. 1 (2012): 224–34. https://doi.org/10.1002/hbm.21207.

Plato. *The Collected Dialogues of Plato*. Edited by Edith Hamilton and Huntington Cairns. Princeton: Princeton University Press, 1961.
Plato. *Plato's Cosmology: The Timaeus of Plato Translated with a Running Commentary*. Translated by Francis Macdonald Cornford. New York: Liberal Arts Press, 1957.
Plato. *Republic*. Translated by Robin Waterfield. Oxford: Oxford University Press, 1993.
Pliny. *Natural History*. Vol. 4. Translated by H. Rackham. Cambridge, MA: Harvard University Press, 1945.
Pollack, Barbara. "Scents & Sensibility." *ARTnews* 110, no. 3 (2011): 92.
Porter, Jess, Brent Craven, Rehan M. Kahn, Shao-Ju Chang, Irene Kang, Benjamn Judkewitz, Jason Volpe, Gary Settles, and Noam Sobel. "Mechanisms of Scent-Tracking in Humans." *Nature Neuroscience* 10, no. 1 (2007): 27–29. https://doi.10.1038/nn1819.
Postrel, Virginia. *The Substance of Style: How the Rise of Aesthetic Value Is Remaking Commerce, Culture and Consciousness*. New York: Harper Collins, 2003.
Prall, D. W. *Aesthetic Judgment*. New York: Thomas Y. Crowell, 1929.
Press, Daniel, and Steven C. Minta. "The Smell of Nature: Olfaction, Knowledge and the Environment." *Philosophy and Geography* 3, no. 2 (2000): 173–86. https://doi.org/10.1080/13668790008573711.
Price, Carolyn. *Emotion*. Cambridge: Polity Press, 2015.
Prinz, Jesse J. "Emotion and Aesthetic Value." In Schellekens and Goldie, *The Aesthetic Mind*, 71–88.
Prinz, Jesse J. *Gut Reactions: A Perceptual Theory of Emotions*. Oxford: Oxford University Press, 2006.
Proust, Marcel. *Remembrance of Things Past*. Translated by C. K. Moncrief and Terence Kilmartin. Vols. I–III. New York: Random Hose, 1981.
Prum, Richard O. *The Evolution of Beauty: How Darwin's Forgotten Theory of Mate Choice Shapes the Animal World—and Us*. New York: Doubleday, 2017.
Pybus, David. *Kodo: The Way of Incense*. Boston: Tuttle Publishing, 2001.
Quilty-Dunn, Jake. "Reid on Olfaction and Secondary Qualities." *Frontiers in Psychology* 4, no. 974 (2013): 1–11. https://doi.org/10.3389/fpsyg.2013.00974.
Reason, Matthew. "Writing the Olfactory in the Live Performance Review." *Performance Research* 8, no. 3 (2003): 73–84.
Reinarz, Jonathan. *Past Scents: Historical Perspectives on Smell*. Urbana: University of Illinois Press, 2014.
Richardson, Louise. "Flavour, Taste and Smell." *Mind and Language* 28, no. 3 (2013): 322–41. https://doi.org/10.1111/mila.12020.
Richardson, Louise. "Sniffing and Smelling." *Philosophical Studies* 162 (2013): 401–19.
Rilke, Rainer Maria. *The Notebooks of Malte Laurids Brigge*. Translated by M. D. Herter. New York: Norton, 1949.
Rindisbacher, Hans. *The Smell of Books: A Cultural-Historical Study of Olfactory Perception in Literature*. Ann Arbor: University of Michigan Press, 1992.
Robbins, Tom. *Jitterbug Perfume*. New York: Bantam, 1990.
Roberts, David. *The Total Work of Art in European Modernism*. Ithaca: Cornell University Press, 2011.
Robertson, David. "Incensed over Incense: Incense and Community in Seventeenth Century Literature." In *Writing and Religion in England, 1558–1689: Studies in Community Making and Cultural Memory*, edited by Roger D. Sell and Anthony W. Johnson, 389–410. Farnham, UK: Ashgate, 2009.

Robinson, Jenifer. *Deeper Than Reason: Emotion and Its Role in Literature, Music and Art.* Oxford: Oxford University Press, 2005.

Robinson, Jenifer. "On Being Moved by Architecture." *Journal of Aesthetics and Art Criticism* 70, no. 4 (2012): 337–53.

Robinson, Michael D., Edward R. Watkins, and Eddie-Harmon-Jones, eds. *Handbook of Cognition and Emotion.* New York: Guilford Publications, 2013.

Ropion, Dominique. *Aphorisms of a Perfumer.* Translated by Annie Tate-Harte. Paris: NEZlittérature, 2018.

Rosenberg, Karen. "Scent of 100 Women: Artist Anicka Yi on Her New Viral Feminism Campaign at the Kitchen." *Artspace*, March 12, 2015. https://www.artspace.com.

Ross, Stephanie. *What Gardens Mean.* Chicago: University of Chicago Press, 1998.

Rothenberg, David. *Survival of the Beautiful: Art, Science, and Evolution.* New York: Bloomsbury, 2011.

Rouby, Catherine, Benoist Schaal, Danièle Dubois, Rémi Gervais, and André Holley, eds. *Olfaction, Taste, and Cognition.* Cambridge: Cambridge University Press, 2002.

Roudnitska, Edmund. *L'Ésthétique en question: Introduction à une esthétique de l'odorat.* Paris: Presses Universitaires de France, 1977.

Royet, Jean-Pierre, Jane Plailly, Anne-Lise Saive, Alexandra Veyrac, and Chantal Delon-Martin. "The Impact of Expertise in Olfaction." *Frontiers in Psychology* 4, no. 928 (2013): 1–11. http://doi:3389/fpsyg.2-13.00928.

Ruckel, Terri Smith. "The Scent of a New World Novel: Translating the Olfactory Language of Faulkner and Garcia Márquez." PhD diss., Louisiana State University, 2006.

Ryan, Michael J. *A Taste for the Beautiful: The Evolution of Attraction.* Princeton: Princeton University Press, 2018.

Ryan, Zoë. *Bless, Boudicca and Sandra Backlund.* Chicago: Art Institute of Chicago, 2012.

Sacks, Oliver. *Hallucinations.* London: Picador, 2013.

Sacks, Oliver. *The Man Who Mistook His Wife for a Hat.* New York: Summit Books, 1985.

Sacks, Oliver. *On the Move: A Life.* New York: Alfred A. Knopf, 2015.

Saito, Yuriko. *The Aesthetics of Everyday Life.* Oxford: Oxford University Press, 2009.

Saito, Yuriko. *Aesthetics of the Familiar: Everyday Life and World-Making.* Oxford: Oxford University Press, 2017.

Saito, Yuriko. "Everyday Aesthetics in the Japanese Tradition." In Yuedi and Carter, *Aesthetics of Everyday Life*, 164.

Santayana, George. *The Sense of Beauty.* New York: Random House, 1955.

Sapolsky, Robert M. *Behave: The Biology of Humans at Our Best and Worst.* Harmondsworth, UK: Penguin, 2017.

Sarkissian, Marie-Anouch. "Trois oeuvres d'art olfactive mettant en pratique le concept d'interface Parfum-Musique." In Jaquet, *L'Art olfactif*, 155–69.

Scarantino, Andrea and Ronald de Sousa. "Emotion." In Edward N. Zalta, *The Stanford Encyclopedia of Philosophy* (Winter 2018 Edition), https://plato.stanford.edu/archives/win2018/entries/emotion/.

Schaal, Benoist. "Emerging Chemosensory Preferences: Another Playground for the Innate-Acquired Dichotomy in Human Cognition." In Zucco et al., *Olfactory Cognition*, 237–63.

Schaal, Benoist. "Prenatal and Postnatal Human Olfactory Development: Influences on Cognition and Behavior." In Doty, *Handbook of Olfaction*, 305–36.

Schafer, Edward H. *The Golden Peaches of Samarkand.* Berkeley: University of California Press, 1985.

Schellekens, Elisabeth. "The Aesthetic Value of Ideas." In *Philosophy and Conceptual Art*, edited by Peter Goldie and Elisabeth Schellekens, 71–91. Oxford: Oxford University Press, 2007.
Schellekens, Elisabeth. *Aesthetics and Morality*. London: Continuum, 2007.
Schellekens, Elisabeth, and Peter Goldie, eds. *The Aesthetic Mind*. Oxford: Oxford University Press, 2011.
Schouwenberg, Louise, ed. *Hella Jongerius: Misfit*. London: Phaidon Press, 2010.
Schumacher, Claude, ed. *Naturalism and Symbolism in European Theatre, 1850–1918*. Cambridge: Cambridge University Press, 1996.
Schusterman, Richard. "Somatic Style." *Journal of Aesthetics and Art Criticism* 69, no. 2 (2011): 147–59.
Scruton, Roger. *The Aesthetics of Architecture*. Princeton: Princeton University Press, 1979.
Scruton, Roger. *I Drink Therefore I Am*. London: Continuum, 2009.
Sebag-Montefiore, Clarissa. "Scent of Memory: Smell and Classical Music Prove an Intoxicating Combination." *The Guardian*, November 11, 2016. https://www.theguardian.com/.../2016/nov/11/scent-of-memory-smell-and-classical-m....
Sela, Lee, and Noam Sobel. "Human Olfaction: A Constant State of Change-Blindness." *Experimental Brain Research* 205, no. 1 (2010): 13–29. https://doi.org/10.1007/s00221-010-2348-6.
Sell, Charles S., and David H. Pybus, eds. *The Chemistry of Fragrances: From Perfumer to Consumer*. Cambridge: Royal Chemical Society Publishing, 2006.
Sezille, Caroline, Arnaud Fournel, Catherine Rouby, Fanny Rinck, and Moustafa Bensafi. "Hedonic Appreciation and Verbal Description of Pleasant and Unpleasant Odors in Untrained, Trainee Cooks, Flavorists, and Perfumers." *Frontiers in Psychology* 5, no. 12 (2014): 1–8. https://doi.3389/fpsyg.2014.00012.
Sharp, Chris. "Anicka Yi's Allegorical Bouquets." *Cura* 22 (2016). https://www.curamagazine.com/chris-sharp/.
Shaw, Barney. *The Smell of Fresh Rain: The Unexpected Pleasures of Our Most Elusive Sense*. London: Icon Books, 2017.
Shelley, James. "The Aesthetic." In *The Routledge Companion to Aesthetics*, edited by Berys Gaut and Dominic McIver Lopes, 246–56. London: Routledge, 2018.
Shelley, James. "Aesthetics and the World at Large." *British Journal of Aesthetics* 47, no. 2 (2007): 169–83.
Shelley, James. "The Concept of the Aesthetic." In *The Stanford Encyclopedia of Philosophy*, edited by Edward N. Zalta. Winter 2017, https://plato.stanford.edu/archives/win2017/entries/aesthetic-concept/.
Shepherd, Gordon M. *Neurogastronomy: How the Brain Creates Flavor and Why It Matters*. New York: Columbia University Press, 2012.
Shiner, Larry. "Art Scents: Perfume, Design and Olfactory Art." *British Journal of Aesthetics* 55, no. 3 (2015): 375–92.
Shiner, Larry. *The Invention of Art: A Cultural History*. Chicago: University of Chicago Press, 2001.
Shiner, Larry, and Yulia Kriskovets. "The Aesthetics of Smelly Art." *Journal of Aesthetics and Art Criticism* 65, no. 1 (2007): 273–86.
Shulman, David. "The Scent of Memory in Hindu South India." In Drobnick, *Smell Culture Reader*, 411–26.
Sibley, Frank. *Approaches to Aesthetics: Collected Papers on Philosophical Aesthetics*. Edited by J. Benson, B. Redfern, and J. Roxbee Cox. Oxford: Oxford University Press, 2001.

Skinner, Jeffery. "The Bookshelf of the God of Infinite Space." *Poetry Magazine* 207, no. 3 (2015): 230.
Smith, Barry C. "The Aesthetics of Wine." In *The Encyclopedia of Aesthetics*, 2nd ed., edited by Michael Kelly, 286–87. Oxford: Oxford University Press, 2014.
Smith, Barry C. "The Chemical Senses." In *Oxford Handbook of Philosophy of Perception*, edited by Mohan Matten, 314–52. Oxford: Oxford University Press, 2015.
Smith, Barry C., ed. *Questions of Taste: The Philosophy of Wine*. Oxford: Oxford University Press, 2007.
Smith, Mark M. *How Race Is Made: Slavery, Segregation, and the Senses*. Chapel Hill: University of North Carolina Press, 2006.
Smith, Mark M., ed. *Smell and History: A Reader*. Morgantown: West Virginia University Press, 2019.
Smith, Mark M. "Smelling the Past." In Smith, *Smell and History*, ix–xxv.
Smith, Michael Oakley, Alison Kubler, and Daphne Guiness, eds. *Art/Fashion in the 21st Century*. London: Thames & Hudson, 2013.
Smith, Murray. *Film, Art, and the Third Culture: A Naturalized Aesthetics of Film*. Oxford: Oxford University Press, 2017.
Sobchack, Vivian. *Carnal Thoughts: Embodiment and Moving Image Culture*. Berkeley: University of California Press, 2004.
Sobel, Noam. "Revisiting the Revisit: Added Evidence for a Social Chemosignal in Human Emotional Tears." *Cognition and Emotion* 31, no. 1 (2017): 151–57.
Soutter, Lucy. *Why Art Photography?* London: Routledge, 2013.
Spence, C., F. P. McGlone, B. Kettenmann, and G. Kobal. "Attention to Olfaction: A Psychophysical Investigation." *Experimental Brain Research* 138, no. 4 (2001): 432–37. https://doi.10.1007/s002210100713.
Spence, Charles. *Gastrophysics: The New Science of Eating*. New York: Penguin Random House, 2017.
Spence Charles, and Betina Piqueroa-Fiszman. *The Perfect Meal: The Multisensory Science of Food and Drink*. Hoboken, NJ: Wiley Blackwell, 2014.
Spence, Charles, Nancy M. Puccinelli, Dhruv Grewal, and Anne L. Roggeveen. "Store Atmospherics: A Multisensory Perspective." *Psychology and Marketing* 31, no. 7 (2014): 472–88. https://doi.org/10.1002/mar.20709.
Spinoza. *Ethics*. Edited and translated by G. H. R. Parkinson. Oxford: Oxford University Press, 2000.
Stamelman, Richard. *Perfume: Joy, Obsession, Scandal, Sin. A Cultural History of Fragrance from 1750 to the Present*. New York: Rizzoli International Publications, 2006.
Starr, G. Gabrielle. *Feeling Beauty: The Neuroscience of Aesthetic Experience*. Cambridge, MA: MIT Press, 2013.
Stecker, Robert. "Defining Art." In *The Oxford Handbook of Aesthetics*, edited by Jarrold Levinson, 136–54. Oxford: Oxford University Press, 2003.
Stern, Tiffany. "Taking Part: Actors and Audiences on the Stage at Blackfriars." In *Inside Shakespeare: Essays on the Blackfriars Stage*, edited by Paul Menzer, 35–53. Selinsgrove: Susquehanna University Press, 2006.
Stevenson, Richard J. "An Initial Evaluation of the Functions of Human Olfaction." *Chemical Senses* 35, no. 1 (2010): 3–20.
Stevenson, Richard J. "Olfactory Perception." In Zucco et al., *Olfactory Cognition*, 73–94.
Stevenson, Richard J., and Tuki Attuquayefio. "Human Olfactory Consciousness and Cognition: Its Unusual Features May Not Result from Unusual Function but

from Limited Neocortical Processing Resources." *Frontiers in Psychology* 4, no. 819 (2013): 1–12. https://doi.org/10.3389/fpsyg.2013.00819.
Stieglitz, Alfred. "Pictorial Photography." In *Photography: Essays and Images*, edited by Beaumont Newhall, 63–66. New York: Museum of Modern Art, 1980.
Stoddart, Michael. *Adam's Nose, and the Making of Humankind*. London: Imperial College Press, 2015.
Sulmont-Rosse, Claire. "Odor Naming Methodology: Correct Identification with Multiple-Choice versus Repeatable Identification in a Free Task." *Chemical Senses* 30, no. 1 (2005): 23–27. https://doi.org/10.1093/chemse/bjh252.
Summers, David. *The Judgment of Sense: Renaissance Naturalism and the Rise of Aesthetics*. Cambridge: Cambridge University Press, 1990.
Süskind, Patrick. *Perfume: The Story of a Murderer*. Translated by John E. Woods. New York: Random House, 1988.
Sweeney, Kevin W. *The Aesthetics of Food: The Philosophical Debate about What We Eat and Drink*. Lanham, MD: Rowman and Littlefield, 2017.
Sweeney, Kevin W. "Medium." In *The Routledge Companion to Philosophy and Film*, edited by Paisley Livingston and Carl Plantinga, 181–83. New York: Routledge, 2009.
Tafalla, Marta. "Anosmic Aesthetics." *Estetika: The Central European Journal of Aesthetics* 56, no. 1 (2013): 53–80.
Tafalla, Marta. "Smell and Anosmia in the Aesthetic Appreciation of Gardens." *Contemporary Aesthetics* 12 (2014). https://www.contempaesthetics.org/.
Tafalla, Marta. "A World without the Olfactory Dimension." *Anatomical Record* 2 (2013): 1287–96.
Thomas-Danguin, Thierry, Charlotte Sinding, Sébastien Romagny, Fouzia El Mountassir, Boriana Atanasova, Elodie Le Berre, Anne-Marie Le Bon, and Gérard Coureaud. "The Perception of Odor Objects in Everyday Life: A Review on the Processing of Odor Mixtures." *Frontiers in Psychology* 5, no. 504 (2014): 1–18. https://doi.10.3389/fpsyg.2014.00504.
Thurlkill, Mary. *Sacred Scents in Early Christianity and Islam*. Lanham, MD: Lexington Books, 2017.
Todd, Cain. *The Philosophy of Wine: A Case of Truth, Beauty and Intoxication*. Montreal: McGill-Queens University Press, 2010.
Totelin, Laurence. "Smell as Sign and Cure in Ancient Medicine." In Bradley, *Ancient Senses*, 17–29.
Troscianko, Emily T. "Cognitive Realism and Memory in Proust's Madeleine Episode." *Memory Studies* 6, no. 4 (2013): 437–56.
Turin, Luca. *The Secret of Scent: Adventures in Perfume and the Science of Smell*. New York: HarperCollins, 2006.
Turin, Luca, and Tania Sanchez. *Perfumes: The Guide*. New York: Viking Penguin, 2008.
Ullmann, Stephen. *Language and Style: Collected Papers*. Oxford: Basil Blackwell, 1966.
Ullmann, Stephen. *The Principles of Semantics*. Oxford: Oxford University Press, 1957.
Upitis, Alise, ed. *Anicka Yi, 6, 070, 430 K of Digital Spit*. Cambridge, MA: MIT List Visual Arts Center, 2015.
Van den Bergh, Omer, Richard J. Brown, Sibylle Pertersen, and Michael Witthöft. "Idiopathic Environmental Intolerance: A Comprehensive Model." *Clinical Psychological Science* 5, no. 3 (2017): 551–67. https://doi.10. 1177/2167702617693327.
Veramendi, Mily, Pilar Herencia, and Gastón Ares. "Perfume Odor Categorization." *Journal of Sensory Studies* 28, no. 1 (2013): 76–89. https://doi.org/10.1111/joss.12025.

Verbeek, Caro. "Inhaling Futurism: On the Use of Olfaction in Futurism and Olfactory (Re)constructions." In Henshaw et al., *Designing with Smell*, 201–3.

Verbeek, Caro. "In Search of Lost Scents: The Role of Olfaction in Futurism." In van Brakel et al., *Sense of Smell*, 59–69.

Verbeek, Caro. "Surreal Aromas: (Re) constructing the Volatile Heritage of Marcel Duchamp." In Ahlers et al., *Belle Haleine*, 115–25.

Vermetten, Eric, and J. Douglas Bremner. "Olfaction as a Traumatic Reminder in Posttraumatic Stress Disorder: Case Reports and Review." *Journal of Clinical Psychiatry* 64 (2003): 202–7.

Vernant, J.-P. "Introduction." In *The Gardens of Adonis: Spices in Greek Mythology*, edited by Marcel Detienne, vii–xl. London: Harvester Press, 1977.

Vogel, Wendy. "What's That Smell in the Kitchen? Art's Olfactory Turn." *Art in America*, April 8, 2015. https://www.artinamericamagazine.com/.../whats-that-smell-in-the-kitchen-arts-olfacto.

Walter, Frédéric. 2003. *Extraits de parfums: Une anthologie de Platon à Colette*. Paris: Editions du Regard.

Warren, Samantha, and Kathleen Riach. "Olfactory Control, Aroma Power and Organizational Smellscapes." In Henshaw et al., *Designing with Smell*, 148–56.

Wassmann, Jürg, and Katharina Stockhaus, eds. *Experiencing New Worlds*. New York: Berghahan Books, 2007.

Weil, Jennifer. "Album Mixes Music with Scent." *WWD*, May 7, 2015. https://www.wwd.com.

Weiss, Taio, Lavo Secundo, and Noam Sobel. "Human Olfaction: A Typical Yet Special Mammalian Olfactory System." In *The Olfactory System: From Odor Molecules to Motivational Behaviors*, edited by Kensaku Mori, 177–202. Tokyo: Springer, 2014.

Wernung, Markus, Jens Fleischhauer, and Hakan Beseoglu. "The Cognitive Accessibility of Synaesthetic Metaphors." In *Proceedings of the 28th Annual Conference of the Cognitive Science Society*, edited by R. Sun, 2365–78. Hillsdale, NJ: Lawrence Erlbaum Associates, 2006.

Wilson, Daniel. "The Japanese Tea Ceremony and Pancultural Definitions of Art." *Journal of Aesthetics and Art Criticism* 76, no. 1 (2018): 33–42.

Wilson, Donald A., Julie Chapuis, and Regina M. Sullivan. "Cortical Olfactory Anatomy and Physiology." In Doty, *Handbook of Olfaction*, 209–24.

Wilson, Donald A., and Richard J. Stevenson. *Learning to Smell: Olfactory Perception from Neurobiology to Behavior*. Baltimore: Johns Hopkins University Press, 2006.

Wilson, Donald A., Wenjin Xu, Benjamin Sadrian, Emmanuelle Courtio, Yaniv Cohen, and Dylan C. Barnes. "Cortical Odor Processing in Health and Disease." *Odor Memory and Perception: Progress in Brain Research* 208 (2014): 275–305. https://doi:10.1016 /B978-0-444-63350-7.00011-5.

Wilson, Edward O. *The Origins of Creativity*. New York: Norton, 2017.

Wisman, Arnaud, and Ilan Shrira. "The Smell of Death: Evidence That Putrescine Elicits Threat Management Mechanisms." *Frontiers in Psychology* 6, no. 1274 (2015): 1–11. https://doi.org/10.3389/fpsyg.2015.01274.

Wnuk, Ewelina, and Asifa Majid. "Revisiting the Limits of Language: The Odor Lexicon of Maniq." *Cognition* 131 (2014): 125–38. https://doi.10.1016/j.cognition.2013.12.008.

Wolterstorff, Nicholas. *Art Rethought: The Social Practices of Art*. Oxford: Oxford University Press, 2015.

Woolf, Virginia. *Flush: A Biography*. New York: Harcourt, Brace & World, 1933.

Wrangham, Richard W. *Catching Fire: How Cooking Made Us Human*. New York: Basic Books, 2010.
Xenophon. *Symposium*. Translated by A. J. Bowen. Warminster, UK: Aris & Phillips, 1998.
Yamagata, Hiroshi. "Pour une esthétique du troisième sens: L'odorat." *Revue d'esthétique* 21 (1992): 81–87.
Yeshurun, Yaara, and Noam Sobel. "An Odor Is Not Worth a Thousand Words: From Multidimensional Odors to Unidimensional Odor Objects." *Annual Review of Psychology* 61 (2010): 219–41. https://doi: 10.1146/annurev.psych.60.110707.163639.
Yuedi, Liu, and Curtis L. Carter, eds. *Aesthetics of Everyday Life: East and West*. Newcastle upon Tyne, UK: Cambridge Scholars Publishing, 2014.
Zangwill, Nick. "Are There Counterexamples to Aesthetic Theories of Art?" *Journal of Aesthetics and Art Criticism* 60, no. 2 (2002): 111–18.
Zangwill, Nick. *The Metaphysics of Beauty*. Ithaca: Cornell University Press, 2001.
Zarzo, Manuel, and David T. Stanton. "Understanding the Underlying Dimensions in Perfumers' Odor Perception of Space as a Basis for Developing Meaningful Odor Maps." *Attention, Perception and Psychophysics* 71, no. 2 (2009): 225–47. https://doi.org/10.3758/APP.71.2.225.
Zhao, Qingqing, and Chu-Ren Huang, "A Corpus-Based Study on Synesthetic Adjectives in Modern Chinese." In *Chinese Lexical Semantics: 16th Workshop*, edited by Qin Lu and Helena Hong Gao, 535–42. Cham, Switzerland: Springer International, 2016. https://www.springer.com/cn/book/9783319271934.
Zelano, C., M. Bensafi, J. Porter, J. Mainland, B. Johnson, E. Bremner, C. Telles, R. Khan, and N. Sobel. "Attentional Modulation in the Human Primary Olfactory Cortex." *Nature Neuroscience* 8, no. 1 (2005): 114–20. https://doi.10.1038/nn1368.
Zucco, Gesualdo M., Rachel S. Herz, and Benoist Schaal, eds. *Olfactory Cognition: From Perception and Memory to Environmental Odours and Neuroscience*. Amsterdam: John Benjamins Publishing, 2012.
Zumthor, Peter. *Atmospheres: Architectural Environments, Surrounding Objects*. Translated by Ian Galbraith. Basel: Birkhäusaer, 2006.

Index

accords in perfume design, 212–13, 212–13n15
Achaz, Grant, 299–300
adornment, 87–88, 89–90, 91, 92n28, 99, 100–1, 243, 265, 265n3, 273. *See also* bodies and embodiment; odors; perfumes and incenses
Adorno, Theodor, 133
   Dialectic of Enlightenment: Philosophical Fragments (with Horkheimer), 24
Adrià, Farran, 299–300
advertising, 186n12, 217, 257, 265, 274–75, 304–5
Aeschylus, 158
aesthetic theories of fine art, 211
aesthetics, 3, 4, 211n11, 316–18. *See also* olfactory aesthetics
   of atmosphere, 257–58, 278–79
   beauty and, 14–15, 71–72, 145–46, 196, 210–11, 211n12
   cognition and, 5, 14–15, 51–52, 71–73, 210, 212–13, 257–58n13
   critical practice and, 193
   of cuisine, 297–99, 300–2
   deodorization and, 93–94
   disinterestedness, 211n12
   emotion and, 68, 71–75
   engagement, 211n12
   of environment, 307–8, 309
   ethics and, 255, 258–59, 258n20
   everyday, 256–58, 291, 301–2
   evolution and, 84–86
   fine art and, 211–12, 217–18, 231–34, 248
   kodo and, 100–1, 202, 204–7, 205n17
   pluralism and, 257–58, 257n12, 257–58n13, 301
   sciences and, 30–31
affective aspects. *See* emotions
Africans and African Americans, 23–24, 46–47n40
Aftel, Mandy, 117
Ahlers, Lisa Anette, 148
AI (artificial intelligence), 44–45, 44–45n34, 247–48, 247–48n18

Albright-Knox, 189–90
Alexanderplatz station, *U-deur* installation at, 145–46
Alinsky, Saul, 46–47, 46–47n40
Allégret, Marina Jung, 212–13n15
ambient scenting, 1, 2–3, 282–83, 285, 286–90
Ambrose of Milan, 268
American Medical Association (AMA), 2, 28
amygdala, 39, 56–57, 69
ancient Greeks and perfumes, 89, 91n23, 266–67. *See also specific Greek philosophers*
Anglican Church, 91
animals
   pheromones and, 51–52
   preservation and, 309
   smell and, 1–2, 20n5, 23, 24–25, 28, 43, 52, 58, 82–84, 94, 182–83, 300–1
anosmia, 28–29, 29n49, 43–45, 54–55, 180, 213, 297, 310–11n17
anthropology, 5–6
antiaesthetic art, 145–46, 196
Antunes, Luis Rocha, 168–69n32
   The Multi-sensory Film Experience, 168–69
apocrine glands, 23–24
Aquinas, Thomas, 20–21
Arabic odor terms, 111
architecture
   aesthetics and, 255–56
   ambient scenting in, 285, 286–90
   as fine art, 234–35
   smell and, 283–85
Aristophanes: *Lysistrata,* 266–67
Aristotle
   on aesthetics, 204
   on cuisine, 20n8
   on emotions, 68–69, 70
   on perfume, 267, 267n11
   Poetics, 72
   on smell, 19–20, 23, 288–89, 292
   on theater, 163
AromaRama, 165–66
aromas. *See* foods and food scents; odors; perfumes and incenses
aromatherapy, 8, 289–92

art, 4. *See also* design and design art; fine art; specific arts
  aesthetic theories of, 211
  contextual/historical theories of, 218
  definitions and boundaries of, 229–31
  fine art vs., 14–15, 229–32, 234, 241, 243–44, 249
  hierarchy in, 248–49
  institutional theories of, 219, 219n45
  practice model of, 218, 218n40
artificial intelligence (AI), 44–45, 44–45n34, 247–48, 247–48n17
The Art of Scent (exhibition), 13, 88n6, 148, 210, 210n6, 210–11n9, 230–31, 236
art perfumes, 234–41, 247–48. *See also* perfumes and incenses
art photography, 235–36, 237
art quilts, 238
art scents. *See* olfactory arts
Ascherson, Neal, 279–80
Ashikaga (emperor), 202
Asians and Asian societies. *See also* kodo
  body odors and, 23–24
  deodorization in, 2, 98, 99, 101, 280
  everyday aesthetics in, 258n20
  perfumes and incenses in, 5, 98–101, 201–7, 270
Assoulen, Laurent: *Sentire*, 171–72
atmospheres, 186, 186n12, 204, 257–59, 278–79, 291. *See also* ambient scenting
attention, 63–64
Attuquayefio, Tuki, 57–58
Auschwitz, 134–36
Australian Art Quartet: "Scent of Memory," 171–72
Autocomplete: Synosmy (Georgsdorf), 179–80
avant-garde art, 163–64, 169–70, 194, 217, 232–34, 299–301
Axel, Richard, 37

Baars, Bernard, 40
Bacci, Francesca, 149–50, 183
Balinese performing arts, 158, 170
Balla, Giacomo, 246n9
Balti Kings (play), 160, 163–64
Banes, Sally, 159–60, 160n12
bans/ordinances on scents, 2, 93–94, 292, 292n48, 292n49
Barkat-Defrades, 111n18
Barnett, Cynthia, 307
Barthes, Roland, 147, 156
Battles, Ford Lewis, 269n19
Batty, Clare, 25–27, 42–43, 64–65, 160–61

Baudelaire, Charles, 6, 147, 153–54, 159–60, 170
  Correspondences, 118–20, 133–34, 215
  "The Perfume," 209, 209n1
Baumgarten, Alexander, 71–72
Beardsley, Monroe, 21–22, 192–93
  on odors, 29–30, 174, 212
beauty
  aesthetics and, 14–15, 71–72, 145–46, 196, 210–11, 211n12
  functional, 8, 255–58
  kodo and, 205
  smell and, 19, 20–21, 27
Becker, Calice, 247n11
Behrend, Jelena, 220
Belasco, David, 159–60
Belle Haleine: The Scent of Art (exhibition), 148, 152, 189–90
Berleant, Arnold, 211n12, 255, 308
Berlin, Germany, 279–80
Berlin Improvisers Orchestra: "Orchestral Whifftracks," 179
Berntsen, Dorthe, 128
Bhagavad Gita, 98
B-H (Bohlens-Haring) database, 105–6
Bible, 88–89, 88–89n10, 90, 152–53
Big Sur Backpacker (perfume), 216
bioculturalism, 79–81, 103, 315
birth and smell communication, 46
Bisch, Quentin, 171–72
Blodgett, Bonnie: *Remembering Smell*, 297
blogs on perfume, 59, 109–10n12, 149–50, 206–7, 216
Blumenthal, Hans, 299–300
bodies and embodiment
  architecture and, 283–84
  chemosignaling and, 45–47
  emotions and, 69, 70
  Kant on, 21
  odor of, 23–24, 23–24n27, 35–36, 52–53, 111, 134
  olfactory arts and, 182–83, 197
  Paradise and, 312
  perfume and, 8, 90, 248, 272
  smell and, 2, 17–18, 21, 25–26
  Yi on, 194–95
Bohlens-Haring (B-H) database, 105–6
Böhme, Gernot, 186n12, 257, 258, 274–75, 278–79, 283–84, 286
Boholano (people of the Philippines), 111–12n20
Boisson, Charles, 110–12
Boisvert, Raymond, 301
Bolton, Andrew, 246n9

Bororo (people of Brazil), 111–12
Bouasse, Sarah, 247n11
Bourgeois, Louise: *The Smell of the Feet,* 152
Bradford, Kevin, 286–87, 288–89
Bradley, Mark, 268
Brady, Emily, 196, 196n46, 278n1, 308, 309, 309n9
Brady, Michael S.: *Emotional Insight: The Epistemic Role of Emotional Experience,* 68–69
brains. *See also* neuroscience and neuroscientists
  damage to, 69–70
  evolution and, 83, 84
  odor treatments for, 161, 180n11
  plasticity of, 44–45, 60, 61–62, 63–64, 104, 213
Bremner, James Douglas, 128
Breuer, Marcel, 214
briefs for perfume production, 221–22
Brillat-Savarin, Jean Antheleme, 297–98
Broca, Paul, 82–83, 83n5
Brodland, Gene, 128n5
Brooks, Isis, 309, 309n9
Brouwer, Adriaen: *Interior of a Tavern,* 152–53
Brueghel, Jan: *Allegory of the Five Senses,* 152
Bruguera, Tanya: *Tatlin's Whisper #6 (Havana Version),* 217
Buck, Linda, 37
Buckle, Jane: *Clinical aromatherapy,* 290–91
Buddhism, 99
Burenhult, Niclas, 112–13
Burke, Edmund, 196
Burr, Chandler, 13, 13n1, 15, 148, 210, 210–11n9, 218–19

Cage, John, 187–88, 217, 232–33
Calkin, Robert, 210
Calvin, John, 91, 269
Calvino, Italo: "The Name, the Nose," 4, 35–36, 41, 52, 55, 315
Carlson, Allen, 21, 255–57, 255–56n1, 256–57n8, 258, 308
Carné, Violaine de: *The Scents of the Soul,* 148–49, 161–62, 163–65, 169–70
Carra, Carlo, 154, 188
Carroll, Noël, 167–68, 169–70, 218, 218n39, 230, 230n1
Castellucci, Romeo: *On the Concept of the Face of the Son of God,* 162–63, 164
catharsis, 72, 163
Cather, Willa: *Death Comes for the Archbishop,* 120

ceramics, 247n14
chado, 100–1, 202, 205, 205n17, 206n23, 257n12
Chadwick, Edwin, 93
Chalayan, Hussein, 246
characters and smell in literature, 122–25
chemosignals, 45–47, 51–52
Chicago Art Institute, 229
China, 99–100, 100n13, 110, 258n20, 270
Choux, Patricia, 216n34
Christianity, 89–91, 265, 267, 268, 311–13
Chu, Simon, 129n7
Cinnabun Company, 302–3
cities. *See also* deodorization; smellwalks
  smell and, 279–81
  urban design and, 281–83
Classen, Constance, 87–88n3
classification of smells, 104–6
Clement of Alexandria, 268, 268n15
Clements, Ashley, 89
clinical aromatherapy, 290–91
cluster theories, 231–32
coffee and coffee aromas, 1, 2, 37–38, 93–94, 197, 305
cognition, 315
  aesthetics and, 5, 14–15, 51–52, 71–73, 210, 212–13, 257–58n13
  ambient scenting and, 287
  emotions and, 68–73, 132–33
  odor terms and, 109–10
  Proust and, 132–33
  smell and, 4, 24–25, 26–27, 29–30, 39–40, 41, 43–44, 45–47, 54–66
  taste and, 297–99
cognitive affect, 164–65
cognitive penetration, 56, 63–64
cognitive theory, 70
Coles, Alex, 245–46
Collingwood, R. G., 73, 132
colonialism, 1, 99, 197
comics, 155–56, 249
concentration camps, 134–36
conceptual art. *See* fine art; olfactory arts; scent art; *specific artists*
Condillac, 20–21
conservation of olfactory art, 191–93
contemporary art, 210–11, 221
contextual/historical theories of fine art, 218–19, 218n41, 231–34
cooking, 83–84. *See also* cuisine
Cooper, David: *A Philosophy of Gardens,* 311, 311n18
Cooper-Hewitt, 189–90
Corbin, Alain, 93, 269

Cox, Christophe, 188
critics and criticism, 59, 193–96, 210–11, 257–58, 317n2
Crowe, Stephen, 179
cuisine, 316–17. *See also* flavors; taste
    aesthetics of, 297–99, 300–2
    Aristotle on, 20n8
    as fine art, 232, 299–301
    olfactory arts and, 160

Damasio, Antonio: *Descartes' Error: Emotion, Reason, and the Human Brain,* 69–70
Danto, Arthur, 219n45
    What Art Is, 230–31
Darwin, Charles, 58, 82–86, 83n5
    on smell, 5, 28, 80–81, 101, 127–28, 315
Dary, Michael, 280–81
David, Sophie, 111n18
Davies, David, 145, 218n40, 218–19, 233–34
Davies, Stephen, 87, 232–33, 241, 257n12, 265n3
Debussy, Claude, 170–71
deception and smell, 25, 268, 271–76
de Cupere, Peter, 74
    Black Beauty Smell Happening, 185
    The Deflowering, 184
    Invisible Scent Paintings, 155
    Olfactiano, 22, 178
    "Olfactory Art Manifest," 190
    Soap Paintings, 154–55
    Tree Virus, 38, 149–50, 194
deficient smell, 24. *See also* cognition
Deigh, John, 68–69
Dejean, Antoine, 269
Delaunay, Sonia, 246n9
Delon-Martin, Chantal, 61–62, 64
Delvoye, Wim: *Cloaca Professional,* 149–50, 194
Democritus, 305
deodorization
    aromatherapy and, 289–90
    in Asian societies, 2, 98, 99, 101, 280
    in cities, 280
    history of, 87–88, 261
    perfume bans and, 2, 93–94, 292, 292n48, 292n49
    smell and, 8, 289–90
    in Western societies, 2, 91–94, 103, 281
De Profundis (perfume), 216
Descartes, René, 20–21
design and design art, 1, 7–8, 316. *See also* architecture
    aesthetics of, 8, 255–58
    critics and, 58–59
    cuisine and, 300
    emotions and, 74
    ethics of, 258
    as fine art, 217, 218, 232, 233–34
    free art vs., 243–49
    perfume and, 214, 218–19, 223–24, 231, 240, 241, 243–49
    urban planning, 281–83
Desrochers, Debra, 286–87, 288–89
Detienne, Marcel, 266
Devlin-Glass, Frances, 123
Dewey, John, 73
DIA Foundation, 186
Diaconu, Mădălina, 178–79n5, 283–84
Di Benedetto, Stephen, 160–61
Dickie, George, 218n40, 219n45
Dickinson, Emily, 203n9, 209
diffusion of odors
    in architecture, 285
    fast food and, 302–3
    in hotels, 289
    in olfactory art, 159–60, 161–63, 165–66, 171–72, 173, 178–79, 182, 186
Digonnet, Rémi, 117–18n2
Diller, Elizabeth, 283
dispensable smell, 28
disrepute and smell, 23–24n27, 23
dogs, 23, 43, 83–84, 121–22, 265, 318
Donne, John: "The Perfume," 117–18
Dostoyevsky: *The Brothers Karamazov,* 90
Downes, John Joseph, 129n7
Dream Air (perfume company), 191n33, 222, 247
Drobnick, Jim, 59, 278n1
    on Gauguin, 153–54, 153–54n4
    interpretive framework of, 195–96
    on olfactory art, 148, 149–50, 182–83, 184, 186–87, 188
    "Reveries, Assaults, and Evaporating Presences," 189
    on Tolaas, 279
    on Ursitti, 190n27
D.S. & Durga (perfumers), 147–48
Dubois, Danièle, 106–7
Duccio, 152–53
Duchamp, Marcel, 188, 189, 217, 232–33
    Belle Haleine, Eau de Voilette, 189n24
Dugan, Holly: *The Ephemeral History of Perfume: Scent and Sense in Early Modern England,* 92, 92n29
Dumas, Alexander, 269
Dutton, Dennis, 1, 21, 29–30, 215

Eau d'Issy (perfume), 13
Edwards, David, 305
Edwards, Michael: "Fragrance Wheel,"
    105–6, 119–20
Egyptians and perfumes, 88–89, 309–10
electro-olfactogram, 37
Ellena, Jean-Claude, 210, 214, 215, 216, 247
    Un jardin en méditerranée, 223–24, 233–34
Ellsworth, Angela: *Actual Odor*, 185
embodiment. *See* bodies and embodiment
emotions
    aesthetics and, 68, 71–75, 138, 211–12
    cognitive affect and, 164–65, 287
    film scores and, 167–68
    intelligence of, 68–71
    memory and, 127–28
    perfume and, 275
    smell and, 55, 56–63, 68, 71, 132
engineering, 243–44, 283
environment, 316–17
    aesthetics of, 307–8, 309
    ethics of, 258, 258n20, 273–74, 291, 309
Ephraim Syrus: *Hymns on Paradise,* 311–12
Epicurus, 309–10
epithelium, 37–39
Erasmus, 91
Erwine, Barbara, 283
Essence of Perfume (Sanskrit treatise), 98
Etat Libre d'Orange (perfume company), 216
ethics, 8, 316–17
    aesthetics and, 255, 258–59, 258n20
    ambient scenting and, 282–83, 285, 286–90
    Aristotle on, 68–69
    aromatherapy and, 290–92
    environmental, 258, 258n20, 273–74, 291, 309
    perfume and, 272, 273–74
Euripides: *Hippolytus,* 89
evolution, 5
    bioculturalism and, 80
    cooking and, 83–84
    olfactory aesthetics and, 138–39
    paleocortex and, 39, 57–58
    pheromones and, 51–52
    smell and, 5, 40, 41, 57–58, 79, 82–86
exhaust systems. *See* diffusion of odors
exhibition of olfactory art, 191–93
expertise. *See* perfumers and flavorists
expressivity of perfume and wine, 214, 215–17, 224, 238–39

Fanuel, Laurence, 161n19
farts and fecal odors, 46–47, 162–63, 164, 168–69, 196

fashion design, 7–8, 214, 230–31, 243, 246, 246n9
fast food, 281, 302–5
Faulkner, William: *The Sound and the Fury,* 123, 129
Feagin, Susan, 160–61
fear, 24, 44, 45–46, 47
Ficino, Marsilio, 91, 91n23, 305
film, 147, 153–54, 165–70
fine art, 316
    aesthetic theories of, 211–17
    art vs., 229–32, 234, 241, 243–44, 249
    contextual/historical theories of, 218–19, 218n41
    composite theories of, 231–34
    cuisine as, 232, 299–301
    gardens as, 310–11
    institutions and, 211, 218–19, 218n40, 219n45, 220–21, 223–24, 232–34, 237–38, 240–41, 300
    intention in, 145–46, 186–87, 215–16, 218n43, 218–21, 299–300
    kodo and, 203, 205–7
    local analogies approach to, 234–35
    olfactory arts and, 145–46, 182–83, 317
    practice model for, 218–19
    perfumes as, 198, 206–7, 209–11, 216–24, 230–31, 232–34, 235–41, 243–49
    pluralism and, 206, 210–11, 231–32, 249, 301
    postmedium turn of, 145, 189–90, 210–11, 218
Fläck-Wood (ventilation firm), 173, 178–79
flavorists. *See* perfumers and flavorists
flavors, 8. *See also* cuisine; foods and food scents; taste
    expression through, 215
    fast food and, 303–4
    manufacture of, 247–48n17
    smell and, 28, 38, 83–84, 129–30, 296–99
Fleischer, Mary, 159–60
Fliess, Wilhelm, 28n42
fMRI imaging, 37, 40, 42, 56, 62
foods and food scents. *See also* cuisine; flavors; taste
    classification of, 8, 105
    ethics of, 302–5
    manufacture of, 2–3, 247–48n17
    perception of, 168–69, 258–59, 298–99
formalism, 194–96, 210–12, 217
Forsey, Jane, 21, 243, 256–57n8
Foster, Hal, 145–46
fourth wall, 162, 163, 165
fragrances. *See* odors; perfumes and incenses

Fragrance Wheel, 105–6
frame problem, 69
Fran Bodkin, 280–81
Frasnelli, Johannes, 61–62
free art, 244–45, 248, 249. *See also* fine art
Freeland, Cynthia, 168–69, 168–69n31
Freeman, Eleanor, 296
Freud, Sigmund, 28, 28n42, 82–83, 133
Frost, Laura, 129
Futurism, 154, 187–88

Gagosian Gallery, 182
gardens, 309–11
Gardner, Howard, 28
Gauguin, Paul, 153–54, 153–54n4, 273
Gefter, Amanda, 173–74
Gell, Alfred, 275
Gell, Ernest, 112
Gelstein, S., 46n39
gender and chemosignaling, 46–47
gene-culture coevolution, 80
Georgsdorf, Wolfgang, 7
    Smeller 2.0 and, 7, 178–80, 191
germ theory of disease, 93
Gesamtkunstwerk ideal, 158, 172, 299
Gilbert, Avery, 43, 130–31, 165–66, 173–74, 210–11n9
Gill, A. A., 281–82
Giotto, 152–53
Girdle of Hara (Sanskrit treatise), 98
Givaudan, 223–24
Goeltzenleuchter, Brian, 182–83
    Institutional Wellbeing, 191
    Olfactory Memoirs, 147–48
    Sillage, 185
Gottfried, Jay, 37, 39–40, 42, 44–45, 55, 63, 64, 79–80
Gottlieb, Ann, 274
Grande Musée du Parfum, 239, 239n17
graphic novels, 155–56, 249
The Great Wall (film), 165
Green, Deborah, 88–89, 269
Green Aria: A Scent Opera, 1, 6, 22, 167–68, 172–74, 178–79, 180, 218–19, 315–16
Greenberg, Clement, 155, 194–96
Grosjman, Sophia, 214, 274
Guyer, Paul: *History of Modern Aesthetics*, 71–72, 73

habituation, 56, 92, 191–92, 302, 307–8, 317
Halifax, Nova Scotia, 93–94, 292
Harmon-Jones, Eddie, 69
Harvey, Susan, 90, 311–12

Haug, Helgard: *U-deur,* 145–47, 185, 196, 279–80
Hawaiian odor terms, 111
Hazlitt, William, 152–53
headspace technology, 182, 215, 215n26
healing. *See* medical uses of perfumes and incenses
Heaney, Seamus: "Digging," 120
hearing
    bias toward, 2, 139–40
    emotions and, 70–71
    historical perspectives on, 20–21
    neuroscience of, 39, 39–40n10, 168–69
    perception models based on, 29–30
    sound art and, 187
hedonic responses, 57–64, 71, 74, 75, 106–7
Hegel, G. W. F.
    on aesthetics, 71–72, 73
    on art, 73n13
    on odors, 21, 192–93
    on smell, 1, 21, 133
Heldke, Lisa, 301
Henshaw, Victoria, 278, 278n2, 280, 281–83, 292
Hepburn, Ronald, 255
Hermès, 223–24
Hershey, Pennsylvania, 280, 280n8
Herz, Rachel, 38n3, 55, 56–57, 68n1, 127–28n2, 271, 272–73, 274
Hinduism, 98–99, 270
hippocampus, 39
Hirst, Damien: *A Thousand Years,* 189
HLA/MHC (human leukocyte antigen / major histocompatibility complex), 52, 84–85, 272–73
Hobbes, 23
Holl, Steven, 283–84
Höller, Carsten: *Double Club* (restaurant), 300, 300n15
Holley, André, 60, 107, 216–17
Horace, 73
Horkheimer, Max: *Dialectic of Enlightenment: Philosophical Fragments* (with Adorno), 24
Hornetz, Christoph, 222
Horowitz, Alexandra, 265
    Being a Dog: Following the Dog into a World of Smell, 317–18
hotels and ambient scents, 1, 2–3, 286, 289
household products, 2–3
Howes, David, 28n42, 87–88n3, 111–12, 132
Hsu, Husan L., 191–92n35
Huang, Chu-Ren, 110, 111–12

Huber, Carlos: "Scent of Memory," 171–72
Hugo Boss Prize, 189–90
Hume, David, 58–59, 107
Humiecki & Graef (perfume company), 222
Hummel, Thomas, 180
Huxley, Aldous: *Brave New World,* 178
Huysmans, J.-K.: *Against Nature,* 8, 153, 209–10, 209–10n3, 261–62
hybridity
  in design art, 245
  in olfactory art, 195–96, 218–19, 236, 317
  in sound art, 188

identifying odors, 5–6, 15, 43n23, 43–45, 44n27, 47, 55–56, 57, 58–63, 64–65, 87, 104–8
identity, 8, 182–83, 192–93, 274–75, 292
idiopathic environmental intolerance (IEI), 291–92, 291–92n42
immigrants and odors, 23–24, 281–82
imperialism, 101
improvisational art, 22
incenses. *See* perfumes and incenses
India, 98–99, 101, 270, 307
indole, 117
insects, 51
installation art, 184, 217, 219n47, 220–21. *See also* fine art; *specific installations*
Institute for Art and Olfaction, 239
institutions and fine art, 211, 218–19, 218n40, 219n45, 220–21, 223–24, 232–34, 237–38, 240–41, 300
intention in fine art, 145–46, 186–87, 215–16, 218n43, 218–21, 299–300
International Flavors and Fragrances, 223–24
Irvin, Sherri, 192–93, 218n43, 256n4, 257–58n13
Iseminger, Gary, 217, 218n40, 232, 233–34, 241
Islam, 111, 270, 284–85, 309–10, 311–13
Iwasaki, Yoko, 193–94, 195–96, 204–5

Jacobs, Tim, 39–40
J'adore (perfume), 274–75
Jahai language, 112–14
James, William, 56
Japan, 100–1, 201, 270. *See also* kodo
Japanese Ministry of Environment, 308–9
Jaquet, Chantal
  on Debussy, 171
  on Gauguin, 153–54
  on historical perspectives on smell, 20, 270
  on kodo, 205, 205n17, 206–7
  on Nietzsche, 17n3
  on olfactory art, 192–93

on perfume, 148
Philosophie de l'odorat, 3
jasmine, 117
Jellinek, J. Stephan, 105–6n3, 106, 210
Jellinek, Peter: "Odor Effects Diagram," 105–6, 119–20
Jenner, Mark S. R., 87–88n4
Jews, 24, 89–90n22
Jha, Radhika: *Smell,* 124–25
John Chrysostom, 268, 272
Johnstone, Sam, 179n6
Jones, Caroline
  Eyesight Alone: Clement Greenberg's Modernism and the Bureaucratization of the Senses, 194
  Yi and, 194–96
Jones, Cat: *Scent of Sydney,* 280–81
Jongerius, Hella, 245n7
Jonson, Ben, 158–59
Joyce, James, 6
  Ulysses, 122–23, 129, 133–34
Judaism, 90n18, 90n22, 265, 269, 311–12
Juniper Ridge (perfume company), 216

Kac, Eduardo: *Aromapoetry,* 147–48
Kaeppler, Kathrin, 105
Kames, Henry Home, 20–21
Kant, Immanuel
  on aesthetics, 14–15, 14–15n2, 21, 71–72, 73, 204, 255, 298, 315
  on gardens, 310–11
  on smell, 1, 5, 21, 23, 25, 26–27, 28, 58, 124, 133
  on the sublime, 196
Kaprow, Alan, 194
Kapsiki (people of Cameroon), 111–12
Kaufmann, Walter, 38–39n2
Kawakubo, Rei, 246
Keller, Andreas, 25–26, 58, 63–64, 71
  Philosophy of Olfactory Perception, 42–43
Kessler, David, 303–4
Kiechle, Melanie A., 93n36
Kienholz, Edward: *The State Hospital,* 189, 189n24
Kiki (perfume), 219–21
Kim Jeong A: *Before the Rain,* 186
Kincaid, Jamaica: *Autobiography of My Mother,* 124
King, J. B., 290
Kirk, Lisa, 216n34, 220–21, 220n46, 223–24, 236, 240–41, 245–46
Kissina, Julia, 179
Koch, Robert, 93

kodo, 7, 100–1, 201–7, 205n17, 239–40, 249, 257–58, 276, 316
Koran, 312. *See also* Islam
Korsmeyer, Carolyn, 19, 244–45, 248
   Making Sense of Taste, 230–31, 300–2
   Savouring Disgust, 196, 216n33
   on Spence, 296n1
   on taste, 297
Köster, Peter, 57, 58, 79–80, 105, 187, 262, 279
Kraft, Philip, 247
kumikoh, 202, 203, 205
Kurkdjian, Francis, 247

LaBelle, Brandon: *Background Noise: Perspectives on Sound Art*, 187–88
Lacher, Allison, 186n13
Lady and the Unicorn tapestries, 152
Laib, Wolfgang, 186
   Milkstone, 191–92
Laing, David G., 60
Lang, Ulrich, 216n34, 220, 245–46
language
   chemosignaling and, 46
   odor terms and, 108–14, 109–10n11
   olfactory experts and, 60–61, 63, 106
   smell and, 19, 55–56, 64, 65–66, 79, 95, 101, 103–14, 121–22, 140
   of smell in literature, 117–25, 133–34, 135–36, 139–40
Larmarque, Peter, 218n40
Laudamiel, Christophe
   art and design practices of, 247
   collectible scent art and, 192
   Dream Air and, 191n33
   Green Aria: A Scent Opera, 172–73, 178–79
   Liberté, Egalité, Fragrancité, 191, 286
   as olfactory artist, 191
   Over 21, 1, 238–39
   production of *Trust* and, 222
   Scent Parabols, 237–38, 240–41, 246
   Scent Squares, 191, 192, 240–41, 246
   Süskind's *Perfume: The Story of a Murderer* and, 167–70
laundry hanging, 256
Lavoisier, Antoine-Laurent, 93
lawns, 258
Lazarus, 152–53
learning smells, 44–45, 54–55, 64, 104–8, 127–28, 180. *See also* perfumers and flavorists
Le Breton, David, 275, 276
Leddy, Tom, 275

The Extraordinary in the Ordinary, 256
Lee, Barry, 216n33
Lengyel, Olga: *Five Chimneys: The Story of Auschwitz*, 134
Lennon, J. Michael, 124n21
Les Métaboles (choral group), 171–72
Let the Fancy (exhibition), 186n13
Levi, Michael, 166–67
Levi, Primo, 134–35
Levinson, Jerrold, 218, 218–19n43, 257n12
Le Whiff, 305
Licht, Alan, 188, 188n20
limbic system, 39, 40
Lingua: or, the Combat of the Tongue and the Five Senses for Superiority (play), 158–59
linguistics, 5–6
L'Interdit (perfume), 13
literature
   kodo and, 203, 204–5
   smell in, 117–25, 129–36, 139, 147–48, 168–69, 261–64
Livermore, Andrew, 60
local analogies, 234–40
lock and key theory of smell, 38–39, 38–39n4
Locke, John, 20–21
Lopes, Dominic, 80, 196–97n49, 206n24, 218n40, 218–19, 234–35, 249, 257–58
López-Mascaraque, Laura, 278n1
Los Angeles Institute for Art and Olfaction, 167
Lotze, Martin, 62–63
Low, Kelvin E. Y., 258n20
Lucretius, 268, 269, 275
Lutens, Serge, 216
Luther, Martin, 91
Lycan, William, 24–27, 42–43, 64–65
Lynn, Gwenn-Aël: *Audiofactory Creolization*, 184–85

Mace, John H., 129n7
Macia, Oswaldo
   The Library of Cynicism, 190
   "Manifesto for Olfactory-Acoustic Sculpture," 190
   Transition, 184, 194
MacIntyre, Alasdair, 218n40
Magnus, Albertus, 309–10
Mag Uidhir, Christy, 249, 257
Mahabharata, 98
Mailer, Norman: *An American Dream*, 124
Majid, Asifa, 112–13
Malick, Terrence: *The New World*, 166–67

Malle, Frédéric, 222–23
manipulation through smell, 271, 272, 286–88, 304–5
Maniq language, 112–13
Manzoni, Piero: *Merda d'Artista,* 156
Marcus, P. D., 249, 257
Margiella, Martin, 246
Marketou, Jenny, 279
  Smell It, 185
Marrakech Medina, 2
Marriot hotels, 286, 289
Marvick, Andrew, 154, 154n7, 156
Matthew, Stewart, 172, 178–79
McDonald's, 303–4, 303n20
McGann, John P., 82–83
McHugh, James, 270
  Sandalwood and Carrion: Smell in Indian Religion and Culture, 98–99, 101
McLean, Kate, 156, 279, 280, 318
McQueen, Alexander, 246
MCS. *See* multiple chemical sensitivity
media for artworks, 218–19. *See also* fine art; perfumes and incenses; *specific media*
medical uses of perfumes and incenses, 89–90, 91, 92–94, 99, 290–91
Meighan, Michael, 281–82
  Glasgow Smells: A Nostalgic Tour of the City, 280
Meireles, Cildo: *Volátil,* 189
memory
  accuracy of, 127–28n2
  Nkanga on, 197
  odor treatments for, 161, 180n11
  perfume and, 213
  Proust and, 129–36, 130–31n11
  voluntary vs. involuntary, 127–29, 130–31, 133–34
Menardo, Annick, 212–13
Mercier, Louis-Sébastien, 92
Merleau-Ponty, Maurice, 283–84, 311n18
Merrick, Christine, 40
Métailié, Georges, 99
metaphor, 109–10, 111–12, 117–18, 117–18n2, 119–21
Middle East and perfume, 270
Miller, Mara: *The Garden as Art,* 310–11, 310–11n14
mindfulness, 195–96, 203, 204
Minneapolis Theater in the Round: *Treasure Island,* 160
Minta, Steven C., 307–8n3
misfit theory of olfaction, 57

Montaigne, Michel de, 91, 159n4, 268, 269, 272
Montesquiou, Robert de, 209–10n3
morals. *See* ethics
Moreau, Gustave, 152–53
Morita, Kiyoko, 202, 203, 203n9
Morris, Matt, 186, 186n13
Morrison, Toni: *Sula,* 124
Morrison, Rachel: *Smelling the Books,* 185
Morse, Margaret, 145n1
Motte-Florac, Elizabeth, 111n18
mouth capture, 296
Mueller, Friedrich, 105
Muhammad, 270
Müller-Alsbach, Annja, 148
multiple chemical sensitivity (MCS), 8, 289–90, 291–92, 292n45
music, 6
  evolution and, 80, 82
  expertise and, 59, 61–63, 64
  film and, 167–68
  fine art and, 14, 15, 22, 234–35
  olfactory arts and, 147, 170–74
  perfume and, 212–13, 214, 236–37
  sound art and, 187–88
Muslims. *See* Islam

Nagel, Thomas, 84–85
Nalls, Gayil: *World Sensorium,* 185
Nanay, Bence, 24–25n29
nasopharynx, 38
natural selection. *See* evolution
nature. *See* environment
needs versus wants, 281
Negrin, Llewellyn, 248
Nejinsky, Rosina, 118n5
neocortex, 39–40, 57–58, 84
neonates, 46
Neto, Ernesto, 184, 196
  Mother Body Densities, 194
Neuhaus, Max, 187–88n19
  Times Square, 187–88
neuroscience and neuroscientists, 30–31, 37
  literature and, 147
  on olfactory experts, 58–63, 138, 213
  Proust and, 129
  of smell and odors, 37–40, 54–55
  on smell's powers of detection, discrimination, and learning, 43–45
  Smeller 2.0 and, 180n11
  on smell's cognitive limitations, 54–58, 138
  on taste, 168–69
The New World (film), 166–67

New York Museum of Arts and Design, 13
Nietzsche, 2, 15, 17–18, 73, 88–89, 300–1
Nkanga, Otobong: *Anamnesis,* 1, 197
Noble, Ann C.: "Wine Aroma Wheel," 106, 107
nose. *See* odors; orthonasal smell; retronasal smell; smell
novels and novelistic devices, 113–14, 120–25, 179
nudge theory, 304–5
Nussbaum, Martha, 70

objects and objectivity
　aesthetics and, 211–13, 255–56n1, 256–57
　odor object theory, 41–43
　of odors, 38–39, 41–43, 47, 56, 106–7, 213
　perception models for, 25–27
Obringer, Frédéric, 100
Odor Effects Diagram, 105–6
odors, 315–16. *See also* ambient scenting; deodorization; diffusion of odors; foods and food scents; olfactory aesthetics; olfactory arts; perfumes and incenses; smell
　as artistic medium, 21–22
　of bodies, 23–24, 23–24n27, 35–36, 52–53, 111, 134
　in Calvino's "The Name, the Nose," 35–36
　classification of, 104–6
　disreputability and, 23
　fine art and, 13–15
　health food and, 304
　identification of, 5–6, 15, 43n23, 43–45, 44n27, 47, 55–56, 57, 58–63, 64–65, 87, 104–8
　in literature, 117–25, 129–36
　memory and, 127–29, 127–28n2
　in nature, 307–8
　object theory of, 38–39, 41–43, 47, 56, 106–7, 213
　ordinances on, 2, 8, 290–92
　sanitation and, 91–94
OFC (orbitofrontal cortex), 39–40, 39–40n10, 42, 44–45, 46, 61–62
Olafsson, Jonas, 55, 61, 63, 79–80
Olfactiano (scent piano), 22, 184
olfaction. *See* odors; smell
olfactory aesthetics, 4, 6, 315–16
　emotions and, 138, 211–12
　everyday, 301–2
　evolution and, 138–39
　flavor perception and, 298
　gardens and, 310–11
　of nature, 308

scent art and, 182–83, 186–87, 196–98, 261
smell and, 13–15, 29–30, 54, 65–66, 71–72, 74, 94n37, 138–40, 300–1
olfactory arts, 7, 9, 95, 315–16
　aesthetics of, 74, 145–47, 163–65, 174
　definitions and criteria for, 145–51, 183–86
　Dutton on, 29–30
　experts and critics of, 58–66, 162
　film and, 165–70
　fine art and, 15, 22, 236, 239, 243
　meaning in, 27
　music and, 170–74
　in non-Western cultures, 158
　painting and, 152–56, 210–11
　perfumes and, 13, 88n6, 147–49, 158, 159–60, 184–85, 186, 191, 247
　Smeller 2.0 and, 7, 178–80
　theater and, 158–65
　Ursitti and, 52–53
olfactory bulbs, 38–39, 41, 42, 82–83
olfactory predella, 154
Ongee (people of Andaman Islands), 111–12
ontology and olfactory art, 192–93
oral referral, 296–97
orbitofrontal cortex (OFC), 39–40, 39–40n10, 42, 44–45, 46, 61–62
orthonasal smell, 8, 316–17
　evolution and, 83–84, 85–86
　food choices and, 305
　neuroanatomy of, 38
　taste and, 129–30, 215–16, 296, 297, 298, 299–300, 305
Orwell, George, 23–24
Osman, Ashraf, 149–50
Osmodrama (exhibition), 7, 178–80
Otherness and smell, 23
Out of Sight! Art of the Senses (exhibition), 189–90
outsider artists, 218–19n44
Oz, Mehmet, 290–91

painting
　fine art and, 234–35
　olfactory arts and, 152–56, 210–11
Pairet, Paul, 299
paleocortex, 39–40, 57–58
Pallasmaa, Juhani, 283–84
Panyembrama (Balinese dance), 158
Paquet, Dominique, 162–63, 168–69
paradigm scenarios, 70
Paradise, 258–59, 270, 308–9, 311–13
Pardo, Carmen, 188, 188n20
Parr, Debora Riley, 147–48n6, 317n1

Parsons, Glenn, 21, 243–44, 247–48, 256–57n8, 258
  Functional Beauty, 255–57, 255–56n1
Participant Inc., 220–21
participatory art, 145–47, 182, 185, 217, 220–21, 299
Pasteur, Louis, 93
Patel, Aniruddh, 80
Pelosi, Paolo, 51
perception, multimodal, 60, 63–65, 84, 168–69, 168–69n32, 278, 298–99. *See also* flavors; odors; *specific senses*
performance art, 158, 185, 206, 217, 220–21
perfumers and flavorists
  art vs. design status of, 245n7, 247, 247n11
  creativity and, 213–14, 215, 223–24, 247n11
  expertise of, 59–63, 65–66, 104, 107, 113–14, 138
  in literature, 261–63
  production and, 221–24
perfumes and incenses, 1–2, 316–17. *See also* kodo
  aesthetics and, 22, 27, 74, 93–95, 209, 211–17, 244–45, 269, 298
  Aristotle on, 20, 20n8
  in Asian societies, 5, 98–101, 201–7
  bans/ordinances on, 2, 93–94, 292, 292n48, 292n49
  classification and labeling of, 105–8, 109–10n12
  deodorization and, 91–94
  as design, 214, 218–19, 223–24, 231, 240, 241, 243–49
  ethics of, 8, 258–59
  expressivity of, 214, 215–17, 224, 238–39
  as fine art, 7–8, 13–14, 15, 198, 206–7, 209–11, 216–24, 230–31, 232–34, 235–41, 243–49
  history of, 87–91, 92n28, 94–95
  in literature, 118–20, 134, 261–64
  meanings of, 265, 271–76
  medical uses of, 89–90, 91, 92–94, 99, 290–91
  from *mitti attar,* 307
  olfactory arts and, 13, 88n6, 147–49, 158, 159–60, 184–85, 186, 191, 247
  production of, 221–24
Perfumes of China: The Culture of Incense in the Imperial Period (exhibition), 100
Perry, Katy, 171–72
petrichor, 307
pheromones, 45, 47, 51–53, 83, 85–86
Philyra (perfume formula algorithm), 247–48, 247–48n18

photography, 234–36, 237, 239–41, 310–11
picturesque, 309n9
piriform cortex (PC), 38–40, 41, 42, 44–45, 56–57, 61–62
plague and sanitation, 92
Plailly, Jane, 62–63, 64
plasticity. *See* brains
Plato, 267
  on reason vs. emotion, 68–69, 70
  on smell, 19–21, 19n2, 57, 288–89, 315
Plautus, 268, 269, 273
pleasure, 8
  aesthetics and, 14–15, 24–25, 28, 40, 73, 75, 131–32, 133, 196, 256
  Aristotle on, 20, 23, 267
  fine art and, 20–21
  Hinduism and, 98
  identity and, 274–75
  Kant on, 14–15, 216–17
  perfumes and, 238–40, 266, 269, 270, 271, 274–75
  Plato on, 19n2, 28
Pliny the Elder: *Natural Histories,* 268, 298–99
pluralism, 4, 7–8, 316
  aesthetic concept of, 257–58, 257n12, 257–58n13, 301
  art concept of, 206, 210–11, 231–32, 249
poetry, 73, 100, 117–20, 147–48, 179, 193–94, 202, 234–35
Poetry Magazine, 147–48
pollution and pollutants, 281
Polyester (film), 165
Porphyry, 89
Postrel, Virginia, 282–83
post-traumatic stress disorder, 128, 128n5
Press, Daniel, 307–8n3
Price, Carolyn, 70–71
Prinz, Jesse, 70, 73
Prior, Jonathan, 309, 309n9
Project KÔDÔ, 206–7
proportional effect, 164, 165
proprioception, 278
Protestant Reformation, 90–92
Proust, Marcel, 120–21, 127–28, 129–36, 130–31n11, 209, 317
  Remembrance of Things Past, 6, 129, 171
Prum, Richard: *The Evolution of Beauty,* 84–86
psychology, 30–31
  bioculturalism and, 79–80
  of emotion, 68–71
  memory and, 127–29
  on olfactory experts, 55–56
  Proust and, 129n7, 129–36

public spaces and ambient scents, 1, 2–3, 282–83, 285, 286–90

quilts, 7, 14, 234–35, 236, 238–40, 238n16

race and odors, 23–24
Rakowitz, Michael: *Rise*, 281
Rasmussen, Susan, 110n13
Raspet, Sean: *Micro-encapsulated Surface Coating*, 155
Ray, Charles: *Hinoki*, 229
Reason, Matthew, 162
Reformation, 90–92
Reid, Thomas, 20–21n11
religion and spirituality. *See also specific religions*
   gardens and, 311
   incense/perfume and, 88–89n10, 89–90, 91–92, 92n29, 94, 99, 111, 133, 201, 206–7
Rembrandt, 152
   reodorization, 94n37
retronasal smell, 8, 316–17
   evolution and, 83–84, 83n5, 85–86
   fast food and, 303, 304, 305
   neuroanatomy of, 38
   taste and, 129–30, 215–16, 297–98, 299–300
Revolution Pipe Bomb (perfume), 216n34, 220–21, 223–24, 233–34, 236, 240–41, 245–46
Riach, Kathleen, 286–87, 288
Richardson, Louise, 25–27, 42–43, 297
Rilke, Rainier Maria: *Notebooks of the Malte Laurids Brigge*, 121
Rindisbacher, Hans, 122n12, 134
   The Smell of Books, 261
ritual incense and perfumes. *See* perfumes and incenses; religion and spirituality
Robbins, Tom: *Jitterbug Perfume*, 124–25
Robertson, David, 91n25, 91n27
Robinson, Jeff, 186n13, 197n49
Robinson, Jenifer, 70–71, 113–14, 283–84
   Deeper Than Reason: Emotion and Its Role in Literature, Music and Art, 73
Robinson, Michael D., 69
Rodchenko, Alexander, 246n9
Romans and perfumes, 89–90, 267–69
Romantics and Romanticism, 117–18, 308–9
Ropion, Dominic, 214–15, 247
   Olfactory Hommage to Francis Poulenc, 216
   Portrait of a Lady, 222–23
Rorty, Amélie, 70
roses, 19, 37–38, 103

Ross, Stephanie: *The Meaning of Gardens*, 310–11, 310–11n14
Rouby, Catherine, 106–7, 111n18
Roudnitska, Edmund, 210, 213, 216
Roudnitska, Michel, 172, 247
Rozin, Paul, 83n5
Rubens, Peter Paul: *Allegory of the Five Senses*, 152
Rubin, David, 128
Rugrats Go Wild (film), 165
Rush University Medical Center, 44
Russolo, Luigi, 187–88
Ryan, Michael: *A Taste for the Beautiful*, 84–86
Ryan, Zoë, 230–31

Sacks, Oliver
   The Man Who Mistook His Wife for a Hat, 74–75
Sadra, I Wayan, 170
Saito, Yuriko, 248, 258, 275, 291
   The Aesthetics of the Familiar, 256
sanitation and sanitary campaigns, 92–94, 281
Santayana, George, 21, 73
Sapolsky, Robert, 40
Scarantino, Andrea, 69n4
scent art, 6, 315–16. *See also* fine art; olfactory arts
   aesthetics of, 182–83, 186–87, 196–98, 261
   bodies and embodiment in, 182–83, 197
   definitions and boundaries of, 149–50, 154
   definitions and criteria for, 183–87
   Drobnick on, 182–83
   exhibition and conservation of, 191–93
   as fine art, 145–46, 182–83, 236, 239, 243, 317
   history of, 188–90
   interpretation of, 193–96
   perfume and, 240, 243
   practitioners of, 190–91
   sound art and, 186–88, 191–92
scenting
   food and, 298–99
   public spaces and, 1, 2–3, 282–83, 285, 286–90
The Scent of Mystery (film), 165–66, 167–68
scent organs, 173, 178
scents. *See* foods and food scents; odors
Scents of the Soul (play), 1
scent tracks. *See* film
Schafer, E. H.: *The Golden Peaches of Samarkand*, 99
Schellekens, Elisabeth, 71–72, 145–46
Schneemann, Carolee, 194
Schopenhauer, Arthur

on aesthetics, 73
on art, 73n13
Schusterman, Richard, 274
Scriabin, Alexander: *Mysterium,* 172
Scruton, Roger, 22, 25–27, 174, 192–93, 215–16
on smell, 1, 215
sculpture
fine art and, 234–35
olfactory arts and, 184
Smeller 2.0 and, 178–79
Secretions Magnifiques (perfume), 216, 216n35
Secundo, Lavo, 40, 45–46
seduction, 8, 206–7, 239–40, 268, 271–72, 274–75, 276. *See also* sexual attraction and selection
senses. *See also specific senses*
aesthetic judgments and, 71–72
historical perspectives on, 19–21, 20–21n12
Jones on, 194
kodo and, 204
multimodal nature of, 60, 63–65, 84, 168–69, 168–69n32, 278, 298–99
in paintings, 152–53
smellwalks and, 278
Yi and, 194–95
The Senses: Design beyond Vision (exhibition), 189–90
Sensorium (exhibition), 189–90
sensory studies, 87
sensus communis, 204
Sereer Ndut (people of Senegal), 111–12
Serrano, Andres, 235–36
sewers and sanitation, 92–94, 281
sexual attraction and selection
in Calvino's "The Name, The Nose," 35–36
perfume and, 92n29, 215, 265–66, 268–69, 271–73, 274–75
pheromones and, 51–53, 83
smell and, 84–85
Sezille, Caroline, 60–61, 64
Shakespeare, William, 19, 159
Shaw, Barney, 317–18
The Smell of Fresh Rain: The Unexpected Pleasures of Our Most Elusive Sense, 140, 140n3
Shepher-Barr, Kristin, 159–60
Shepherd, Gordon, 39–40, 83–84, 83n5, 303n20, 304
Shereikis, Judy, 44n29
Shereikis, Rebecca, 110n13
Shiner, Larry, 218n40, 234–35n12
Shioya, Nobi: *7S,* 184–85
Shiseido (scent firm), 286

Shoyeido Incense Company, 203
Sibley, Frank, 22, 29–30, 58–59, 71–72, 206–7, 211n12, 215, 317
sick building syndrome, 285, 285n28
Skinner, Jeffrey, 147–48
smell, 1–3, 317–18. *See also* odors; olfactory aesthetics
age and, 43–44
in Asian societies, 98–101
communication through, 45–47, 83, 164–65
cultivating sense of, 22, 58, 140, 317–18
emotions and, 55, 56–63, 68, 71, 132
gender and, 43–44
human vs. animal, 20n5, 43, 52, 82–84, 94, 182–83, 300–1
impairment/loss of, 28–29, 29n49, 43–45, 54–55, 180, 213, 297, 310–11n17
language and, 19, 55–56, 64, 65–66, 79, 95, 101, 103–14, 121–22, 140
neuroanatomy of, 37–40, 54–55
perception models for, 25–27
prejudice against, 1–2, 19–21, 23–28, 39, 65, 138, 182–83, 190, 239–40, 281
terms for, 108–14, 109–10n11
Smeller 2.0, 7, 178–80, 184, 191, 315–16
smellmaps, 156, 280–81
smell of the "other," 23–24
Smell-O-Vision, 165–66
smellwalks, 8, 156, 278–81, 278n1, 318
Smith, Barry C., 27, 38, 61, 74, 278n1
Smith, Kiki, 219–21
Smith, Mark M., 87–88
How Race Is Made, 23–24
Smith, Murray, 80, 80n4, 139–40
Film, Art and the Third Culture, 30–31, 30–31n54
Snow, C. P., 30–31n54
Sobchack, Vivian, 168–69, 168–69n32
Sobel, Noam, 40, 45–46, 46n39, 57, 61, 79–80, 112–13n24
social sciences and humanities, 80, 87, 138–39
Solomon, Robert, 70
Song of Songs (Bible), 88–89, 88–89n10, 159–60, 266
sound art, 150, 186–88, 187–88n18, 188n19, 189–90, 191–92
sounds in film, 166–69
Sousa, Ronald de, 69n4, 70
Souter, Lucy, 235–36
Spence, Charles, 40, 296, 298, 298n9, 299, 303n20, 304, 305n26
Spinoza, 20–21, 68–69, 270, 276, 291
spirituality. *See* religion and spirituality

squibs, 159
Stamelman, Richard: *Perfume: Joy, Obsession, Scandal,* 274
Stanton, David T., 105–6
Starck, Philippe: *Juicy Salif,* 245
Starr, G. Gabrielle, 147
   Feeling Beauty: The Neuroscience of Aesthetic Experience, 73
Stecker, Robert, 232–34, 241
Sterbak, Janna
   Meat Dress for an Albino Anorexic, 189
   Perspiration: An Olfactory Portrait, 186
Stevenson, Richard J., 29n49, 55, 57–58, 63
   Learning to Smell (with Wilson), 41
Stieglitz, Alfred, 237
Stoddart, Michael, 68, 83, 84–86
   Adam's Nose, and the Making of Humankind, 52
Stone, Carl, 179
Stylites, Simeon, 90
sublime, 196
Suchetelen, Anna van, 147–48
Summers, David, 20n8
Süskind, Patrick: *Perfume: The Story of a Murderer,* 8, 167, 168–69, 261, 262–64
Sweeney, Kevin, 27, 74, 107–8, 169–70, 189n25, 215–16
   The Aesthetics of Food, 297–98, 299–301
Symrise scent production company, 216n34, 220–21, 247–48n17
synesthetic metaphor, 109–10, 111–12, 117–18, 119–21
Synnott, Anthony, 87–88n3

Tacoma, Washington, 280, 280n8
Tafalla, Marta, 28–29, 278, 310–11n17
The Tale of Genji, 100–1, 202
taste. *See also* orthonasal smell; retronasal smell
   aesthetics and, 29–30, 71–72, 139–40, 300–1
   cognition and, 297–99
   historical perspectives on, 20–21, 24–25, 297–98
   meaning in, 27
   multimodal perception of, 168–69
   philosophy of, 3, 25
   Proust and, 129–30, 133
   science of, 38, 168–69
   smell and, 83–84, 296–97
Tate Britain, 189–90
temporality
   in fine cuisine, 297–98
   in gardens, 310–11
   in perfumes and wines, 212–13, 215–16, 236–37, 255–56
thalamus, 39, 39–40n10
Thames Embankment, 93
theater, 72, 158–65
Theophrastus, 266
Theresa of Avila, 90
Thierry Mugler perfume house, 167
Thommasini, Anthony, 172–73
Thurlkill, Mary, 312–13
Til Birnam Wood (play), 160–61
Tinguely Museum, 1, 191–92
Tiravanija, Rirkrit, 300
Todd, Cain, 74, 215–16, 217, 278n1
   Philosophy of Wine, 230–31
Todd, Mike, 165–66
To Dig a Hole and Fill It Up Again (exhibition), 197n49
Tolaas, Sissel
   The Fear of Smell and the Smell of Fear, 13–14, 15, 45, 74, 155, 196–97, 218–19
   smellwalks and, 279–80
   Valentinelli compared to, 188
Toporowicz, Maciej: *Lure,* 186
Totelin, Laurence, 89
touch
   aesthetic judgments and, 71–72, 139–40
   neuroscience of, 39
   olfactory arts and, 155, 182–83
   role in flavor, 38
transcendence, 129–36, 274–76
trigeminal nerve, 38, 38n3
Trust (perfume), 222
Turin, Luca, 38–39n4
21 Drops Essential Oil Therapy, 290
Tykwer, Tom: *Perfume: The Story of a Murderer* (film), 167, 168–69
type/token problem, 214

U-deur (scent installation), 145
Ueda, Maki: *Invisible White,* 186, 194
Ullmann, Stephen, 117–18
Umeda (people of Papua), 112, 275–76
University of Düsseldorf, 109–10
Un jardin en méditerranée (perfume), 223–24, 233–34
urban planning, 281–83. *See also* architecture
Ursitti, Clara, 74, 190
   Pheromone Link™ Scent Library, 52–53
   Poison Ladies, 190
   Self-Portraits in Scent, 146–47, 186, 194, 218–19, 240–41

Valentinelli, Ennio, 188
Van Beirendonck, Walter, 246
ventilation. *See* diffusion of odors
Verbeek, Caro, 188
Vermetten, Eric, 128
Vernant, Jean-Pierre, 89, 266
vibration theory of smell, 38–39, 38–39n4
Viola, Bill, 189
vision
　AI and, 44–45, 44–45n34
　architecture and, 283–84
　bias toward, 19, 24–25, 139–40, 279, 310–11
　emotions and, 70–71
　evolution and, 82–83, 84
　high status of, 2
　historical perspectives on, 20–21
　Jones on, 194
　language and, 55
　Lopes on, 196–97n48
　neuroscience of, 39, 39–40n10, 54–55, 168–69
　perception models based on, 25–26, 27, 29–30
　Yi on, 182
visual arts. *See also* painting
　criticism and, 195–96
　fine art and, 14
　olfactory arts compared to, 182–83, 186–87, 189
　postmedium turn of, 145, 189–90
　sound art and, 187–88
Volatile! A Poetry and Scent Exhibition, 147–48
vomeronasal organs, 51–52, 83, 84–86
von Meiss, Pierre: *The Elements of Architecture,* 284–85

Wagner, Richard, 158
Waitress (musical), 148–49, 171–72
Walker, Kara: *Marvelous Sugar Baby,* 300
Wall, Jeff, 235–36
Walpole, Horace, 310–11n13
Walton, Kendall, 255–56n1
Ware, Chris, 155–56
Warhol, Andy: *Brillo Box,* 230–31
Warren, Samantha, 286–87, 288
Waters, John: *Polyester,* 1–2
Watkins, Edward R., 69
Wawrzynik, Martynka: *Smell Me,* 186
Webster, John: *White Devil,* 158–59
Webster, Meg: *Moss Bed, Queen,* 194
Weiss, Taio, 40, 45–46

Wenz, Peter, 273–74n32
Westerners and Western societies. *See also* deodorization
　art practice and, 206–7
　deodorization in, 2, 91–94, 103, 281
　gardens in, 309–10
　identifying smells in, 104, 112–14, 162–63
　odor terms in, 108–10, 113–14
　sensory pleasure and, 270
Westin hotels, 286, 289
Whitney Museum of American Art, 238
wilderness, 308–9
Williams, Tennessee: *The Glass Menagerie,* 160–61
Wilson, Daniel, 206n23, 257n12
Wilson, Donald, 38–39
　Learning to Smell (with Stevenson), 41
Wilson, Edward O., 51, 307–8
wine and wine tasting, 27, 38, 59–61, 64
　aesthetics and, 74, 106–7, 215–16
　classifications in, 105, 106, 107–8, 138
　expressivity of, 214, 215–17, 224, 238–39
　as fine art, 230–31
Wine Aroma Wheel, 105, 106, 107
Wollheim, Richard, 219n45
Wolterstorff, Nicholas, 218n40, 248–49
Woolf, Virginia, 6
　Flush: A Biography, 121–22, 133–34
Wrangham, Richard: *Catching Fire: How Cooking Made Us Human,* 83, 84
Wu, Keng Nei, 44–45, 64

Xenophon: *Symposium,* 267

Yamagata, Hiroshi, 204–5
Yeshurun, Yaara, 56–57, 61, 112–13n24
Yi, Anicka
　Auras, Orgasms, and Nervous Peaches, 184–85
　Divorce, 182, 194–95
　Hugo Boss prize award to, 189–90
　Jones and, 194–96
　You Can Call Me F, 182–84, 194–95, 196
Youssef, Jozef, 298–300

Zangwill, Nick, 217
Zarzo, Manuel, 105–6
Zhao, Qingqing, 110, 111–12
Zittel, Andrea, 245
Zumthor, Peter, 284, 284–85n27

www.ingramcontent.com/pod-product-compliance
Ingram Content Group UK Ltd.
Pitfield, Milton Keynes, MK11 3LW, UK
UKHW022152230426
12049UKWH00003BA/58

9 780190 089818